Deep Time of the Media

ELECTRONIC CULTURE: HISTORY, THEORY, AND PRACTICE

Ars Electronica: Facing the Future: A Survey of Two Decades edited by Timothy Druckrey

net_condition: art and global media edited by Peter Weibel and Timothy Druckrey

Dark Fiber: Tracking Critical Internet Culture by Geert Lovink

Future Cinema: The Cinematic Imaginary after Film edited by Jeffrey Shaw and Peter Weibel

Stelarc: The Monograph edited by Marquard Smith

Deep Time of the Media: Toward an Archaeology of Hearing and Seeing by Technical Means by Siegfried Zielinski

Deep Time of the Media

Toward an Archaeology of Hearing and Seeing by Technical Means

Siegfried Zielinski

TRANSLATED BY GLORIA CUSTANCE

The MIT Press
Cambridge, Massachusetts
London, England

Originally published as *Archäologie der Medien: Zur Tiefenzeit des technischen Hörens und Sehens,* © Rowohlt Taschenbuch Verlag, Reinbek bei Hamburg, 2002

The publication of this work was supported by a grant from the Goethe-Institut.

MIT Press books may be purchased at special quantity discounts for business or sales promotional use. For information, please e-mail special_sales@mitpress.mit.edu or write to Special Sales Department, The MIT Press, 55 Hayward Street, Cambridge, MA 02142.

This book was set in Bell Gothic and Garamond 3 by Graphic Composition, Inc., Athens, Georgia. Printed and bound in the United States of America.

Library of Congress Cataloging-in-Publication Data
Zielinski, Siegfried.
[Archäologie der Medien. English]
Deep time of the media : toward an archaeology of hearing and seeing by technical means / Siegfried Zielinski ; translated by Gloria Custance.
    p.   cm.—(Electronic culture—history, theory, practice)
Includes bibliographical references and index.
ISBN 978-0-262-24049-9 (hardcover: alk. paper), 978-0-262-74032-6 (paperback)
1. Mass media—Historiography. 2. Mass media—Philosophy. I. Title. II. Series.
P91.Z53813   302.23'0722—dc22   2005047856

10   9   8   7   6   5   4   3   2

Contents

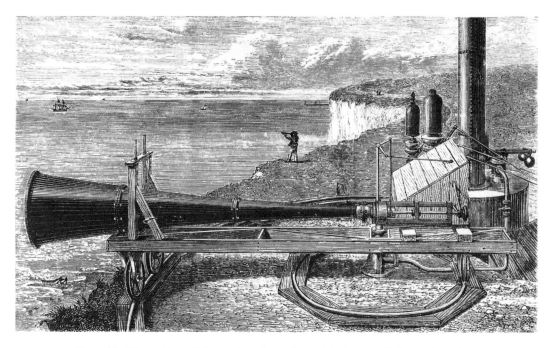

Figure I.i   "The problem: At times, when a fact is thrown into the smoothly flowing river of scientific development that completely contradicts earlier conceptions, one of the strangest transformations takes place. What is slightly new is either dissolved and assimilated or, if it is too deviant in the present situation, it sinks to the bottom as a foreign body where the deposits of time cover it — it either has an effect much later or never at all. That which is significantly new, however, rapidly has a conspicuous influence on the entire state [of things]. A violent perturbation of ideas about and over this commences. . . ." (Text: Ostwald 1896, p. 1f; illustration: Tyndall, 1883, frontispiece)

Foreword

The sense of present which we live each day, as a conflict between the representatives of ideas having different systematic ages and all competing for possession of the future, can be grafted upon the most inexpressive archaeological record. Every shred mutely testifies to the presence of the same conflicts. Each material remnant is like the reminder of the lost causes whose only record is the successful outcome among simultaneous sequences.

—GEORGE KUBLER, *THE SHAPE OF TIME*

An anemic and evolutionary model has come to dominate many studies in the so-called media. Trapped in progressive trajectories, their evidence so often retrieves a technological past already incorporated into the staging of the contemporary as the mere outcome of history. These awkward histories have reinforced teleologies that simplify historical research and attempt to expound an evolutionary model unhinged from much more than vague (or eccentric) readings of either the available canon or its most obvious examples. Anecdotal, reflexive, idiosyncratic, synthetic, the equilibrium supported by lazy linearity has comfortably subsumed the media by cataloguing its forms, its apparatuses, its predictability, its necessity. Ingrained in this model is a flawed notion of survivability of the fittest, the slow assimilation of the most efficient mutation, the perfectibility of the unadapted, and perhaps, a reactionary avant-gardism. In this model there is less failure than dopey momentum and fewer ruptures than can be easily accounted for. As a historiography it provides an orthodox itinerary uncluttered by speculation or dissent, unfettered by difference, disconnected from the archive, averse to heterogeneity.

This laissez-faire historiography dominates American writing concerned with the histories of media and has fueled both oversimplification and imprecision. History is, after all, not merely the accumulation of fact, but an active revisioning, a necessary corrective discourse, and fundamentally an act of interrogation—not just of the facts, but of the displaced, the forgotten, the disregarded.

For some in the media, "archaeology" has come to supplant basic history, replacing it with a form of material retrieval—as if the preservation of materiality was tantamount to preserving history itself. This has led to an archaeology (really more a mere cataloging) of the apparatus itself, rather than an investigation of the scenes in which the apparatus found its way into the spheres of research and experience.

Michel Foucault's *The Archaeology of Knowledge* is defiant in distinguishing archaeology from other forms of historiography. Archaeology is "the systematic description of a discourse-object," (139) it "tries to establish the system of transformations that constitute change." (173), it "does not have a unifying, but a diversifying effect," (160) it "is not supposed to carry any suggestion of anticipation." (206)

It is the analysis of silent births, or distant correspondences, of permanences that persist . . . of slow formations that profit from the innumerable blind complicities. . . . Genesis, continuity, totalization: these are the themes of the history of ideas.

But archaeological description is precisely such an abandonment of the history of ideas, a systematic rejection of its postulates and procedures . . . (Foucault, 138)

As such, archaeology is not a substitute for "the history of ideas," not a proxy for iconography, not an alternative for eccentric discovery or collecting, not a surrogate for rigorous research. With this in mind, it seems imperative to delineate an approach to "media archaeology" that, on the one hand, avoids idiosyncrasies or subjectivities, and, on the other, doesn't lull itself into isolating media history as a specialized discipline insulated from its discursive historical role.

There's little doubt that the multithreaded developments of media have numerous unresolved histories and that an enormous task of retrieval and conceptualization has yet to be achieved. How a media archaeology can constitute itself against self-legitimation or self-reflexivity is crucial if it is to circumvent the reinvention of unifying, progressive, cyclical, or "anticipatory" history—even as it is challenged to constitute these very vague histories as an antidote to the gaping lapses in traditional historiography. Indeed it is this very problem that afflicts media archaeology. The mere rediscovery of the forgotten, the establishment of oddball paleontologies, of idiosyncratic genealogies, uncertain lineages, the excavation of antique technologies or images, the account of erratic technical developments, are, in themselves, insufficient to the building of a coherent discursive methodology.

In this sense the notion of resurrecting dead media could prove farcical, futile, or more hopefully, deeply fertile. A broad accounting of the evolution of the apparatus, of the media image, of the history of the media effect, of excavating the embedded intellectual history, and so on, is surely the precursor of what will be an invaluable reconfiguration of a history largely focused on the device and its illusory images. Similarly, the rediscovery of uncommon or singular apparatuses, novel and fantastic as they might be, is neither decisive nor fully adequate to formulate an inclusive approach that distinguishes it from connoisseurship, or worse, antiquarianism. Merely reconstituting or retrofitting "old" media into "new" contexts could, in this sense, only emerge as techno-retro-kitsch.

What is most necessary for the field of media archaeology is to both distinguish it as a nascent discipline and to set some boundaries in order to avoid its trivialization. Archaeology, as Foucault writes, "is not a return to the innermost secret of the origin," rather it "describes discourses as practices specified in the element of the archive" (p. 138 from same source.) Without evolving coherences that are neither reductive nor dogmatic, media archaeology faces numerous issues: to evolve histories of technologies, apparatuses, effects, images, iconographies, and so forth, within a larger scheme of reintegration in order to expand a largely ignored aspect of conventional history.

Already some useful examples of this exist, from Siegfried Giedion's *Mechanization Takes Command* or E. J. Dijksterhuis's *Mechanization of the World Picture* to Friedrich Kittler's *Gramophone, Film, Typewriter* or Wolfgang Schivelbusch's *Railway Journey* or *Disenchanted Night: The Industrialization of Light in the 19th Century*, Michel Foucault's *Archaeology of Knowledge*, Laurent Mannoni's *The Great Art of Light and Shadow: Archaeology of the Cinema*, Norman Klein's *The Vatican to Vegas: A History of Special Effects*. Each tackles the apparatus (or its "effects") as integral to the substantive changes they wrought as modernity emerged. Not under the spell of linearity, these books stand as guidebooks (among many others), for the establishment of diversified approaches to a media history and, more specifically, a media archaeology that stands as a decisive field if it can develop forms that extrapolate more than missing links.

Siegfried Zielinski's *Deep Time of the Media* intensifies and extends these studies with a wide range of scholarship from Stephen Jay Gould's "punctuated equilibrium" to Georges Bataille's "general economy," and, more deeply, into the original volumes of Athanasius Kircher, Giovan Battista della Porta, and Giuseppe Mazzolari. Instead of tracking the reverberations, *Deep Time of the Media* situates the effect in the midst of its own milieu. Though particular approaches may represent harbingers, augurs, precursors, they are purposefully rooted and serve particular goals.

It is in this context that Zielinski's *Deep Time of the Media* comes as a pivotal work challenging the field in a number of ways. In rebridging (perhaps demolishing) the widening gulf between *tekhne* and *episteme, Deep Time of the Media* refuses the mere instrumentalization of technology as meticulously as it integrates the responsibilities of knowledge. Riding through the stratifications has revealed far more than the unearthing of new "species" of media, but is leading toward a rethinking of the bleak search for origins by imagining (exposing) intricate topologies that link movement and coincidence, failure and possibility, obscurity and revelation. This move through and across the "tectonic" flows suggests a sweeping remapping of the hitherto centralized nodes of learning and that traces the decentralized currents of time, space, and communication as a kind of historical formation in which routes replace nodes and in which east meets west meets north meets south. In this the epistemic centers in the Eurocentric canon just don't hold and nor does a singular rationalistic scientific *logos*.

In its "case studies" *Deep Time of the Media* provides both a rigorous methodology and a reconceptualization of media studies. For Zielinski only full primary sources provide adequate evidence. So in tandem with a rigorous and dedicated teaching and lecturing schedule, his peripatetic research has taken him on the nomadic circuits of his subjects. Here he constructs the new cartography, seizes on the crossed path, the forgotten archive. His lectures, always laden with the trade-mark overhead projector, always trace an adventure into some new facet of the journey—with an obscure archive a decisive discovery.

Abandoning historical convention in favor of historical acuity, *Deep Time of the Media* travels into deep time and discovers not just more remains, but instead neglected constellations. Within these are towering figures of scientific and philosophical investigation—della Porta, Kircher, Ritter, Hutton, Lombroso, among many others. These bold personalities demand our attention not because they outdid their times, but rather because they embodied them.

With them come the shifting objects of study—less and less material—light and shadow, electricity and conduction, sound and transmission, magic and illusion, vision and stimuli—in short, conditional phenomena. Fleeting and contingent, the phenomenal world was lured into visibility by instruments whose ingenuity often eclipsed their discoveries. At least we had been convinced that this is so. Zielinski proves us wrong. Through their instruments the sphere of representation exploded. Its fragments resonate in every future media apparatus. Through their instruments the interface emerged, through their instruments a fragile imaginary was brought to light, through their instruments time, sound, reflex, could be seen, through their instruments the world was no longer a paltry given, it was a moving target, a dynamic presence, it was, to put it bluntly, alive.

Ever since, our machines have aspired to the "real" and, luckily, have fallen short of their phony virtual utopias. This surely explains why the last chapter of *Deep Time* focuses on the "artistic, scientific, technical, and magical challenges" that persist in contemporary media praxis. Zielinski's tenacious role as a historian has never restrained his enormous commitment to colleagues and students. His unyielding charge is to relentlessly cultivate "dramaturgies of difference," to "intervene" into the omnivorous systems from the periphery, to refuse centralization, to seize the imagination back from its grim and superfluous engineers, and to construct an art worthy of its "deep time." As Deleuze writes:

It is not enough to disturb the sensory-motor connections. It is necessary to combine the optical-sound image with the enormous forces that are not those of simply intellectual consciousness, nor of the social one, but of a profound, vital intuition.

—Timothy Druckrey

Acknowledgments

My grateful thanks are due to Gloria Custance for her untiring and exceptional work in translating this book. She also translated all quotations from the German, unless noted otherwise.

Nils Röller read the greater part of the original manuscript and I thank him for the many fruitful discussions and constructive suggestions, including in the earlier years of our collaboration. I am indebted to Timothy Druckrey, Keith Griffiths, Dietmar Kamper, Anthony Moore, Miklós Peternàk, The Brothers Quay, and Otto E. Roessler for their generous intellectual support and encouragement, which played an essential role in the realization of the project. Werner Nekes I thank for his hospitality and the many visits to his unique archive. Wolfgang Ernst, Thomas Hensel, Angela Huemer, Christine Karralus, Friedrich Kittler, Jürgen Klauke, David Link, Alla Mitrofanova, Morgane, Peter Pancke, Hans Ulrich Reck, Elisabeth von Samsonow, Silvia Wagnermaier, and Sigrid Weigel listened patiently to my expedition reports and provided invaluable aid in the form of material, questions, commentaries, and suggestions. Nadine Minkwitz and Juan Orozco gave generously of her time for the digital processing of illustrations, and I am grateful to Heiko Diekmeier and Claudia Trekel for their skill with reproductions. For his great help with the texts in Latin, I thank Franz Fischer; for translations, Angela Huemer and Rosa Barba (Italian), Peter Frucht and Adèle Eisenstein (Hungarian), Lioudmila Voropai (Russian), and Gloria Custance (English and French), whose assistance with the German edition was, in many respects, completely indispensable. Anke Simon, Daniela Behne, Uschi Buechel, Andrea Lindner, and Birgit Trogemann were tireless in their efforts to procure all the books and media I required. Patricia Nocera I thank for generously giving me access to the treasures of the Biblioteca Nazionale di Napoli and guiding me through its labyrinth. In the Herzogin Anna Amalia library in Weimar, Katrin Lehmann provided exceptional assistance for my research. The Staatsbibliothek Berlin and the university libraries in Cologne and Salzburg were very helpful in connection with work on Dee, Fludd, Kircher, Llull, Porta, and Schott. For their help in matters of organization, I thank Suse Pachale and Heidrun Hertell.

Many thanks to Roger Conover for preparing and overseeing this edition for MIT Press and to Lisa Reeve for her editorial support.

My research and the writing of the original text were possible within such a short time period only because the Ministry of Science, North Rhine-Westphalia, granted me an additional sabbatical semester. My special thanks go to Burkhard König at Rowohlt for his faith in the project, and his constant and unfailing support for my endeavors.

1

Introduction: The Idea of a Deep Time of the Media

Our sexuality . . . belongs to a different stage of evolution than our state of mind.
—BRUNO SCHULZ,"AN WITOLD GOMBROWICZ." IN: *DIE REPUBLIK DER TRÄUME*

In the early 1980s, the Texan science-fiction author Bruce Sterling invented the phenomenon of cyberpunk, together with the sci-fi writers William Gibson from Canada and Samuel R. Delany of New York, an ex-boxer and professor of literature. Their creation married clean high-tech and dirty rubbish, order and anarchy, eternal artificial life and decomposing matter. Techno- and necro-romanticism came together to create a new *Lebensgefühl.* The inspired collaboration of Ridley Scott, film director, and Douglas Trumball, designer and set decorator, translated this feeling into cinema in the brilliant *Bladerunner* (1982). *The Matrix* (1999), directed by Andy and Larry Wachowski, fulfilled a similar function at the end of the 1990s for the now computer-literate fans of cyberculture, who by then were all linked via worldwide data networks. The horror that stalks the film *Matrix* is no longer an individual, amoral machine that operates locally and has taken on human form, as in *Bladerunner,* but, instead, is a data network that spans the entire globe and controls each and every action, emotion, and expression.

When one generation of computer hardware and software began to follow the next at ever shorter intervals, Sterling initiated "The Dead Media Project." There, he exchanged his wanderings through an imaginary everyday life in the future for an energetic movement that traversed the past to arrive in the present. Together with like-minded people, in 1995 he started a mailing list (at that time, still an attractive option on the Internet) to collect obsolete software. This list was soon expanded to include dead ideas or discarded artifacts and systems from the history of technical media: inventions that appeared suddenly and disappeared just as quickly, which dead-ended and were never developed further; models that never left the drawing board; or actual products that were bought and used and subsequently vanished into thin air.[1] Sterling's project confronted burgeoning fantasies about the immortality of machines with the simple facticity of a continuously growing list of things that have become defunct. Machines can die.[2] Once again, romantic notions of technology and of death were closely intertwined in "The Dead Media Project."

Media are special cases within the history of civilization. They have contributed their share to the gigantic rubbish heaps that cover the face of our planet or to the mobile junk that zips through outer space. While the USSR was falling apart, the cameraman of Tarkovsky's legendary *Solaris,* Vadim Yusov, was teaching astronauts from the MIR space station to take pictures of Earth for Andrei Ujica's *Out of the Present* (1995). The 35mm camera they used is probably still orbiting up there over our heads. After the rolls of film had been shot and stunning pictures of the blue planet were in the can, the camera was simply thrown out of the escape hatch. Taking it back to Earth would have been too expensive, and it was not considered worthwhile to develop a special program just to destroy a few kilograms of media technology.

The stories and histories that have been written on the evolution of media had the opportunity—at least theoretically—to do some recycling, in line with the rubbish theory proposed by Michael Thompson:[3] they might have searched through the heaps of refuse and uncovered some shining jewels from what has been discarded or forgotten. Nothing endures in the culture of technology; however, we do have the ability to influence how long ideas and concepts retain their radiance and luminescence. Up to now, media historians have neglected to do anything of the kind, mainly on ideological grounds, and this has also had methodological repercussions. In the extensive literature on the genealogies of telematics (from antiquity's metal speaking-tube to the telephone; from Aeneas's water telegraph to the Integrated Service Data Network [ISDN]), or cinema archaeology (from the cave paintings of Lascaux to the immersive IMAX), or the history of computers (from Wilhelm Schickard's mechanical calculating apparatus to the universal Turing machine), one thing above all others is refined and expanded: the idea of inexorable, quasi-natural, technical progress. It is related to other basic assumptions, such as the history of political hegemony developing from the strictly hierarchical to strictly democratic organization of systems, the rationale of economic expediency, the absolute necessity for simple technical artifacts to develop into complex technological systems, or the continual perfecting of the illusionizing potential of media. In essence, such genealogies are comforting fables about a bright future, where everything that ever existed is subjugated to the notion of technology as a power to "banish fear" and a "universal driving force."[4]

Michelangelo's ceiling paintings in the Sistine Chapel in Rome do not anticipate that which today goes by the name of virtual reality and is produced on outrageously expensive computer systems, like the CAVE. What would this genius, master of two-dimensional illusions using painted images, colors, and geometry, have found of interest in such an idea, weak and already backward a couple of years after its "invention"? Having said that, there is something akin to a topicality of what has passed. However, if we are to understand history as being present not only when it demands to be accepted as a responsibility and a heavy burden, but also when there is value in allowing it to develop as a special attraction, we will need a different perspective from that which is only able to seek the old in the new. In the latter perspective, history is the promise of continuity and a celebration of the continual march of progress in the name of humankind. Everything has always been around, only in a less elaborate form; one needs only to look. Past centuries were there only to polish and perfect the great archaic ideas. This view is primitive pedagogy that is boring and saps the energy to work for the changes that are so desperately needed. Now, if we deliberately alter the emphasis, turn it around, and experiment, the result is worthwhile: do not seek the old in the new, but find something new in the old. If we are lucky and find it, we shall have to say goodbye to much that is familiar in a variety of respects. In this book, I shall attempt to describe this approach in the form of an (an)archaeological expedition or quest.

For Isaac Newton, the great world-mechanic, and his contemporaries, what we call "our" planet was still thought to be not much more than six thousand years old. God's representatives here below, men like the Anglican prelate James Ussher, had "proved" that this was so in the mid-seventeenth century, and that was that. As more and more evidence of immense qualitative geological changes piled up, their only resort was the trick of compressing the time periods in which the deposits had accreted. In the seventeenth century, Athanasius Kircher used the same theoretical crutch in his description of the subterranean world. In the eighteenth century, doubts were increasingly voiced about this extremely short chronology, and by the nineteenth century, geologists were calculating in millions of years. It was only in the twentieth century that there was absolute certainty that the history of the Earth spans billions of years. Such numbers surpass our powers of imagination, just as it is almost impossible to imagine the existence of infinite parallel universes or the coexistence of different space-times.

At the turn of the eighteenth to the nineteenth century, the idea that the Earth was far older than previously supposed became a fashionable topic in the academies and bourgeois salons, just as electrical impulses in the bodies of organisms or between heterogeneous materials already were. Time structures on the large scale began to arouse interest, as well as their peculiarities on the small scale. In addition, the solidity of territories began to lose its dependability and comfortable familiarity as national boundaries were redrawn at ever decreasing intervals and traditional hierarchies were questioned. In Germany, Abraham Gottlob Werner, a mining engineer and lecturer at the famous Bergakademie in Freiburg, pioneered studies on the systematic investigation of minerals and rocks and their origins in the oceans that once covered the Earth. However, he neither could nor wanted to write a history of the Earth. More courageous than the "Neptunist" Werner was the "Vulcanist" James Hutton.[5] Son of a wealthy Scottish merchant, Hutton supplemented his already ample income by producing useful chemical compounds. His wealth provided him with a comfortable lifestyle in Edinburgh and the means to travel, conduct research, and undertake geological fieldwork for his own intellectual pleasure, entirely independent of any institutions. What is more, he had the time to write up and illustrate his observations. Hutton's *Theory of the Earth* of 1778, one thousand pages long, and the two-volume edition published in 1795 no longer explained the history of the Earth in terms of the old theological dogma. Hutton asserted that Earth's history could be explained exactly and scientifically from the actual state of the "natural bodies" at a given moment in time, which became known as the doctrine of uniformitarianism. Further, Hutton did not describe the Earth's evolution as a linear and irreversible process but as a dynamic cycle of erosion, deposition, consolidation, and uplifting before erosion starts the cycle anew. At localities in Scotland he observed that granite was not the oldest rock, as Werner and his student Johann Wolfgang von Goethe had assumed. Underneath the granite were deep vertical strata of slate, which were much older. These conclusions were presented in a powerful illustration that adorned the second edition of Hutton's *Theory of the Earth.* Underneath the familiar horizontal line depicting the Earth's surface, the slate deposits plunge into the depths, exceeding by far the strata lying above them. John McPhee's *Basin and Range* (1980), which first introduced the concept of "deep time," displays Hutton's illustration on the cover. This discovery must have been as stunning and important for geology as were the first depictions of the Copernican view of the solar system, which firmly dislodged the Earth from the center of the universe.

Hutton's illustration also introduces the chapter devoted to the Scotsman in Stephen Jay Gould's *Time's Arrow, Time's Cycle,* his important work on the history of the Earth and organic life.[6] Gould, the Harvard geologist and zoologist who regarded himself primarily as a paleontologist, says that the idea of geological deep time is so foreign to us that we can understand it only as a metaphor. Imagine the age of the Earth as represented by one Old English yard, "the distance from the king's nose to the tip of his outstretched hand. One stroke of a nail file on his middle finger erases human history."[7] Hutton's concept of Earth as a cyclic self-renewing machine,[8] without beginning or end, is in stark contrast to the time reckoning instituted by humans. Gould takes this concept a step further when, for his field, he rejects all ideas of divine plans or visions of progress. In a specific continuation of uniformitarianism, Gould's studies on the long chronology are marked by a contemporary concern for the ongoing loss of diversity. In *Wonderful Life,* which came after *Time's Arrow, Time's Cycle,* he introduces a new category that runs contrary to linear thinking: "excellence," which should be measured with reference to diversification events and the spread of diversity.[9] Thus, Gould adds to the idea of deep time a quantitative dimension as well as a qualitative one that addresses the density of differences and their distributions. Taken together, these ideas result in a very different picture of what has hitherto been called progress. The notion of continuous progress from lower to higher, from simple to complex, must be abandoned, together with all the images, metaphors, and iconography that have been—and still are—used to describe progress. Tree structures, steps and stairs, ladders, or cones with the point facing downwards (very similar to the ancient mythological symbol for the female, which is a triangle with the base above and the point directed toward the Earth) are, from a paleontological point of view, misleading and should therefore be discarded.[10] From this deep perspective, looking back over the time that nature has taken to evolve on Earth, even at our current level of knowledge we can recognize past events where a considerable reduction in diversity occurred. Now, if we make a horizontal cut across such events when represented as a tree structure, for example, branching diversity will be far greater below the cut—that is, in the Earth's more distant past—than above. In this paleontological perspective, humankind is no longer the hub and pivot of the world in which we live but, instead, a tiny accident that occurred in one of evolution's side branches. Genetically, the human brain has changed little during the last ten thousand years—a mere blink in geological terms that can hardly even be measured. Humans share the same stasis in their biological development with other successful species. The price that they pay for this is a relatively short life span and a narrow range of variations in their specific biological traits. At the other end of the scale are the bacteria, with their enormous variety and capacity for survival. It was Gould's own existential experience of illness—in 1982 he was diagnosed with a rare form of cancer and the statistical mean predicted he had only months to live—that made him deeply distrustful of any interpretation of living organisms that is based on considerations of the average. In reality, there was no mean for Gould. He took individual variations to be the only trustworthy value and punctuated equilibrium as the mode in which change takes place.[11]

The paradigm of technology as an organ was a crutch used in the development of mechanics; similarly, the organic becoming technology is now a poor prosthesis in the age of electronics and computation. Technology is not human; in a specific sense, it is deeply inhuman. The best, fully functioning technology can be created only in opposition to the traditional image of what is human and living, seldom as its extension or expansion. All of the great inventions that form the basis of technology, such as clockwork, rotation in mechanics, fixed wings in aeronautics, or digital calculators in electronics, were developed within a relationship of tension to the relative inertia of the organic and what is possible for humans. The development of geological and biological evolution on the one hand and that of civilization on the other are fundamentally different. Evolution, which is counted in billions of years, progresses very slowly. The changes that have taken place within the short time span of what we call civilization have occurred quickly by comparison and now occur at ever shorter intervals. In Gould's view, this difference is demonstrated by two particular traits, which influence cultural development decisively. The first is topological. Humans are nomadic animals; and our migrations lead to productive mixes of different situations and traditions, which often find expression in subsequent periods of rapid development. The second trait that has influenced the development of civilization is the culturally acquired ability to collect and store knowledge and experience and to pass these on to others. This ability can also lead to periods where qualitative developments are extremely concentrated: these could not possibly be achieved via the mechanisms of biological evolution.[12]

An investigation of the deep time of media attractions must provide more than a simple analogy between the findings of research on the history of Earth and its organisms and the evolution of technical media. I use certain conceptual premises from paleontology, which are illuminating for my own specific field of inquiry—the archaeology of the media—as orientations: the history of civilization does not follow a divine plan, nor do I accept that, under a layer of granite, there are no further strata of intriguing discoveries to be made. The history of the media is not the product of a predictable and necessary advance from primitive to complex apparatus. The current state of the art does not necessarily represent the best possible state, in the sense of Gould's excellence. Media are spaces of action for constructed attempts to connect what is separated. There have been periods of particularly intensive and necessary work on this effort, not the least in order to stop people from going crazy, among other reasons. It is in such periods that I make my cuts. If the interface of my method and the following story are positioned correctly, then the exposed surfaces of my cuts should reveal great diversity, which either has been lost because of the genealogical way of looking at things or was ignored by this view. Instead of looking for obligatory trends, master media, or imperative vanishing points, one should be able to discover individual variations. Possibly, one will discover fractures or turning points in historical master plans that provide useful ideas for navigating the labyrinth of what is currently firmly established. In the longer term, the body of individual anarchaeological studies should form a *variantology* of the media.

The idea for this book originated in the late 1980s, while I was writing *Audiovisions: Cinema and Television as Entr'actes in History* for Rowohlt's Encyclopaedia book series. *Audiovisions* attempted to locate the two most popular audiovisual media of the twentieth century and their parallel development within a wider context of the history of the development of technology and culture. My intention was to make cinema and television comprehensible as two particular media events and structures whose hegemonial power is historically limited. At the time of writing, there were already hectic signs heralding a technological and cultural transition centered on the digital and computers. I sought to offer a more considered and calm perspective, but by no means a complacent one. This overhasty orientation on a new master medium toward which all signifying praxis would be directed for a time—until the next one is defined—demanded the delineation of an independent and constructive way of dealing with this new phenomenon as a different possibility. In my understanding, *Audiovisions* was a plea for the heterogeneity of the arts of image and sound and against the beginning *psychopathia medialis.*[13]

Certain attitudes, which one already encountered on a daily basis in the late 1980s, became even more pronounced during the course of the 1990s. The shifts, which had become standard practice, were judged to be a revolution, entirely comparable in significance to the Industrial Revolution. Hailed as the beginning of the information society and new economy, where people would no longer have to earn a living by the sweat of their brow, the proclaimed revolution stood wholly under the sign of the present, and it was assumed that the new would lose its terrors. Every last digital phenomenon and data network was celebrated as a brilliant and dramatic innovation. It was this vociferous audacity, found not only in the daily fare served up by the media but also in theoretical reflections, that provoked me to undertake a far-ranging quest. In the beginning, it was patchy, with considerable time lapses, and dependent upon the places where I worked.

At the University of Salzburg I found a fine stock of books from an excellent Jesuit library. For the first time ever, I held in my hands original books and manuscripts by Giovan Battista della Porta, Athanasius Kircher, Caspar Schott, Christoph Scheiner, and other authors of the sixteenth and seventeenth centuries. A key experience was when I chanced upon a copy of John Dee's *Monas Hieroglyphica* of 1591, which had been bound together in one volume with a treatise on alchemy dating from the thirteenth century by Roger Bacon. This discovery coincided with a workshop on John Dee and Edward Kelley, to which I had invited the British filmmaker and producer Keith Griffiths. He encouraged me to delve into the rare texts by Dee, court mathematician to Elizabeth I, to explore the Prague of Rudolf II, and to appreciate as truly exciting texts the alchemists' writings with their strange worlds of images. Helmut Birkhan, a classical scholar from Vienna who, on his own testimony, is one of the half-dozen people in the world to have actually read the unpublished fifteenth-century *Buch der Heiligen Dreifaltigkeit* by the Franciscan monk Ulmannus, introduced me to the special hermeticism of alchemistic texts. He is able to interpret this strange material in the way that I "read" films by Jean-Luc Godard or Alain Robbe-Grillet with my students and, moreover, with the same enthusiasm. It was from

Birkhan that I first learned that a crucial characteristic of alchemistic writings, in contrast to the published findings of modern science, is the *private* nature of the elaborated treatises; for this reason, they are replete with cover-up strategies and practices to preserve their secrets. Words conceal one meaning behind others: for example, "a young boy's urine" can also stand for what we call vinegar—one of the easier examples to decipher. The special language employed by alchemists was regarded by some adepts as "destructive to discourse." In one of the earliest texts, *Turba philosophorum,* a meeting of alchemists was convened for the purpose of standardizing linguistic signs to facilitate mutual comprehension. However, "it failed utterly in its goal, for the various participants . . . Greek natural philosophers, such as Anaximenes and Pythagoras, with arabicised and distorted forms of names . . . scarcely referred to what others had said and contented themselves with making general statements or ones couched in singular language. It did not result in norms for the language of alchemy nor must this ever come about!" Heaven forbid, then anyone could make the *lapis* and, as Birkhan once made unmistakably clear to his audience during a lecture, for this we lack all the prerequisites.

Parallel to studying advanced media technologies, I began to develop a deep affection for several of the early dreamers and modelers. I had never encountered them in the course of my university education, and they have been left out of the discourse of media studies almost entirely. These two fields of interest were virtually inseparable: forays into forgotten or hitherto invisible layers and events in the historical development of the media, and the fascination exuded by my professional setting, filled with Unix and Macintosh computers, PCs, networks, analogue and digital studios for producing and processing images and sound, and including attempts by artists and scientists to coax new languages from this world of machines or to teach them laughter and tears. During the 1990s, this close mesh of media theory and artistic praxis led me to define two areas that, in my view, represented a pressing challenge:

• After a brief period of confusion and fierce competition between various systems of hardware and software, there emerged a strong trend toward standardization and uniformity among the competing electronic and digital technologies. The workings of this contradiction became abundantly clear to those involved with the new technical systems in the example of the international data networks. Telematic media were incorporated very quickly in the globalization strategies of transnational corporations and their political administrators and thus became extremely dependent on existing power structures. At the other end of the scale, there were individuals, or comparatively small groups, who projected great hopes onto these networks as a testing ground for cultural, artistic, and political models that would give greater prominence and weight to diversity and plurality. This goal of facilitating heterogeneity as before, or even developing it further with the aid of advanced media systems, was in direct contradiction to the trend toward universalization being demanded by the centers of technological and political power.

▪   As so often before, the tension between calculation and imagination, between certainty and unpredictability, proved to be an inexhaustible fount of discussion about cultural techniques and technological culture. It is a debate where no consensus is possible, and any dogmatic opting for one side or the other can lead only to stasis. However, it is possible to explore the options in experiments that are, in turn, a source of fresh insights. Radical experiments, which aim to push the limits of what can be formalized as far as possible in the direction of the incalculable and, vice versa, to assist the forces of imagination to penetrate the world of algorithms as far as is possible, are potentially invaluable for shedding light on a culture that is strongly influenced by media and for opening up new spaces for maneuvering. A most important arena where the two sides engaged, both theoretically and practically, proved to be a specific area of media praxis and theory, namely, the handling and design of the *interfaces* between artifacts and systems and their users. Cutting-edge media theory and praxis became action at the interface between media people and media machines.

My quest in researching the deep time of media constellations is not a contemplative retrospective nor an invitation to cultural pessimists to indulge in nostalgia. On the contrary, we shall encounter past situations where things and situations were still in a state of flux, where the options for development in various directions were still wide open, where the future was conceivable as holding multifarious possibilities of technical and cultural solutions for constructing media worlds. We shall encounter people who loved to experiment and take risks. In media, we move in the realm of illusions. Dietmar Kamper, philosopher and sociologist, used to insist in public debates that the verb *illudere* not only means to feign or simulate something, but also includes the sense of risking something, perhaps even one's own position or convictions: I think that this is of crucial importance for engaging with media.

If we are to learn from artists who have opted to play the risky game of seeking to sensitize us for the other through and with advanced technology, then gradually we must begin to turn around what is familiar. When the spaces for action become ever smaller for all that is unwieldy or does not entirely fit in, that is unfamiliar and foreign, then we must attempt to confront the possible with its own impossibilities, thus rendering it more inspiring and worth experiencing. We must also seek a reversal with respect to time, which—in an era characterized by high-speed technologies and their permeation of teaching, research, and design—has arguably become the most prized commodity of all. These excursions into the deep time of the media do not make any attempt to expand the present nor do they contain any plea for slowing the pace. The goal is to uncover dynamic moments in the media-archaeological record that abound and revel in heterogeneity and, in this way, to enter into a relationship of tension with various present-day moments, relativize them, and render them more decisive.

"Another place, another time"[14]—I developed an awareness of different periods that we often experience with regard to places: for example, to discover Kraków in Palermo, to come across Rome in New York, or to see cities like Prague, Florence, or Jena converge in Wrocław. At times, I was not certain where I actually was. Phases, moments, or periods that sported particular data as labels began to overlap in their meanings and valencies. Wasn't Petrograd's early techno-scene in the 1910s and 1920s more relevant and faster than that of London, Detroit, or Cologne at the turn of the last century? Did the Secret Academy in the heart of Naples necessarily have to be a sixteenth-century foundation, or wouldn't it have flourished better if founded under new conditions in the future? Don't we need more scientists with eyes as sharp as lynxes and hearing as acute as locusts, and more artists who are prepared to run risks instead of merely moderating social progress by using aesthetic devices?

Fortuitous Finds instead of Searching in Vain: Methodological Borrowings and Affinities for an Anarchaeology of Seeing and Hearing by Technical Means

> Satie bought seven identical velvet suits complete with matching
> hats that he wore uninterruptedly for seven years.
> —VOLTA, *ERIK SATIE*

On Things That Emit Their Own Light

Bioluminescence is a curious phenomenon: it is the ability of certain plants and animals, independent of all sources of artificial and natural light in their vicinity, to emit short flashes of light or to glow over a longer period of time without any increase in the organism's temperature. For this reason, it is also known as cold luminescence. Pliny the Elder was the first to approach it analytically in the first century A.D., and it has continued to fascinate scientists and philosophers of nature ever since. Although there are many intriguing speculations, thus far biological research has not offered a fully satisfactory theoretical explanation for the phenomenon of living organisms that emit their own light. It has been established that biochemical reactions are involved, oxidation processes. In order for organisms to bioluminesce, oxygen has to react with at least two groups of molecules, one of which are luciferins. These light-producing organic substances react very fast with oxygen and release energy in the form of photons. However, this process would be destructive for the luciferins—the molecules would immediately disappear after contact with oxygen and their power to emit light would be too weak to be visible—were it not for the presence of their catalyzing partner, luciferase. This enzyme coordinates the reaction of luciferin with oxygen so that a large number react at the same time and thus, in concert, produce light.[1]

In nature, bioluminescence has a number of different functions. Fireflies produce their soft intermittent light especially for the purpose of courtship, whereas certain species of fish use light to lure their prey. There is also the unicellular *Pyrocystis noctiluca,* one of a group of microscopic marine organisms, dinoflagellates, which belong to marine plankton. The action of luciferin and luciferase can generate many light flashes in their single cells. In warm and quiet summer weather, mass propagation, or blooms, of *P. noctiluca* can occur. Then, all the light flashes that they produce—only at night—are so strong that the sea glows. Although the cellular mechanisms are understood, little is known about why *P. noctiluca* puts on light shows. The same applies to the marine fireflies, which the Japanese call *umibotaru,* that are found in great numbers at the coastlines of their islands. The insects are only two or three millimeters long, yet they produce a strong blue light.

A favorite laboratory workhorse of marine biologists is the jellyfish *Aequorea Victoria,* a coelenterate, of which particularly good specimens are found in the deeper sections of the Bay of Naples at the foot of Vesuvius.[2] At the end of the twentieth century, Belgian scientists working on *A. victoria* discovered a new substance called coelenterazine, which is a submolecule of luciferin. Genetically, its function is twofold. First, it acts to guard the cell against superoxides and hydrogen peroxide, so-called free radicals. These molecules are so energetic that the slightest contact is sufficient to destroy the fragile double helices of DNA and cell membranes.[3] However, its role as protector against these dangerous invaders is not enough for the enterprising coelenterazine. It uses its considerable excess energy to produce aesthetic surplus value. In periods when their microworld is not under threat from any quarter, these submolecules of the luciferins enable the bioluminescing invertebrates in the darkness of the ocean to stage a quasi-poetic release of accumulated energy: a phenomenal economy of squanderous expenditure.

Georges Bataille understood his provocative "general economy" as a critique of the productivity mania of the capitalist system that, in principle, communism would also perpetuate. As an alternative to this paradigm, he proposes a truly luxurious concept of economy, formulated as a metaphor in *An Economy within the Framework of the Universe.* In Bataille's thinking, wealth is equated with energy—"Energy is the reason for and purpose of production"—and the issue is how surplus energy, which results from all production, is used. The purpose of a poetic form of expenditure, which he sees as a possible way out of the compulsion to accumulate, he describes in a comparison with the energy of the sun: "The Sun's rays, *which we are,* ultimately find nature and the meaning of the Sun again: it has to expend itself, *lose itself without calculating the cost.* A living system either grows, or it expends itself *for no reason.*"[4]

Physica Sacrorum

The anthropologist Gotthilf Heinrich von Schubert initially studied theology in Leipzig and later turned to science and theoretical and practical medicine in Jena before gaining his doctorate in medicine in 1803. His dissertation was entitled "On the Use of Galvanism to Treat Persons Born Deaf." He set up a general practice in the idyllic small town of Altenburg and, at first, flourished. When the paying patients stayed away, he turned to writing to make a living and, in a matter of weeks, produced a lengthy novel in two volumes, *Die Kirche und die Götter* [The Church and the Gods] (1804). A young physicist and expert on Galvanism, Johann Wilhelm Ritter, arranged for the work to be printed but pocketed Schubert's advance fee because he needed money urgently for his own experiments.[5] Schubert became the editor of the journal *Altenburger medizinische Annalen* but decided to return to university to qualify as a general science teacher. 1805 found him studying in Freiburg with Werner, a famous mineralogist and geologist of the period. The year after, he went to Dresden to complete his studies. While in Jena, Schubert had attended Friedrich W. J. Schelling's lectures, which at that time were a popular social event that provided the philosopher with a good supplementary income. Schubert was also keen to start teaching. The University of Jena invited him, in the winter term of 1807, to lecture to the "educated upper classes" on a subject that was "of highest general interest: on the expressions of inner mental life in specific states where the physical disposition is constrained, which are elicited by animal magnetism or manifested without it in dreams, in premonitions of the future, in mental visionings, etc."[6] In the spring of 1808, Schubert published these lectures under the title *Ansichten von der Nachtseite der Naturwissenschaft* [Views from the Night Side of the Natural Sciences].

In this way, Schubert wanted to draw people's attention to those natural phenomena that, as a rule, were excluded from close examination or analysis. However, "the Other" to which he refers is revealed in the course of his lecture texts as not so much a difference in the objects of his study (these belong to the standard repertoire of natural philosophy of the period) but rather as his development of a method, which characterizes the specific approach and perception of the investigator. Citing contemporary astronomers, Schubert defines the "night side" as "that half of a planet, which, as a result of it revolving on its own axis, is turned away from the Sun and, instead of being illuminated by the Sun's light, an infinite number of stars shine upon it." This phosphorescent light, which Schubert wanted to distinguish from the brilliant "rose-light" of the sun,[7] has the quality of "allowing us to see everything around us only in rather broad and large outlines." This light addresses, "with the particular terrors that attend it, above all that kindred part of our being, which exists in semi-dark feelings rather than clear and calm understanding; its shimmer always has something ambivalent and indefinable about it."[8]

Schubert was by no means an obscurantist or mystic, although he was often labeled as such in later years[9] and, for this reason, virtually banished from the history of science. After publishing the anthropological *Ahndungen einer allgemeinen Geschichte des Lebens* [Presages of a General History of Life] (1806–1807), Schubert wrote introductory texts on specialist fields of research, such as *Handbuch der Geognosie und Bergbaukunde* [Handbook of Geology and Mining] (1813) and *Handbuch zur Mineralogie* [Handbook of Mineralogy] (1816), and also lectured regularly on the history of the natural sciences and geology. In essence, however, he did not accept that any hard and fast divisions existed between different areas of intellectual activity. For Schubert, clear judgment and scientific analysis are just as capable of leading to understanding and expression as dreaming, somnambulism, clairvoyance, or ecstatic trance. These are merely different modes among which the pursuit of an understanding of nature alternates. He also wrote a book on the dark side of the psyche that was far ahead of its time: when Sigmund Freud's *Traumdeutung* [The Interpretation of Dreams] was first published in 1900, Schubert's *Symbolik des Traums* [The Symbolism of Dreams], with a section on "The Language of the Waking State," was being reprinted for the fifth time. The book was written in 1814. "The language of dreams,"[10] he was convinced, could be understood only within the context of its close relationship to mythology, poetry, and physical and mental experience of nature and natural bodies. On the relationship of sexuality, pain, and death, he writes: "This strange, close union appears to have been well understood by former ages, when they placed a phallus or its colossal symbol, the pyramid, on graves as a memento, or celebrated the secret rites of the God of Death by carrying a phallus in procession; although sacrificing to the instrument of carnal lust may have been the primitive expression of a different, deeper insight. In the

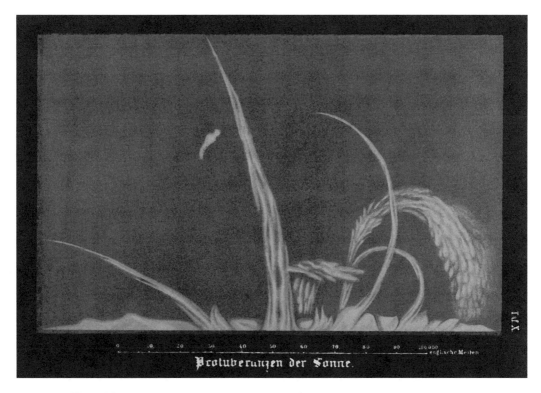

Figure 2.1 In astrophysics, protuberances are the masses of burning hydrogen, which flare up from the sun's surface at a speed of ca. 6 miles/sec and reach a height of up to 30,000 miles. Seen through a telescope, at the edges of this extravagantly wasteful star dynamic forms glow against the blackness of space: slender fountains, shapes reminiscent of plants. These phenomena can be observed especially well during a total eclipse of the sun, when the moon shuts out the light from the fiery ball. W. Denker drew this sketch to record his observations of the sun's eclipse in the summer of 1887.

midst of the obsequies and laments of the mysteries, there rang out the sound of . . . laughter."[11]

Schubert had planned to collect his many individual studies on anthropological themes into an all-embracing "physica sacra," or sacred physics,[12] but he did not manage to complete it, though he lived to be eighty. Nevertheless, this man, who had studied with Herder, Schelling, and Werner and was the close friend of the physico-chemist Ritter, at least came close to realizing his project of defining anthropology as a physics of the sacred, in fragments. His strange books and essays can be read as expressions of a single endeavor to write poetry specific to nature from the perspective of the latest scientific discoveries in the era of romanticism. The French translation of his lectures on the night side of the natural sciences was published under the title *Esprits des choses.* In the volume of Novalis's fragments entitled *Blütenstaub* [Pollen], the poet laments bitterly that in our ardent search for the absolute we find only things. Schubert had begun to turn around his contemporary's complaint in a direction that does not of necessity lead to despair. Untiringly, he sought the diversity of things and sometimes found in them the absolute, hidden or expressed in a language that we have yet to learn. Although this is a journey that can be full of tricks and difficulties, it does enable a passionate relationship with the world rather than one that is characterized chiefly by lamentations.

In the 1840 edition of his lectures, Schubert tucked away in the appendix cursory reflections on the progress he had made in his field. He compensates the reader for this disappointing brevity by adding a new preface. There he characterizes the thirty-year-old lectures as "tents" that have become riddled with holes and are now no more than "stopovers and resting places" during the brisk hike through "the vast area that the contemplation of nature represents," which is how he understands his own teaching and research. "The wanderer cannot have any possessions; if you own property, you are not free to wander," said Massimo Cacciari in his study of the philosopher of wandering, Edmond Jabès.[13] And Dietmar Kamper wrote at the end of his history of the imagination, "The true location where reflection takes place is no longer the writing desk or the professorial chair but while on the move, in time. Those who embark on such travels are not able to contribute much to the *state* of the art and they must needs develop a precarious relationship to knowledge as *property*. . . . The demand that is currently raised because of the contemporary level of complexity of social developments, namely, that any sociological theory must be able to apply the rules it establishes to itself, cannot be met with the mobility that sitting permits."[14]

Inverted Astronomy

In 1637, Athanasius Kircher was given the unexpected opportunity of going on a journey that was, for the period, a long one. At the time, he had a professorial post in Rome with a heavy workload and commitments. The Landgrave of Hesse-Darmstadt, who was going to Malta, invited Kircher to accompany him as his father confessor. Kircher accepted immediately, knowing that these light clerical duties would leave him ample time for studies and research. Malta interested him because of the fossils that had been found there and the opportunity for speleological expeditions. The island has many deep caves, which Kircher explored for their geology. When the Landgrave no longer required his services, Kircher fulfilled a long-standing private wish and, on his way back to Rome, visited southern Italy and Sicily. In the ancient ruins of Syracuse, he checked the legend of the listening system known as "the Ear of Dionysus," and was particularly keen to investigate the veracity of another legend. It was said that when the Roman army under Marcellus attacked Syracuse (214–212 B.C.), to defend the city, Archimedes set some of their galleys on fire with the aid of mirrors. All the foremost writers on theoretical optics, including Ibn al-Haytham, Roger Bacon, and Giovan Battista della Porta, had looked into this legend and confirmed its probable truth through calculations involving various mirrors and their focal points. Then, in 1637, Descartes in his *Dioptrique* flatly denied that the story had any basis in reality. Descartes's arguments were theoretical. Surprisingly, he linked them to his calculation of the sun's size in relation to the distance of its rays to Earth: a hundredfold focal length in relation to the radius of a mirror would not produce more heat at the focal point than the sun's rays would produce unaided by any reflecting mirror. Even a great number of mirrors would not make any difference; the temperature of the reflected sun rays would remain constant.[15] Kircher did not correct Descartes's position theoretically, but empirically and experimentally. He inspected the fortifications of Syracuse harbor, calculated the probable distance to the Roman galleys, and concluded that the distance was considerably less and, therefore, the focal length of the reflected sunrays would be much shorter, than commentators had previously assumed. Additionally, he experimented with different mirrors and proved that rays reflected by several mirrors and concentrated on the same point would indeed produce much more heat than one flat or parabolic mirror; moreover, they were capable of igniting wood.[16]

Kircher's main interest, however, centered on the volcanoes in the area: the geological triangle of Etna, Stromboli, and Vesuvius, which exerts such an overwhelming influence on the whole of southern Italy and the island of Sicily. He was convinced that there were subterranean connections between the three fire-spewing mountains. During his stay in Sicily, Kircher extensively studied Mount Etna, which had been active continuously since the end of 1634. From there, he made a trip to the Aeoliae Insulae, or Lipari Islands, where he explored both Vulcano and Stromboli. He planned to climb the volcano on Stromboli but was denied access for safety reasons.[17] On the way back to Rome from Messina, Kircher had planned to visit a number of Jesuit stations in Calabria before

Figure 2.2 Kircher's treatise on the legend of Archimedes of Syracuse and his burning mirrors. *Bottom right:* Kircher's diagram demonstrates the transmission of writing using a parabolic mirror. The device can be used both to destroy and to communicate; in this case it was used for destruction, but it could also have been used for prevention. (Kircher, *Ars magna lucis et umbrae,* 1671, p. 764).

traveling on to Naples, where he intended to study Vesuvius. However, the voyage turned out to be a nightmare experience that had a lasting impact on his thinking. Two results were his works *Iter extaticum II* [Ecstatic Journey], published in 1657 as a geological sequel to his fictitious journey into space of 1656, and the two-volume *Mundus subterraneus* [Subterranean World] in 1664–1665, in which the entire second chapter of the introduction is devoted to describing this journey.[18] The significance of the experience for Kircher can be gauged by the fact that this text appears again, word for word, in his autobiography.[19]

The journey began on March 24, 1638. The weather was unsettled but, initially, without particular incident. Three days into the voyage, however, heavy seas slowed progress considerably. Both Etna and Stromboli had begun to erupt, sending out massive clouds of smoke and ash, and in the north, Vesuvius had also become active. From port to port, the situation worsened. Wherever the ship put in, they were forced to leave again quickly because of violent earth tremors that sent parts of the coastline plunging into the sea, such as the clifftop village of St. Eusémia on the southwest Calabrian coast. This massive volcanic activity caused the sea's temperature to rise sharply; in places, it seemed to boil. Kircher described his situation in highly dramatic terms: "I was convinced that I had reached the end of my days and commended my soul to God unceasingly. Ah! In my distress, how contemptible all worldly pleasures seemed to me. Honour, high office, influential positions, learning—all these disappeared instantly at that time, like smoke or bubbles." His prayers were heard: miraculously, the party survived the eruptions and earthquakes of March and April 1638 and eventually reached Naples. The very same evening, Kircher engaged a guide, who needed considerable persuading and demanded a high fee, and climbed Vesuvius. He wanted to retrace the footsteps of Pliny the Elder (Secundus) and inspect the volcano at close quarters, but without sharing the same fate, for Pliny had died near there on August 24, 79 A.D., suffocated by Vesuvius's poisonous gases. On reaching the crater, Kircher was confronted by "a terrible sight. The eerie crater was entirely lit up by fire and gave off an unbearable smell of sulphur and pitch. It seemed as though Kircher had reached the abode of the underworld, the dwelling place of evil spirits." Nevertheless, his curiosity proved stronger than his fear. In the early hours of the next morning, he had himself let down on a rope to a rock ledge in the crater to examine the "underground workshop" at close quarters: "This wonderful natural phenomenon strengthened our conviction still further that the interior of the earth is in a molten state. Thus, we regard all volcanoes as mere safety valves for the subterranean fire source."[20]

In the foreword to *Mundus subterraneus,* Kircher notes with regret that there is a dearth of writing on such wonderful works of God that are hidden from the eyes of most people. It was his ambition to help remedy this state of affairs. For this reason, he had dared to take the steps necessary to research the Earth's interior. In the twelve books that comprise *Mundus subterraneus,* Kircher undertakes a colossal *tour d'horizon* of what he terms the "geocosmos," beginning with a geometrical and philosophical-theological concept of the gravitational center of the earth, which he calls "centrosophia." In the following twenty chapters he covers the composition of the earth, provides a special treatise on water with reflections on tides, and discusses meteorology, the roots of plants, minerals, and metals. In the final book, he gives a detailed account of alchemy, which finishes with a scathing critique of the forms that the Catholic Church had anathematized. However, the heart of the work is to be found in the fourth book of the first volume, where Kircher sets down his observations made at the volcano. In the Earth's interior, a fire burns at the center (*"ignis centralis"*), from which all things come and to which all return. This fire is usually hidden from view, "something truly wondrous, which seeks to emulate the Divinity (*"divinitatis aemulus"*) as it were, wherein the greatest almost coincides with the smallest, which joins together all radiant things into the diversity and variety of the whole world, absorbs everything into itself and knows it and develops everything, which is outside."[21] For Kircher, the fiery core of the earth has become *the* central phenomenon; it is to geology what the sun is to astronomy. The moon he assigns to water. The myriad forms of interplay between the two, the inner fire and water, give rise to everything that we call nature and life.

Baron Georg Philipp Friedrich von Hardenberg was also no stranger to the world below ground: he earned his living as an administrative assistant in the Saxony salt works. As a poet, he called himself Novalis. In chapter 5 of his unfinished novel *Heinrich von Ofterdingen (1802),* his alter ego in a twofold sense, Friedrich von Hohenzollern, who is an aristocrat and a miner, meets with a hermit in the course of his travels. At one point in their dialogue the Count says: "Our art rather requires us to familiarize ourselves closely with the earth; it is almost as though a subterranean fire drives the miner on." The hermit replies, "You are almost inverted astrologers. Astrologers observe the heavens and their immeasurable spaces; you turn your gaze toward the ground and explore its construction. They study the power and influence of the stars, and you examine the

Figure 2.3 Frontispiece of Athanasius Kircher's *Mundus subterraneus,* 1665.

Figure 2.4 *Top:* Two-page illustration at the end of the second book in the preface to Kircher's *Mundus subterraneus* (1665). For the engraving, a wash drawing was used of which Kircher had done most himself (Strasser 1982, p. 364). The original gives a stronger impression than this reproduction of the drastic impact that climbing Mount Vesuvius had on Kircher. Out of the black interior of the volcano, deep red and sulfurous yellow flames leap high into the sky. At the top, they become white, then dirty gray smoke. *Bottom:* A similar, not quite so expressive drawing of Mount Etna follows page 186 in Book 7. The drawing is based on Kircher's observations in 1637. Morello (2001) includes color reproductions of this phylum of illustrations.

powers of the rocks and mountains and the many and diverse actions of soil and rock strata. For astrologers the heavens are the book of the future, whereas the earth shows you monuments of the primeval world."[22]

"Mittel und Meere"

The writer and literary critic Édouard Glissant from Martinique believes that European intellectuals all suffer from a fundamental problem. The lands, which have been constituent for their identity, are all grouped around a single great sea that lies at the center, exuding warmth and light, promising leisure and happiness. Since classical antiquity, all desires and movements have been directed toward this center, which has also been the driver of conquest. It is from the greater Mediterranean area that all technical inventions and all scientific, philosophical, aesthetic, and political models have come, which continue to influence our culture through the present day.[23] The compelling need to construct universal worldviews and theories, which have had devastating effects in our history, can only be understood with this in mind: one sea in the middle, one God, one ideology, one truth, which must be binding for all. The old empires, such as ancient Greece and the Imperium Romanum, and the various forms of colonialism must be understood in the light of this central perspective. The entire gamut of social models, theories, and worldviews that seek to universalize have arisen from this notion of the center: the modern nation-state and democracy, capitalism and communism, Christendom, the notion of the world as a harmonic organism or as a single gigantic mechanism. In late medieval times and the Renaissance, with courageous thinkers like Raimundus Lullus from Majorca, the Englishman Roger Bacon, or the later proponents of a magical conception of nature, whose ideas ran at odds to conventional wisdom, there existed theoretically a chance of a radical new departure. However, the compulsion to standardize thought that was exerted by the Catholic Church discriminated against these men and others like them, which made it impossible to realize any alternatives. As Édouard Glissant writes, "What the West will spread around the world, what it will force upon it, are not heresies but systems of thought. . . . After thinking in systems has triumphed, the Universal—initially as Christian and later as rational—will spread and represent the special achievement of the West."[24]

According to Glissant, such compulsion to establish the principle of universalization would be unthinkable for the inhabitants of the Caribbean. They do not live on territory that is enclosed but on fragments of land separated by the waters of the Atlantic Ocean and Caribbean Sea. The absence of something that could unify the islands and their peoples is not felt to be a lack. On the contrary, the only unifying, or standardizing, factor they have ever experienced is an invisible trace running along the sea floor—the chains of the slave trade. The cultural and economic activities of the islanders are characterized by instituting flexible relations between the land fragments. Attempts to impose universalization via the language of the colonizers the islanders have countered with creolization, in which the semantics of French, for example, is fractured and subverted through introduction of the speakers' own rhythms and rule-breaking syntax. Their musical expression is song with highly disparate voices. By contrast, the European invention of polyphony is "the uniform and complete dissolution of all differences in tone and voice for these are viewed as being inadequately distinctive in themselves."[25]

Rather than be defined by "identity machines," Glissant opts for the potential power of a "poetry of relationships." For Glissant, magic and poetry are inherently similar and are extensions of creolization and heresies; they are forces that work against globalization's abolition of potent diversity: "Only heresy keeps the cry of what is special going forcefully, the accumulation of non-reducible differences, and, ultimately, the obsession not to understand the unknown in order to generalize it in formulas and systems."[26] A poet, playwright, novelist, and critic from Martinique, Glissant teaches in New York but lives mainly in Paris, where he attended university. The main thrust of his critique is directed toward the entirety of European thought, which has given rise to its hegemonial position in the West and Northern Hemisphere. His ideas link him with the work of all thinkers, particularly French intellectuals, who, during the last century of uniformities and terrible destruction, did not abandon the attempt to give all that is heterologous a chance: Georges Bataille, Maurice Blanchot, Gilles Deleuze, Jacques Derrida, Michel Foucault. As an answer to the strategy of globalization, Glissant introduces the concept of *mondialité,* in which the players come from the periphery, the niches, and the margins of the territories of the world powers: "Those who are gathered here, always come from 'over there,' from faraway, and they have decided to bring their uncertain knowledge, which they acquired There, to Here." By concretizing the type of knowledge that he is concerned about, Glissant takes up one of the most fruitful thoughts from Derrida's *Grammatologie* [Of Grammatology]: "Fragmentary knowledge is not mandatory science. We sense things, we follow a trail."[27]

The idea is enticing: to see the activity of tracking as something that defies all systematic order. However, trails are not simple phenomena. They are impregnations of events and movements, and even prehistoric hunter-gatherers needed to learn much in order to decode, read, and classify the signs.[28] The same applies to an even greater degree when we consider history, with its evolved and constructed civilizations, and particularly the history of the media. What can be found there, analogous to spores, broken twigs, feces, or lost fur and feathers, was produced entirely by cultural and technical means. By seeking, collecting, and sorting, the archaeologist attaches meanings; and these meanings may be entirely different from the ones the objects had originally. The paradox that arises when engaged in this work is that one is dependent upon the instruments of cultural techniques for ordering and classifying, while, at the same time, one's goal is to respect diversity and specialness. The only resolution of this dilemma is to reject the notion that this work is ground-breaking: to renounce power, which one could easily grasp, is much more difficult than to attain a position where it is possible to wield it.[29]

Reality as a Mere Shadow of What Is Possible

The concept of *archaiologia,* stories from history, comprises not only the old, the original (*archaios*), but also the act of governing, of ruling (*archein*) and its substantive *archos* (leader). A *narchos* is the *nomen agentis* to *archein,* and it means "the absence of a leader," also "the lack of restraint or discipline."[30] Discussing Foucault's concept of an archaeology of knowledge, Rudi Visker used the term "anarchéologie" more than ten years ago to describe a method that evades the potential of identifying a "standardized object of an original experience."[31] A history that entails envisioning, listening, and the art of combining by using technical devices, which privileges a sense of their multifarious possibilities over their realities in the form of products, cannot be written with avant-gardist pretensions or with a mindset of leading the way. Such a history must reserve the option to gallop off at a tangent, to be wildly enthusiastic, and, at the same time, to criticize what needs to be criticized. This method describes a pattern of searching, and delights in any gifts of true surprises. In his critique of Hitler's brand of fascism, Bertolt Brecht frequently pointed out that order is a sign of lack, not of abundance. This idea does not apply only to the extreme sociopolitical situation under fascism. For example, the most exciting libraries are those with such abundant resources that it is impossible to organize them without employing armies of staff who would ultimately engineer the loss of this cornucopia. The London Library in St. James Square, founded in 1841 as a private club, is such a library. There, you are less likely to find the book you have long been looking for without success and more likely, in the course of your explorations of the labyrinthine gangways with their floors of iron gratings, to chance upon a book that you did not even know existed and that is of far greater value than the one you were actually looking for. Of far greater value, because your find opens up other paths and vistas that you did not even entertain during your focused search. This is a possible course to take: within a clearly defined context, the unsuccessful search for something is balanced by a fortuitous find, and this discovery is acknowledged as a possibility of equal worth. One simply has to try it out. However, it must be stressed that this method has absolutely nothing to do with aimless wandering and meandering.

In the first volume of his epic novel *The Man without Qualities,* Robert Musil wrote:

To get through open doors successfully, it is necessary to respect the fact that they have solid frames. This principle, by which the old professor had always lived, is simply a requisite of the sense of reality. However, if there is a sense of reality—and no one doubts its justification for existing—then there must also be something we might call a sense of possibility.

Whoever has it does not say, for example, this or that has happened, will happen, or must happen here; instead, they invent: this or that might, could, or ought to happen in this case. If they are told that something is the way it is, they think: Well, it could just as well be otherwise. Thus, the sense of possibility can be defined as the ability to conceive of everything there might be just as well and to attach no more importance to what is than to what is not.[32]

In his posthumously published *Notes on Philosophy,* Wittgenstein—a contemporary of Musil and, like him, a trained engineer—states that "one of the most deeply rooted errors of philosophy" is that it understands possibility as a "shadow of reality."[33] For the people, ideas, concepts, and models that I encountered in the course of this anarchaeological search trajectory, this view is reversed: their place of abode is the possible, and reality, which has actually happened, becomes a shadow by comparison.

Duration and Moment

"Who owns the world?" This was the provocative question asked by the many activists fighting for a better life for the majority after World War I. Bertolt Brecht asked the same question and included it in the title of the film *Kuhle Wampe,* which he made in 1932 with Slatan Dudow. The question refers to rights over property and territory in the broad sense—ownership of factories, machines, land, even entire countries or continents. It still needs asking today; however, another question is gradually taking over, which will be decisive in the coming decades: Who owns time?[34] Between the beginning of the twentieth century and the beginning of the twenty-first century, there was a marked shift in the quality of political and economic power relations that both involved the media and drove their development: away from rights of disposal over territories and toward rights of disposal over time; less with regard to quantity, and more in connection with refining its structure, rhythm, and the design of its intensity. This shift is not immediately apparent in global relationships, but if one scrutinizes the microstructures of the most technologically advanced nations and their corporations, it is quite apparent.

Karl Marx wrote for posterity. Thanks to his care in citing sources, the remark of an anonymous contemporary (author of a pamphlet) is recorded in his collected works, who, by succinctly summing up his own notion of economy, formulated what later became the touchstone of Marx's critique of the established bourgeois economic system: "A nation is really rich only if no interest is paid for the use of capital; when only six hours instead of twelve hours are worked. . . . "Wealth . . . is *disposable time,* and nothing more."[35] At this historical juncture where time has been declared the most important resource for the economy, technology, and art, we should not pay so much attention to how much or how little time we have. Rather, we should take heed of who or what has power of disposal over our time and the time of others, and in what way. The only efficacious remedy for a melancholy and resigned attitude toward the world is to appropriate, or reappropriate, the power of disposal over the time that life and art need. Only then is the *future* conceivable at all—as a permanent thing of impossibility.

In Greek mythology, *Kronos* stands for duration, time's expanse, which disposes life by using it up. This is the time of history. Chronology fits us into the temporal order of things. Suffering can be chronic, but passion never is. Chronology cripples us because we are not made of enduring stuff and we shall pass. Machines live longer. At the end of the twentieth century, the computer scientist and engineer Danny Hillis, who was one of the codevelopers of the massively parallel architecture of today's supercomputers, presented prototypes of a clock that was to start running in early 2001 and keep time for the next ten thousand years.[36] A group of technology enthusiasts, who call themselves The Long Now Foundation, have ambitions of a time-ecological nature. In reality, the proponents of these ideas merely reveal themselves as infinitely presumptuous: the now, the present, is to be extended far into the future and thus, by implication, preserved for posterity. The idea of preserving the minds of contemporary mortals in artificial and everlasting neural networks for future generations is another example of these rather obscene ideas.

The ancient Greeks understood only too well the dilemma resulting from chronology as the dominant time mode. They attempted to solve it by introducing two more gods of time, *Aion* and *Kairos,* conceived of as antipodes to powerful Kronos, who ultimately devoured his own children. Aion shines at the transcendental dimensions: time that stretches far, far beyond the life span of humans and planet Earth; pure time, like that of machines; or, the fastest way from zero to infinity, as the avant-gard playwright and director Alfred Jarry once defined God. Aion's time is time that we can reckon with. By contrast, Kairos's time is doing the right thing at the right moment: he is the god of the auspicious moment, who in the Greek myth can also prove fatal. He does nothing for us; he challenges us to make a decision. On some ancient reliefs, copies of Lysippus's statues, Kairos is depicted balancing the blade of a knife on his fingertips.[37] The front half of his head is covered in long wavy locks; the back is bald. Once Kairos has passed by, it is too late. One may still be able to catch up again with the unique moment from behind, but from this position, it is no longer possible to seize hold of it. When an opportunity comes along, one must recognize it as auspicious and take it.

Just such a character is the observer in chaos-theoretician Otto E. Roessler's endophysics, which Roessler understands as the physics of the *Now* and which I try to comprehend as the physics of uniqueness. As an actor in the world, Roessler's observer is an activist, not the distanced observer of traditional physics. This observer follows dynamic processes with great presence of mind and visualizes their change from one quality into another. This observer has only the one chance. He or she has absolutely no access to the world's totality and experiences it only in the form of an interface, via which he or she can know and shape it—for example, by simulating the world in computer models. Due to his association with making decisions, the turning-point character of Kairos is also expressed in Greek in the adverb *harmoi* (at this precise time, at the appropriate time), a word that was rarely used. The noun form, *harmós,* means "seam, slit, or joint," and the verb *harmótto* means, among other things, "to submit or comply."[38]

As an activist in the world, the endophysical observer is confronted with two options: contribute to the world's destruction or, for fleeting moments, help to transform it into paradise.[39] This is also the world of media and the art that is produced with and through them. All techniques for reproducing existing worlds and artificially creating new ones are, in a specific sense, time media. Photography froze the time that passed by the camera into a two-dimensional still, not into a moment, for a moment possesses a temporal range that is not calculable. Telegraphy shrank the time that was needed for information to bridge great distances to little more than an instant. Telephony complemented telegraphy with vocal exchanges in real time. The phonograph and records rendered time permanently available in the form of sound recordings. The motion-picture camera presented the illusion of being able to see the bodies in motion that photography had captured as stills. In film, time that had passed technically was rendered repeatable at will; the arrow of time of an event or process could be reversed, stretches of time that had become visual information could be layered, expanded, or speeded-up. Electromechanical television combined all these concepts in a new medium, and electronic television went one step further. Von Braun's cathode ray tube inscribed images dot by dot and line by line. In the electronic camera, a microelement of the image became a unit of time, which in turn could be manipulated. In electromagnetic recordings of image and sound elements, what can be seen and heard can be stored or processed in the smallest particles or in large packages. Cutting, pasting, and replacing, basically invented by the first avant-garde at the beginning of the twentieth century, became advanced cultural techniques.[40] Computers represented a more refined and more effective intervention in time structures, as well as—like television—the synthesis of various existing technologies in a monomedium. In the Internet, all earlier media exist side by side. They also continue to exist independently of the networked machines and programs and, from time to time, come into contact with each other.

For the anarchaeological approach, taking account of the specific character of media with regard to time has two important consequences. The first I touched upon above in relation to the concept of deep time. The field of study cannot encompass the entire process of development; exploring different historical epochs has the aim of allowing qualitative turning points within the development process to emerge clearly. The historical windows that I have selected should be understood as attractive foci, where possible directions for development were tried out and paradigm shifts took place. Changes like these have an ambivalent significance. On the one hand, they support and accelerate economic, political, or desired ideological processes, and on the other, they exclude other alternatives or relegate them to the margins of what is possible. The second consequence involves a heightened alertness to ideas, concepts, and events that can potentially enrich our notions for developing the time arts. Such ideas do not appear frequently, but they are among the most fortunate finds in this quest. They appear in the guise of shifts, as wholly different from the states of inertia or complacency. To cite another idea from Roessler's endophysical universe: the cut through the world, which enables it to be experienced, is similar to Heraclitus's lightning flash, which is the agent of change—often of change that is initially imperceptible. Here the similarity to the concept of *difference,* introduced by Derrida to characterize the linguistic and philosophical operation, is obvious.[41]

In Praise of Curiosities

What media could or might be was defined so often in the course of the 1990s that it is no longer clear what this word, used as a concept, actually describes.[42] This inflation of definitions has to do with the fact that the economic and political powers took the media more and more seriously, and thus the definers found themselves under increasing pressure. *Media* and *future* became synonymous. If you didn't engage with what was then baptized *media,* you were definitely passé. By adding media to their curriculum, institutes, faculties, academies, and universities all hoped to gain access to more staff and new equipment. In the majority of cases, they actually received it—particularly after, in association with the magic word *digital,* media systems were established that the decision makers did not understand. This was another reason they called the process a revolution. The digital became analogous to the alchemists' formula for gold, and it was endowed with infinite powers of transformation. All things digital promised to those who already possessed wealth and power more of the same and, to those who possessed nothing, that they could share in this unbloody revolution without getting their hands dirty. Governments and administrations opened their coffers when the magic word—even better if coupled with the menetekel *Internet*—appeared in grant applications.

In this manner, a shift in focus took place among literary researchers, sociologists, art historians, philosophers, political scientists, psychologists, and also certain "hard" scientists. Over and above studies in their immediate field of research, they increasingly began to develop concepts for media and, in this way, tried to demonstrate to the education policy makers that in fact they were the best in the field of media studies and the right address for competency in media questions. However, the media makers and players continued to concentrate on the business of making money and were not interested in any academic enhancement, or critique, of their praxis.

I write of this in the past tense because I am convinced that this process belonged to the last century, a century that needed media like no other before. It was a century that spawned so many violent caesuras, so much destruction, and so many artificial, that is, humanmade, catastrophes. The twenty-first century will not have the same craving for media. As a matter of course, they will be a part of everyday life, like the railways in the nineteenth century or the introduction of electricity into private households in the twentieth. Thus, it is all the more urgent to undertake field research on the constellations that obtained before media became established as a general phenomenon, when concepts of standardization were apparent but not yet firmly entrenched. This undertaking may be of some help to those who have not given up on Rimbaud's plan to steal the fire and reinvent the worlds of texts, sounds, images, and apparatus each day anew.

My archaeology makes a plea to keep the concept of media as wide open as possible. The case of media is similar to Roessler the endophysicist's relation to consciousness: we swim in it like the fish in the ocean, it is essential for us, and for this reason it is ultimately inaccessible to us. All we can do is to make certain cuts across it to gain operational access. These cuts can be defined as built constructs; in the case of media, as interfaces, devices, programs, technical systems, networks, and media forms of expression and realization, such as film, video, machine installations, books, or websites. We find them located between the one and the other, between the technology and its users, different places and times. In this in-between realm, media process, model, standardize, symbolize, transform, structure, expand, combine, and link. This they perform with the aid of symbols that can be accessed by the human senses: numbers, images, texts, sounds, designs, and choreography. Media worlds are phenomena of the relational. The one or the other may be just as plausible from the way the objects are looked at as the bridges and boundaries that have been constructed between or around them. However, it is not my intention to place a limit on the multitude of possible linkages by pinning them down.

Descartes came in for a lot of criticism because, in his philosophical endeavor to bring more clarity into the world of thought, he made an essential distinction between extension and the indivisible, between substance and spirit. However, Descartes never suggested that there were no connections between the two. He merely said that these connections were not accessible to his system of philosophical thinking in concepts. They belong to other realms, primarily that of experience and that is where he, as a philosopher, will leave them. Gottfried Wilhelm Leibniz, who was both a sharp critic of the Cartesian system and the one to bring it to completion, also returns to this division in his *Monadology*, even going so far as to quantify those parts that are not accessible to philosophical rationalism: "in three-quarters of our actions we are merely empiricists."[43] By not attempting in any way to standardize the found heterogeneous phenomena of the in-between, which play a part in media archaeology, I follow the idea of a tension between a reality that is filed away in concepts and a reality that is experienced. This notion of tension is also understood here, as in the relationship between calculation and imagination, as not opting a priori for one side or the other. At times, it is appropriate to use arguments that generalize, for example, when addressing artifacts or systems from the familiar canon of media history. However, in the course of our journey to visit the attractions, a certain something must be evoked, a sense of what might be termed *media* or *medium* in the various constellations that I describe. Whether it succeeds in this for the reader is the decisive question for the value of my study. It is not a philosophical study—this anarchaeology of media is a collection of curiosities. Slightly disreputable then as now, the word was used by Descartes (who had certainly read his Lullus and Porta)[44] to refer to those areas of knowledge treated in the appendix to his *Discours:* optics, geometry, and meteors.

By curiosities, I mean finds from the rich history of seeing, hearing, and combining using technical means: things in which something sparks or glitters—their bioluminescence—and also points beyond the meaning or function of their immediate context of origin. It is in this sense that I refer to attractions, sensations, events, or phenomena that create a stir and draw our attention; these demand to be portrayed in such a way that their potential to stimulate can develop and flourish. The finds must be approached with respect, care, and goodwill, not disparaged or marginalized. My "deep time" of media is written in a spirit of praise and commendation,[45] not of critique. I am aware that this represents a break with the "proper" approach to history that I was taught at university. At center stage, I shall put people and their works; I shall, on occasion, wander off but always remain close to them. It does not bother me that this type of historiography may be criticized as romantic. We who have chosen to teach, research, and write all have our heroes and heroines. They are not necessarily the teachers who taught us or the masters they followed. The people I am concerned with here are people imbued with an enduring something that interests us

Figure 2.5 The citizens of Syracuse do not appear to care whether the legend of Archimedes' setting fire to Roman galleys with parabolic mirrors is feasible according to the laws of physics and geometry. They erected this monument to their inventive defender at the city's gate. The postcard was printed in Milan.

passionately. I have by no means made a random selection; their work in reflection and experiment in the broad field of media has had enduring, rather than ephemeral, effects.

Empedocles is visited for his early heuristics of the interface, and his expansive and broad-minded approach accompanies us as an inspiration throughout the entire story. Giovan Battista della Porta worked at a time when extremely divergent forces—the beginnings of a new scientific worldview and the traditions of magical and alchemistic experiments with nature—still collided with full momentum. The intellectual openness of certain individuals came into severe conflict with power structures that tried to intervene and regulate free, sometimes delirious, thought. In this constellation, there arose a micro-universe of media concepts and models of the most heterogeneous nature that is without parallel in history. In Robert Fludd's musical monochord, calculation and imagination meet in a special way. His mega-instrument could also be interpreted as an early device of standardization. The tracking movement of our quest leads from Fludd to Athanasius Kircher, whose view of the world is encoded in a strict binary fashion. Kircher's media world is an all-embracing attempt to pacify bipolar opposites in a third. This experiment took place within a network that had powerful ambitions for worldwide expansion, yet, at the same time, the Jesuit's sheer boundless imagination of media evaded being confined through functionalization by the institutions of the Catholic Church. The next chapter focuses on the physicist Johann Wilhelm Ritter, who declared his own body to be a laboratory and a medium, in which he intended to prove experimentally that electrical polarity pervades nature. For many years, Ritter was classified as a romantic natural scientist, but here I focus on him as an indefatigable champion of an artistic and scientific praxis that understands itself as art within time. Joseph Chudy and Jan Evangelista Purkyně [Purkinje] accompany him: the Hungarian as a piano virtuoso who discovered the keyboard as an *interface* for an audiovisual telegraph that worked on the basis of binary codes; the Bohemian doctor and physiologist, who in his research on vision shifted attention away from the representation of external factors to internal ones, including neurological processes, and investigated basic effects for media machines of moving images. The introduction to this section presents the invention of an electrical machine for transmitting written messages over distances in the 1760s at the Jesuit Collegium Romanum in Rome. The development of the media in the nineteenth century has been relatively well researched. Here, with the Italian doctor and psychiatrist Cesare Lombroso, a pivotal figure in a twofold sense, it is again the subject of inquiry. Lombroso carries to the utmost extremes the strategies and methods of measuring and media techniques as an apparatus for providing true representations. Moreover, his argumentation availed itself of media forms that the nineteenth century appeared to have left way behind it. With Aleksej Gastev we reach the first decades of the twentieth century. His ideas of an economy of time, which derive from a binary code of all mechanical operations, also open up the perspective that leads to the twenty-first century.

For the anarchaeologist's quest, mobility is essential. My research entailed traveling to places that seemed to me, schooled as I am in critique of the hegemonial aspect in media history of industrial culture, very remote indeed. I visited all the places where the heroes of my anarchaeology labored. Agrigento, where Empedocles lived, I left rather quickly because, as the administrative center for the valley of ancient temples, it did not seem to have much in common any longer with the place that I had found in his texts. From Catania, I circled (and ascended) Mount Etna and then went on to Syracuse, following in the footsteps of Kircher and Empedocles. The latter I encountered again in Palermo where he has given his name to the gallery of modern art and to myriad other facets of everyday life in the city, like the neon sign of a bar. He is revered there like a Sicilian freedom fighter. In Palermo I also came across completely unexpected presences from the past: Tadeus Kantor's death and love machines in the museum of marionettes that have been so influential in the history of theater and animation; a dilapidated institute for research on human physiology; the Gemellaro Museum of paleontology, whose treasures are lovingly displayed in one cramped little room. After Palermo, I retraced Kircher's movements on his journey through southern Italy, which had inspired him to write of his "subterranean world." His investigations of this world ended in Naples and Vesuvius, also my next port of call: the city of della Porta, where he wrote his *Magia naturalis;* beloved of Goethe, Crowley, Benjamin, Sartre, Pasolini, and Beuys; a city that so many of the masters visited at least once. The Biblioteca Nazionale there proved to be a real treasure trove. To my amazement, I even found works by the English Rosicrucian Robert Fludd and was allowed to turn the pages myself, without wearing white cotton gloves or having any strict supervision. In the winter, in cold and incessant rain, I visited the Jesuits' power center—the Collegium Romanum and the surrounding area in Rome, where Kircher did most of his writing and research—the Roman police's criminological museum, and the main Jesuit church Il Gesù. My movements ended for the time being in Riga, where once Sergei Eisenstein's father had built elegant Jugendstil houses and Aleksej Gastev had published his last book of poems before devoting himself wholeheartedly to the Russian "Time League." Between the stations of Rome and Riga lay many others: Warsaw, Wrocław, Budapest, St. Petersburg, Prague, Weimar, and smaller towns, whose significance will become apparent in the course of my narrative. In this way, a map, a cartography of technical visioning, listening, and—in addition to my original plan—combining came into existence, which is so very different from the geography of media that we are familiar with. It runs through the propositions I advance in the final chapter.

The mythical hero with the gaze that controls is Argus, whose name derives from the Latin *arguere* (to prove, to illuminate). He is the all-seeing one with one hundred eyes, of which only a few ever rest; the others move continually, vigilantly watching and observing. The goddess Hera set Argus to guard her beautiful priestess Io, who was one of Zeus' beloved. Supervision is the gaze that can contain envy, hate, and jealousy. Argus was killed by Hermes, son of Zeus, who made him the messenger of the gods. Soon after his birth, Hermes invented the lyre by stretching strings over a tortoiseshell. The ancient Greeks venerated Hermes for his cunning, inventiveness, and exceptional powers of oratory, but also for his agility and mobility. He was given winged sandals and became the god of traffic and travel, of traders and thieves. Because he could send people to sleep with his caduceus, his wand with serpents twined about it, he was also revered as the god of sleep and dreams. Hermes defies simple definition, as does the slippery field of media. In one of the magnificent frontispieces of his books, Kircher honors him with a special meaning: as god of "the fortuitous find."[46]

Attraction and Repulsion: Empedocles

Pleasure and absence of pleasure are the criteria of what is profitable and what is not.
— DEMOCRITUS[1]

At the beginning of the 1990s, two classical scholars, Alain Martin from Belgium and Oliver Primavesi from Germany, were engaged on an extraordinary project of discovery and decryption. The National Library in Strasbourg had granted the papyrologist Martin permission to select one papyrus for analysis and publication from a collection of around 2,200 unclassified papyri. A combination of excellent knowledge of the characteristics of ancient papyrus documents and intuition led Martin to select two glass frames, which belonged together, containing fifty-two fragments of a papyrus "written in beautiful literary script."[2] Using photographic reproductions of the fragments, Martin spent several years putting the pieces of the puzzle together again. With the help of a computer, he compared the text particles over and over again with ancient Greek texts of known authorship and, in this way, identified the fragments as part of a longer text by Empedocles.[3] Working with Primavesi, a philosopher and authority on Empedocles, he managed to decipher the entire text fragment, which took another three years. In 1997, they presented the results of their labors in Agrigento, Sicily. Because we know the work of the so-called pre-Socratic philosophers primarily through indirect transmission—passages quoted or paraphrased by later authors—the identification of this fragment as being a direct transmission of pieces of Empedoclean text was a tremendous discovery. In 1904 or 1905, an archaeologist representing the Berlin "papyrus cartel" (formed to prevent German museums from bidding against each other when purchasing objects abroad) bought the fragments for £1 sterling from an Egyptian dealer in antiquities. Possibly their significance would have been recognized much earlier had they remained in Berlin where the classical scholar Hermann Diels, an eminent authority on the pre-Socratics, worked. However, the cartel's procedure of distributing acquisitions by drawing lots resulted in the fragments' going to Strasbourg, the capital of Alsace-Lorraine, which belonged to Germany at that time. There, it was carefully preserved but its significance remained undiscovered for nearly ninety years.

For Primavesi, the fragment's content was as spectacular as its discovery. In his view, the fragment demands a radical reappraisal of previous scholarship on Empedocles. In the tradition of Aristotelian interpretation, until now the work of this poet-philosopher has been divided into two areas: his didactic poem on nature *Peri physeos,* and the poem *Karthamoi* [Purifications], which is concerned with the human soul. Primavesi writes, "The papyrus demonstrates that this approach was in error—the physics of the four 'roots' on the one side, and crime and punishment of the soul daimon on the other, are so closely intertwined in the new text that these must be seen as integrated elements of one and the same unified theory."[4]

The twentieth century was a period of disunity, of terrible explosions, murderous political systems, and violent splits, punctuated by phases of economic and cultural prosperity. At the end of the century, we were inundated with concepts of artificial bonding, unifying, and reuniting, as though by way of a conciliatory gesture. Universal machines, globalization, and technological networking of geographical regions and identities that are in reality divided were advanced to counter the de facto divisions that have intruded between individuals and between people and machines because of the unequal distribution of wealth, education, culture, and knowledge. In no way did they serve to diminish the real divisions; they merely created the impression that the real gulfs were easy to bridge using market strategies and technology. At the beginning of the twenty-first century, the situation has escalated again. People who had nothing apart from their bodies, their pride, their ideas of redemption, and their hate used these bodies as weapons against others who have everything but their bodies, pride, and ideas of liberation. These unequal opponents, however, do have something in common: feelings of hatred.

In the sixth and fifth centuries B.C., the region where Empedocles lived and worked was wealthy and prosperous. Not surprisingly, it was a prize fought over by many different invaders. Situated between the territories of Asia Minor, North Africa, and the mainland of Europe, it experienced rapid transitions— from periods of rich prosperity to military campaigns of destruction. From this extraordinary region bordered by the Ionian Sea, which was a kind of dividing line, an interface, between the spheres of influence of the great powers of the age, came a host of exceptional thinkers: Heraclitus of Ephesus, Parmenides of Elea, Anaxagoras of Clazomenae, Democritus of Abdera, and Empedocles of Acragas. Acragas—in Latin, Agrigentum; today called Agrigento—was on the south coast of Sicily, the southernmost outpost of ancient Greek civilization, which faced Carthage in North Africa across the sea. The inhabitants of his city were a rare mix of many cultures. "The men of Agrigentum devote themselves wholly to luxury as if they were to die tomorrow," Empedocles said of his fellow towns-folk, according to Diogenes Laertius, "but they furnish their houses as if they were to live forever."[5] Today Kairos and Kronos are reversed: we construct build-ings that will be ruins in a few decades, or even years, and then demolished; what has become chronic now is *fun,* which has nothing to do with joy, for fun does not require a reason or an occasion.[6]

The German poet Hölderlin despaired while trying to bring together dis-parate things that were poles apart, both in his poetry and in his life. His *Der Tod des Empedokles* presents the drama of a man who was a tragic failure, an Icarus who soared toward the light but flew too near to the sun, which melted the wax that held his wings together. In this moment of failure, Hölderlin's Empedocles plunges into the volcano—a fallen angel, an errant daimon—where, finally, he becomes one with the element that fascinated him the most: fire. For me, it is not important whether there is a grain of truth in this legend or any of the others about the death of the poet-philosopher. What interests me most about Empedocles' fate are his sandals, which, it is claimed, were found at the foot of Mount Etna. They bear witness to his specific and dogged kind of resistance, to the stubbornness of things when confronted by attempts to monopolize and de-stroy them, including historical attempts to interpret them. And more than Empedocles' death, I am interested in the life of this "pilot," as Panthea calls him in Hölderlin's tragedy fragment,[7] and what has survived of his thought, which is also rendered in fascinating verse by Hölderlin. In my understanding, Empe-docles' philosophy is definitely not a concept of failure, but a worldview oriented toward succeeding, precisely because it is aware of the possibility of failure.

At first glance, it may appear somewhat redundant in the age of unlimited reproducibility of things and organisms to study the ideas of a philosopher who formulated his doctrine in fine hexametric poetry two-and-a-half-thousand years ago. Further, it may seem rather anachronistic to place this discussion at the beginning of a quest to examine the relationship of humans and machines from a specific perspective. Yet, at the end of the 1940s, at approximately the same time as Alan Turing was writing his famous essay on intelligent machinery and Norbert Wiener was publishing his book on the reciprocal relationship between control and communication using cybernetics, the eminent physicist Erwin Schroedinger gave a series of lectures in Dublin and London on the relationship of the ancient Greeks to nature, in which he declared the atomist Democritus as his hero. At the time, Schroedinger viewed his own subject, theoretical physics, as in deep crisis, triggered by the theory of relativity, quantum mechanics, the growing strength of biology, plus the historical experience of World War II's destructive violence and force, of which the natural sciences had been co-organizers. For these reasons, Schroedinger thought it appropriate to revisit the origins of systematic thinking about nature. Thus, he took up a committed position that objected strongly to an erroneous understanding of the Enlightenment. Schroedinger cited the opposing position, advanced by the Austrian physicist Ernst Mach, who had claimed in one of his popular lectures that "our culture has gradually acquired full independence, soaring far above that of antiquity. It is following an entirely *new* trend. It centres around mathematical and scientific enlightenment. The traces of ancient ideas, still lingering in philosophy, jurisprudence, art and science constitute impediments rather than assets, and will come to be untenable in the long run in the face of the development of our own views."[8] The "supercilious crudeness" of this view Schroedinger countered by arguing for a reorientation backwards in time toward those points in the history of human thought when the divisions that inform the modern scientific view of nature did not yet exist. Dangerous misconceptions cannot arise "from people knowing too much—but from people believing that they know a good deal more than they do."[9] For his act of backtracking, Schroedinger found a delightful metaphor: "We look back along the wall: could we not pull it down, has it always been there? As we scan its windings over hills and vales back in history we behold a land far, far, away at a space of over two thousand years back, where the wall flattens and disappears and the path was not yet split, but was only *one*. Some of us deem it worth while to walk back and see what can be learnt from the alluring primeval unity."[10]

The suggestion is not to attempt a real or imaginary return to the times before the great divisions came about; obviously, this is not feasible for the public or the private sphere. However, it makes good sense to rethink and re-examine the constellations of the period, which were clearly highly conducive to bold and free thought, in spite of the conflicts of the powerful that dominated everyday life. "It was certainly not the numbers or concentration of socially secure educated persons that was the decisive factor," writes Otto Roessler in a text about Anaxagoras, the founder of chaos theory, who was only a few years older than Empedocles. "It was a wave and general mood of courage and freedom from anxiety. The social trend toward consolidation was outshone, for a time, by the intellectual expansive impetus of the few."[11]

Empedocles was already a legend in his lifetime, and many popular traditions are attached to his figure. Little is known for certain, however. Diodorus of Ephesus writes of his appearance: "swathed in a purple robe, his long flowing hair decorated with garlands and wreaths . . . shod in iron, he walked about the cities with a serious and stern countenance, accompanied by a retinue of slaves."[12] His skills as a physician earned him the reputation of a miracle worker, and he was accredited with a magical relationship with nature. The inhabitants of Selinus worshipped him almost as a god because, at his own expense, he had constructed channels to divert the water from two neighboring rivers into the city's marshy and polluted watercourse. Besides stopping the spread of the plague, this intervention also provided Selinus with wholesome fresh water. As a thinker allied with the Pythagorean tradition, he was greatly involved with music, which he invested with healing powers and is reported to have utilized in therapy. Empedocles was above all a public figure. As a "passionate lover of freedom and suppressor of tyranny,"[13] he was committed to the democratization of the Greek cities in Sicily. Promoting conciliation, he intervened often in the struggle between Syracuse and Agrigento for domination of the island, and he championed the idea of Sicilian political unity. He refused, however, to assume any political office. It is said that he desired to exert influence by virtue of his reputation and not through exercise of power.

Just as Empedocles' political thought is governed by the idea of a peaceful reconciliation of opposites, he developed his concept of the physical world as an attempt to combine incompatible positions. For the older philosopher Parmenides, whose teachings Empedocles studied, "what is" is eternal, uncreated, imperishable, and encapsulated in a homogeneous sphere. On the other hand, Anaxagoras explained all things in the world of phenomena through the principle of mixing: all natural things come into being and pass away through a continual process of mixing elemental substances in varying proportions.

Figure 3.1 Fritz Kahn's five-volume *Das Leben des Menschen* [The Life of Man] is an outstanding example of depicting the human organism as a mechanical system. In this illustration of optical perception, the most important nineteenth-century machines of acceleration — the clock and the train — are brought together with electric warning signals, the objects being perceived here. (Kahn, vol. 4, 1929, plate XXII)

Empedocles attempted to combine these two disparate ideas. His concept of nature is informed by three principles. First, he attributes the plurality of "what is" to four "roots" or elements: fire, earth, water, and air. He also calls them "root clumps," which can be translated as *rhizomata*.[14] All matter is composed of these elements in varying proportions. The second principle concerns the mode of how the composition comes about. For Empedocles, there is no beginning or end to all that is, and therefore neither creation nor destruction. Some thing cannot arise out of no thing, nor can something become nothing. Like Anaxagoras, he conceives all natural processes as types of mixing. The four elements correspond to the properties hot, dry, wet, and cold. These four operate in all existing things and organisms; later, this concept became a basic principle of chemistry. "From them comes all that was and is and will be hereafter—trees have sprung from them, and men and women, and animals and birds and water-nourished fish, and long-lived gods too, highest in honour. For these are the only real things, and as they run through each other they assume different shapes, for the mixing interchanges them."[15] Empedocles does not appear to make any clear distinction between the different kinds of natural life; all are animate and exhibit many similarities. For example, he sees plants as being highly sensitive and having many analogies with humans and animals: leaves are analogous to feathers, hair, or scales.[16] He calls plants nature's "embryos" because they unite both sexes within themselves and are able to propagate without exchanging secretions.

The third principle pervades Empedocles' entire doctrine regarding nature; it is what made his thought so exciting for Plato, Plotinus, and the Neoplatonic philosophers and, later, the magical natural philosophers of the fifteenth and sixteenth centuries. It is probably also what prompted Aristotle to make his dismissive characterization of the poet-philosopher as "stammering, suspecting the truth, but unable to express it in the language of philosophy."[17] The forces that drive the mixing of the elements are attraction and repulsion, or, as Empedocles formulates it in his poetry, Love and Strife. These forces generate all motion. Translated into terms employed by modern science, we speak today of energy, and, with reference to the elements, of matter. With the interplay of energy and matter, which is governed by affinities among the elements, we have arrived at the paradigm, which is regarded today, in both physics and chemistry, as fundamental to the analysis of natural phenomena at the macro- and micro-scale. In Empedocles' cosmology, the degree to which Love or Strife dominates determines the structure of the universe and defines the relationship between the center and the periphery. The ideal form is the dominion of Love. When Love is at the center and commands all motion, the mixtures are distributed equally, the "many" come together into the "one." This is *sphairos,* the state of stillness, peace, and happiness. Its form, the ball or sphere, was also the shape of Parmenides' Entity, the best possible form of what is. However, unlike Parmenides, Empedocles does not view this state as eternal or unchanging; it is subject to constant motion, is in a continual state of flux. When Strife enters the calm and peace in the sphere, there is separation into the "many" (that is, new mixtures) through its agency, and Love migrates toward the outer edge of the circle. Perhaps the language Empedocles uses to describe this state reflects something of his own fate and that of many of his intellectual contemporaries, like Anaxagoras, who were forced into exile. Love is banished to the outer limits of the chaos that is besieging Love.[18] From the periphery, Love then begins the second half of the cycle anew by advancing again; then mixing takes place under its increasing power.

Within this highly flexible framework of the constant motion of elements and their infinite mingling, there is embedded a concept of perception of the one by the other. Empedocles does not make a principal distinction between understanding and sensory perception: both are equal, natural processes. "Fortunate spring-time of the spirit, when Reason still dreamed, and the Dream still thought; when knowledge and poetry were still the two wings of human wisdom."[19] Similarly, the idea of separating all that happens or acts into subjective and objective was foreign to his thought. Empedocles did not see an active agent on the one side, primarily concerned with enjoyment and causing suffering, and a passive body on the other, which mainly suffers and endures: to him, both are active. "Being" in the context of this dynamic mixing process means that there is constant interchange between the one and the other. In order for the others and the other to be active, Empedocles presents all living things with a wonderful gift. He wraps them in a fine skin, or film, which not only protects them but is also permeable in both directions. This is effected by the skin's fine, invisible pores, which have different shapes. Passing back and forth through them is a constant stream of effluences that are not directed at anyone or anything in particular. If there is antipathy, the streams do not meet. When there is sympathy between the one and the other, there is reciprocal contact and they can "pick up" the effluences of each other, which join successfully to create a sensation. In order for this to take place, the requisite pores must correspond in size and shape; there is "symmetry of the pores, each particular object of sense being adapted to some sense [organ]." The senses differ; their pores are different in size and shape, and "we perceive by a fitting into the pores of each sense. So they are not able to discern one another's objects, for the pores of some are too wide and of others too narrow for the object of sensation, so that some things go right through untouched, and others are unable to enter completely."[20]

For Empedocles, the eyes are Aphrodite's work. In the extant fragments, the example of the eyes illustrates most clearly what he means by the work of Love as an essential component for successful perception. One of the finest surviving fragments gives a poetic (and accurate) description of the structure of the eye: "As when a man who intends to make a journey prepares a light for himself, a flame of fire burning through a wintry night; he fits linen screens against all the winds which break the blast of the winds as they blow, but the light that is more diffuse leaps through, and shines across the threshold with unfailing beams. In the same way the elemental fire, wrapped in membranes and delicate tissues, was then concealed in the round pupil—these kept back the surrounding deep water, but let through the more diffuse light."[21]

Here, Empedocles combines poetically the anatomical components of the eye—retina, pupil, vitreous humor—with the most important factor for perceiving the other: the notion of perception as a process of continual flow presupposes the existence of a rich, burning energy within that is inexhaustible. The same holds true for acoustic perception. For Empedocles, hearing is a sensation that takes place inside the ear, at the threshold to the outside world. He describes hearing entirely in physiological terms. The perception of sounds stems from the sounds heard within "when (the air) is set in motion by a sound, there is an echo within." The auricle is "a sprig of flesh." Empedocles likens the hearing organ to the resonating body of a bell, which produces the same sounds within as the noises produced outside by the sounds of things and living creatures. To listen is to hear in sympathy; for Empedocles, this presupposes inner motion: "the impact of wind on the cartilage of the ear, which . . . is hung up inside the ear so as to swing and be struck after the manner of a bell."[22]

Empedocles does not think of the infinite multiplicity of things in terms of any hierarchical order. Nothing is above anything else; everything exists side-by-side, in motion, and with constant interpenetration. Nor does Empedocles propose a hierarchy of the senses; Plato and Aristotle will introduce this idea later. Seeing is not privileged over hearing, nor taste over touch and smell. This latter example provides a further illustration that the Agrigentine poet-philosopher understands perception as an active process: he ascribes "the keenest smell" to

Figure 3.2 "The retina, greatly magnified" (Kahn, *Das Leben des Menschen,* vol. 5, 1931, plate VII). From the top down, the illustration shows component layers of the eye, from the "vitreous humour in front of the retina (Gl.)" and "control cells (Sch.), which connect the optic nerve cells to the vision cells," the "cones" (Z.) and "rods" (Stä.).

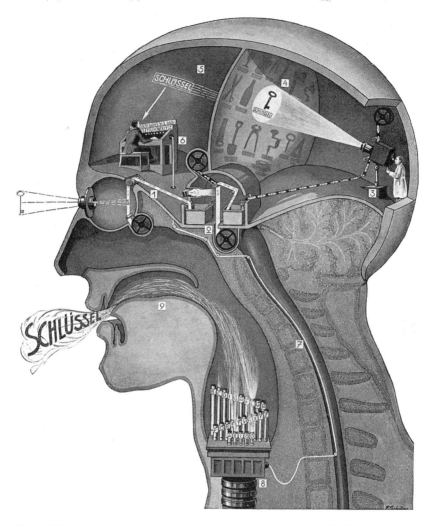

Figure 3.3 The act of seeing and the articulation of what is seen as a functional cycle comprised entirely of media technology: "The image of the key [*Schlüssel*] passes through the eye's lens system to the light-sensitive retina at the back of the eye where it is exposed. . . ." (Kahn, *Das Leben des Menschen,* vol. 4, 1929, plate VIII).

those "whose breath moves most quickly; and the strongest odour arises as an effluence from fine and light bodies."[23] This idea may also represent a transition to the realm of mind, perhaps referring to the *nous* of Anaxagoras, a fine and light substance of which pure mind is composed and which is the driving force behind everything, for "all the old words for soul originally meant air or breath."[24]

Empedocles' theory of pores is a doctrine that does not ascribe a privileged place to humankind. The principles of constantly changing combination and exchange apply to all natural phenomena, including inorganic ones. Just as Empedocles sees vision in organisms with eyes as presupposing an inner fire as the driving force, he also sees this process at work in rocks and metals. He explains the formation of the Earth's crust by volcanism: boulders, rocks, and cliffs are "lifted up and sharpened by the many fires that burn beneath the surface of the Earth,"[25] a process that will continue for as long as the fire burns. He explains reflection as dependent upon fire in a specific way (here it must be remembered that in Empedocles' day, mirrors were made of polished metal, often copper). Reflection occurs because the inner fire of the metal heats the air on the mirror's surface, and the effluences, which stream onto it, become visible. He ascribes special powers to stones, which he also conceives of as wrapped in a porous skin. Magnetism appears to be an impressive confirmation of this conception: the attraction exerted by amber on iron functions because it draws the effluences through the metal's pores toward itself, and the iron follows.[26]

As far as one can judge from the extant fragments and later sources, Empedocles does not appear to have proposed that a third natural or artificial entity should be interposed between the porous skins of the organs of the one and the other during the process of perception. For him, perception comes about when exchange of effluents takes place. The philosopher and student of nature Democritus, who developed his ideas at around the same time as Empedocles but in another, distant city, Abdera in Ionia,[27] suggested a different conception: he gave the effluences a structure and attempted to explain their inner relationship. Democritus conceived of the world as consisting of two opposing entities, which need each other: fullness and emptiness. Fullness is not solid but consists of a multitude of the smallest units, which Democritus named atoms. So small that the human eye cannot see them, atoms are elementary substances composed of the same material but with an infinite number of different sizes and shapes. Because they are in a state of perpetual motion, they need space, or the "void." As substances that cannot be subdivided further, atoms are impenetrable. In their eternal random motion, they collide and move in different di-

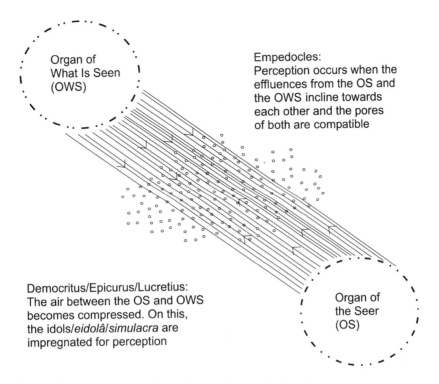

Organ of
What Is Seen
(OWS)

Empedocles:
Perception occurs when the
effluences from the OS and
the OWS incline towards
each other and the pores
of both are compatible

Democritus/Epicurus/Lucretius:
The air between the OS and OWS
becomes compressed. On this,
the idols/*eidolâ*/*simulacra* are
impregnated for perception

Organ of
the Seer
(OS)

Figure 3.4 Empedocles' pore theory of perception as applied to vision.

rections. Everything that exists is composed of these multifarious forms in motion, including the human sensory organs. Thus, Democritus expands Empedocles' theory of pores in two ways: first by introducing a medium, the void, wherein the various configurations can arise, and second by suggesting a concrete—in his terms, material—in-between. The streams that emanate on the one side from the perceiver and on the other from what is perceived compress the air between them. The various constellations of atoms in motion are impregnated on the air and appear there as "idols" (*eidolâ*), images of real objects, which are identified by the sensory organs as different configurations. According to Democritus's theory, perception also arises through successful exchange, via the idols, between the organs and what they encounter. This exchange should be imagined as a kind of balancing, a reciprocal scanning of the many forms via the intervening layer of compressed air, which has the status of an interface between the perceiver and the perceived.

One aspect of Democritus's vision of the universe represents a considerable shift compared to that of Empedocles: in introducing the idea of images in the compressed air, Democritus raises a question that does not occur to the Agrigentine (with the caveat, that is, as far as we know from the extant fragments). It is the issue of whether the idol that appears on the compressed air is true or false. The associations and connotations of the Greek word *eidolon* range from "knowing, recognizing, seeing, and appearance" to "shadow" and "illusion." For Democritus, perception that takes place is not necessarily true. Of the things that are, only the atoms in motion and the void are true—the material elements and the medium: "Sweet exists by convention, bitter by convention, color by convention; atoms and void [alone] exist in reality."[28] Just as the constellations of things that can be perceived change constantly through perpetual motion and collisions, so too do the organs of perception. They are not a consistent and reliable reality but, instead, permanently changing states. Fragment 100 of Democritus's texts, reported by Diogenes Laertius, puts the epistemological crux of the theory in a nutshell: "In reality, we know nothing, for the truth lies in the abyss."[29]

A great number of later thinkers found enormously convincing the idea that emanations from the changing atomic constellations appear on the compressed air as images and are scanned by the sensory organs. In the first century A.D., Lucretius included a paean to atomist philosophy in his poem *De rerum natura* [On the Nature of Things]. There he accentuates the Latin word for *eidolâ,* which came to have central importance for the postmodern discourse on images, located between *Schein* and *Sein* (appearance and reality): "nam si abest quod ames, praesto *simulacra* tamen sunt" [Though she thou lovest now be far away, yet idol-images of her are near].[30] Lucretius has no doubt that the simulacra, of which there are an infinite number and variety, are true in principle. His only reservations in a Democritean sense concern the occasions when the scanning conception of vision clashes with the conceptions of reason.

In his book on classical theories of vision, Gérard Simon elegantly describes the questions debated by the early philosophers when they turned their attention to perception, the complex relations between seeing and what is seen. His critical rereading of the surviving text fragments of the early natural philosophers led him to the following conclusion: the "beam of vision," that fascinating phenomenon referred to so often by the "old geometers" and geometrized by Euclid, should not be understood as a physical quantity. Their object of study was not light and its radiation, but vision. From a science-historical perspective, therefore, the field of classical texts does not belong to physics, nor mathematics and geometry, but rather to the field of a "theory of the soul." The classical philosophers' inquiry was articulated as questions about "the seeing human, his/her relationship to what is visible."[31] While it is certainly correct to insist that we do not do violence to the classical theories by applying our modern categories to them, Simon's unequivocal determination of the competent discipline, which he undertakes in the terminology of modern science, appears to be self-contradictory. Observations of nature, mind, and the soul, as well as the mathematical calculations made by the early philosophers, cannot be separated. Their conception of physiology encompassed it all.[32] This approach had dramatic consequences for Democritus's theory of atomism, for he also applied it to the soul.[33] With the exception of Epicurus, later philosophers did not share this view; particularly, Plato and Aristotle found that it went too far, and they condemned it. Later, the Catholic Church joined in the censure of atomism. They needed the soul as an authority external to matter and the human body, controlled by free will but at the same time in a complex relation of dependence upon Divine Providence and its institutions on Earth. Within such a system of atoms in motion, be it ever so complex, the Fall from Grace is an impossibility; at best, there are only catastrophes for which no one is responsible. Democritus's vision was not taken up again until the magical natural philosophers of the fifteenth and sixteenth centuries had the courage to do so. And in the hearts of the early romantics of the eighteenth century, both Democritus's and Empedocles' unity of nature and soul begin once again to pulse with energy.

The two ancient Greece specialists Alain Martin and Oliver Primavesi have put the jigsaw puzzle of papyrus pieces together to reveal a fragment that, more than two thousand years after the ideas of Empedocles originated, will occasion extensive reinterpretation of the little we know of his work. My montage of text fragments bears absolutely no comparison to their labors. What I have tried to show is how one can arrange some of the extant text particles of Empedocles and Democritus on perception to extract ideas and statements that have some bearing upon the frenetic contemporary sphere of activity that is theory and praxis of media: the interface between the one and the other, which can be defined as the interface between media people and media machines.

Empedocles' theory of pores is a theory of perception both in the simplest and deepest form conceivable. Interpreted technologically, it is a theory of double compatibility: size and relative power of the pores and effluents must match so

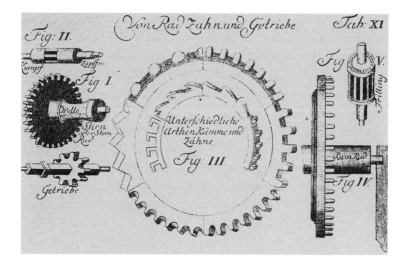

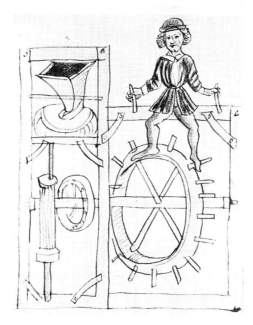

Figure 3.5 In interface theory, engineers distinguish between user-machine interface and machine-machine interface. Beginning in 1724, Jacob Leupold published an eight-volume work, *Theatrum machinarum,* with a total of 1,764 pages on the classical interface of the latter type. *Top:* Illustration of different types of cogwheels, which must mesh exactly in the perfect mechanical interface. Leupold comments: "Wheels and gears are artificial equipment that is most admirable because using only a few wheels and gears, according to the nature of the work, these can be accommodated within a small space and performance can not only be greatly enhanced but, because the motion is continuous, repetition is unnecessary, unlike with levers." *Bottom:* This treadmill, dating from 1430, is a hybrid. Here, the function of the second, compatible cogwheel is taken over by human muscle power. (Mattschoss 1940, pp. 34 and 17)

that exchange can take place. Physically, it is a theory of affinities, which can be described in psychological terms as a concept of reciprocal giving and receiving of attention. Economically, it is a theory of extravagance. In media-heuristics terms, which draw the above aspects together, it is eminently suitable as a theory of a perfect *interface.* Yet, because it is perfect, building it will never be possible. However, precisely because it possesses this potential impossibility, the theory is entirely worthy of consideration for dealing with existing interfaces, which purport to have already established compatibility between the one and the other.

In actual fact, Empedocles' theory of pores renders the construction of any interface superfluous. The porous skins are ubiquitous; they are a material element of all things and people and thus move with them. Every person and every thing has received this gift. Democritus introduced a medium, and thus a third quantity, wherein one can contemplate the "idols," or *simulacra,* including their truth. With Democritus, though, one can imagine that, in the future, more artificial interfaces will have to be constructed in order to bridge the chasm that currently exists between being and appearance.

When Empedocles describes the republic in which he would like to live, in which Love is dominant over Strife, he becomes passionately enthusiastic. It is, in Bruno Schulz's sense, a true republic of dreams. Its beautiful queen is Kypris, to whom lavish gifts are brought: "Her they propitiated with holy images and painted animal figures, with perfumes of subtle fragrance and offerings of distilled myrrh and sweet-smelling frankincense, and pouring on the earth libations of golden honey."[34] Among the few surviving fragments of the *Purifications* that are instructions to his followers is the shortest and most effective formula for a philosophy of succeeding: "Refrain entirely from partaking of the food of woe!"[35]

4

Magic and Experiment: Giovan Battista della Porta

A discovery is premature if its implications cannot be connected by a series of simple logical steps to canonical, or generally accepted, knowledge.

—OLIVER SACKS, "SCOTOMA: FORGETTING AND NEGLECT IN SCIENCE"

Working Untiringly on the World's Multifariousness

The fourteenth book of Giovan Battista della Porta's *Magia naturalis*[1] deals with the good things in life, mainly wine and cooking. The interested reader learns how to induce fowl to lay an egg as large as a fist, which dishes and beverages will drive away undesirable bugs from the table, how to make guests get drunk fast, or how to sober them up again. Chapter 9 contains a recipe for which even Salvador Dali, egomaniacal eccentric and esoteric, was moved to express utmost admiration. As a child, Dali had wanted to be a cook, and he is reputed to have attempted this dish several times.[2] The recipe gives detailed instructions for roasting a goose, a favorite fowl of ancient Roman cuisine, alive. The goose is plucked, apart from the head and neck, and placed at the center of a ring of fires so that it cannot escape. During the cooking process, the bird's head must be kept damp with cold water and its body basted. The goose must be kept supplied with salt water to drink and a mixture of herbs that acts as a laxative, for a goose cooked with full bowels does not taste good. Porta emphasizes that he reports only experiments that he has tried out himself, witnessed as an observer, or has had described to him by absolutely reliable sources. He had cooked this dish for friends; however, they were so ravenous that they pounced on the goose before it was properly cooked.[3]

It sounds paradoxical, but this gruesome cooking recipe expresses in an exemplary fashion Porta's attitude to the world, things, and nature: it is characterized by respect and affection. In everything around him, he discovers marvels that must be tracked down and celebrated. His observations and analyses of living phenomena as well as his physical interventions all have the goal of upholding their attraction and, if possible, enhancing it. From a Pythagorean standpoint, which forbids the eating of anything that has a face, to cook an animal alive is immoral and a crime. However, the most immediate exchange of life for life is also the highest form of dining, prized above all other things, for example, in Japanese cuisine. The paradigm of freshness means precisely this: the difference in time between the preparation of food and its consumption must be kept as small as possible and, at the same time, the boundary between the two is dramatized. The fast cut with a very sharp knife that kills a fish for sashimi, for example, differs only temporally from the bird roasted alive. The latter is a cuisine indebted to Kronos, the god of duration, and the former pays tribute to Kairos, a celebration of the unique moment when one quality changes into another. The relationship of the Japanese to fish can be understood only if one remembers that the inhabitants of those slim incrustations in the ocean are permanently confronted by death—in the ocean depths, from the rumbling volcanoes, or from the earthquakes that periodically shake their islands.

"Even the dull animals of the mainland become weird beasts here," Walter Benjamin wrote of Naples in the section on eating houses in his *Denkbilder* [Thought Figures]: "On the fourth or fifth floors of the tenements, they keep cows. The animals are never taken out and their hooves have grown so long that they can no longer stand."[4] Porta was a Neapolitan. He is said to have been born in Vico Equense, twelve miles to the south of the city, but for the greater part of his life he lived in the port by Vesuvius.[5] The people there are proud of him and he was always proud of his city. Vesuvius, which has impressed so many generations of travelers to Italy, dominates the Gulf of Naples, "this radiant and pleasantly articulated bay";[6] when the weather is bad, it looms threateningly over the landscape. One can look up and see Vesuvius while poring over the magnificent folios and rare manuscripts from the sixteenth and seventeenth centuries in the rare books section of the Biblioteca Nazionale with its splendid but delapidated baroque interior. No other city lives so categorically for the moment as Naples; there is no other place where the quick succession and chaotic whirl of moments seem so precious. A poet, privy councillor, and naturalist from the German province of Thuringia wandering through the streets of Naples for the first time on March 17, 1787 felt the same way: "To walk through such an enormous and restlessly moving crowd is most curious and salutary. How they all mix and mingle, and yet each finds their way and destination!" Two days later, Goethe noted: "One only has to walk along the streets with an open eye to see the most inimitable sights."[7] As if wishing to concur wholeheartedly with Goethe, the young Jean-Paul Sartre wrote an empathic account of his visit in 1936 with classic descriptions of situations, things, and people:

. . . in Naples, chance reigns supreme and its effects are everywhere—from the inspired to the horrible: on Sunday, I encountered a girl walking in the blazing sun. The left side of her face was screwed up against the blinding light. Her left eye was closed and her mouth twisted, but the right side of her face was absolutely immobile and looked dead. Her right eye, wide open, completely blue, completely transparent, sparkled and glittered like a diamond, reflecting the sunbeams with the same non-human indifference as a mirror or a window-pane. It was quite awful but also strangely beautiful—her right eye was made of glass. Only in Naples does chance manage to accomplish this: a dirty girl, blinded and dazzled, with a glittering mineral existing within her poor flesh, almost as though her eye had been torn out deliberately in order to adorn it more splendidly.[8]

Pier Paolo Pasolini, who was originally from Bologna (a city he once described in "Letter to the young Neapolitan Genariello" as "so big and fat" that it could just as easily be a German or French town),[9] liked to compare the denizens of Naples with a tribe of Indians camping defiantly in the middle of a city, who would rather die than submit to the powers that be. It was in Naples that Pasolini made the first film in his trilogy about life, *Il Decamerone* (1970). The city mourned him as one of her own when he was clubbed to death on the night of All Saints' Day, November 1–2, 1975—possibly (the case remains unsolved) by one or more of the *ragazzi di vita,* "the boy prostitutes whom he had portrayed with such loving care."[10]

For Sicilians, Palermo is the Italian city that lies closest to Africa and Naples is the farthest. North of Naples is barbarian territory, which has little or nothing to do with Italy; it is where the exploiters who profit from the south live.

Vesuvius, the volcano that buried Pompeii and Herculaneum under masses of lava and ash, earth tremors, and earthquakes are just the natural catastrophes

Figure 4.1 This Latin edition of *Magia naturalis,* which Goethe worked with in Weimar, was published by Gulielmo Rouillio in Venice three years after the first edition of 1558. The inside cover is richly decorated in black, red, and gold. (Photo: Sigrid Geske)

that permanently threaten Naples. In Porta's time, not only was Naples subjected to the humiliating dictates of ecclesiastical Rome but, from the early sixteenth century, Naples suffered with Sicily for two hundred years under Spanish rule, represented by two viceroys, in The Kingdom of the Two Sicilies.[11] In the heart of this southern city, which grew at an amazing pace in the sixteenth century to become the largest and most densely populated city of what would later be Italy, the spread of venereal disease was equally swift. Indeed, since

1495 syphilis was an ever-present scourge,[12] and epidemics regularly afflicted Neapolitans. In the mid-seventeenth century, these developments peaked in disaster: in 1656, 60 percent of the population died of the plague. In view of the catastrophic hygienic conditions prevailing in the greater part of the city, it is hardly surprising that Porta devotes much attention to the manufacture of sweet-smelling substances in the first edition of *Magia naturalis* (1558) and, in the expanded edition, adds an entire book on perfumes, "De myropoeia." Besides being an early contribution to sexual osphresiology,[13] this text represents just one of many facets within Porta's wide-ranging oeuvre that reflect the close relationship between magical modeling of nature and technology and proactive theory and praxis that aims to heal, not to destroy.

Porta was not an academic in any conventional sense. From the viewpoint of media archaeology, in this respect he is in excellent company, particularly with regard to the twentieth century. Many of the seminal texts that have influenced media theory and studies profoundly were not authored by academics spinning their thoughts on comfortable professorial chairs. Dziga Vertov, for example, developed his radical theory of the all-pervasive "Kino-Eye" as a professional filmmaker. Bertolt Brecht's text fragments that are apostrophized as the first theory of radio are concepts developed by a dramatist and early experimenter with radio plays. Later, the poet, essayist, journalist, translator, dramatist, editor, and publisher Hans Magnus Enzensberger took up these ideas in his *Baukasten zu einer Theorie der Medien* [Construction Kit for a Theory of the Media]. Walter Benjamin wrote his famous essay, "The Work of Art in the Age of Mechanical Reproduction," as an independent scholar and professional writer. In the 1950s, Günther Anders published his provocative ideas on *The Outdatedness of Human Beings* as a freelance writer and activist in the antinuclear movement after returning from exile in the United States, where he had earned his living in soul-destroying industrial jobs. The two most important texts for what came to be known as apparatus theory, which focus on technological and psychoanalytical aspects of film and the media and are today rather ignored (unjustly, to my mind), were written by Jean-Louis Baudry, a Parisian dentist and novelist.[14] Thus, it is clear that not only the media apparatus is a phenomenon of interposition, of the "in-between"; the most fruitful media discourses also move freely between disciplines. Mobility and the state of being in-between are here of equal importance.

Porta was not a disciplined thinker, either by past or present standards. As far as we know, he acquired his wide-ranging fundament of knowledge in his maternal uncle's excellent library, a rich store of books and curiosities that Porta refers to as his "museo" in his book on human physiognomy. This uncle introduced his young nephew to the texts of the ancient Greeks and instructed him in experimental laboratory work. According to Porta's own testimony in later editions of *Magia naturalis,* he performed his first experiments—described in *Magia* I—at the age of fifteen. All his life Porta pursued his studies as a self-taught man. He was proud of being a free spirit, untrammeled by any affiliation to institutions or constraints of a personal nature.[15] He managed to maintain his independence even when the financial situation of his aristocratic family became increasingly straitened as a result of their support for the princes of Salerno over the Spanish viceroy. Porta was obliged to seek paid employment, and earned his living as a doctor, engineer, bookkeeper, astrologer, writer, and winegrower. As a youth, he and his two brothers were accepted at the famous Pythagoras school in Naples, even though the study of music, including playing instruments, was a compulsory part of the curriculum. This "trio of tone-deaf young musicians"[16] won over the school's directors by virtue of their lively intelligence, their unbounded curiosity, and above all their knowledge of mathematics. At that time, music, specifically harmonics, was regarded as the handmaid of mathematics. One of their masters, Domenico Pizzimenti, was the translator of Democritus, whose theory of atoms exerted a profound influence on Porta's thought and, much later, on the physicist Erwin Schroedinger, who declared it "the most advanced epistemological approach" of all the ancient philosophers.[17]

Porta's biographers unanimously testify to the young Neapolitan's anti-authoritarian character, which did not defer to classical authorities of philosophy and science. His greatest strength was a lively "speculative mind that gave small credence to the precepts of the masters unless he had been given tried evidence of their veracity."[18] With certain predecessors, such as Roger Bacon, Ramon Llull, and particularly Giordano Bruno, Porta shared a critical attitude toward the symbiosis of Aristotelian natural philosophy and Christian scholastic dogma as advanced by Thomas Aquinas. Porta's polemic against the self-styled high priests of an abstract truth that was entirely lacking in passion was sharpest in his plays. In the prologue of one of his early comedies, *Duo fratelli rivali* [Two Rival Brothers] of 1601, he writes: "Come hither, Doctor of necessity, you who have failed to devise a law even with six links of the chain; you who claim to know all the sciences, although you know nothing of yourself." Nor did Porta mince words when dealing with his detractors. In the Italian language translation of *Magia naturalis,* he calls one of his English critics a "barbaro Inglese."[19] Sadly, the English barbarian is not mentioned by name. When reading Porta's texts, one cannot fail but get a strong impression of just how much the unwieldy new Latin, the lingua franca of both secular and religious intellectuals, constrained and hampered his expression. In the dialogues of his plays, Porta comes across as a true man of the spoken word and less as one of discursive texts. He had a deep aversion to the obscure, complacent, and exclusive language traditionally used by scholars, a feeling he shared with other contemporaries who were seeking new explanations of the world and its phenomena.[20] Bruno, for example, conceived and wrote his highly polemical *De gli eroici furori* [The Heroic Frenzies] in Italian; it was written in London and published in 1584 to 1585.[21] Thus the intellectual revolution of premodern times, which also involved a fracturing of established conventions of language, received a tremendous boost from the invention of the printing press, which made it possible to produce a great number of copies of any work.

When existing structures hamper and constrain the mind and thought, one must invent new ones or change the old. Porta founded one academy himself and played an important role in another. He named his own society Accademia dei Segreti (or Academia secretorum naturae), the Academy of (Natural) Secrets. It met at a building in the Via Toledo at the Piazza de Carità, at the corner where the once magnificent boulevard today forks with the narrow Via Pignasecca. The academy's premises were likely identical with Porta's living quarters, laboratory, and library. The Accademia dei Segreti is considered the first modern scientific society primarily dedicated to experiment.[22] Aspiring members had to fulfill only one condition before being admitted to the pursuit of study and experiment: they had to have discovered something new about the world and be prepared to share this knowledge with the other members. As a scholar, Porta attached great importance to discussion and cooperation in research as well as fostering the culture of debate. He often quoted in his writing Heraclitus's maxim about the eternal conflict of ideas, which is also a guarantee for the emergence of diversity. For Walter Benjamin, sociability was a characteristic and essential Neapolitan trait: "each and every private attitude or task is permeated by the currents of social life. Existence, which is the most private matter for northern Europeans, is here—like in the Hottentots' kraal—a collective affair."[23]

The Accademia dei Segreti was dedicated to "discovering and investigating those unusual phenomena of nature of which the causes are unknown."[24] This is all—no more, no less—that is meant by the title *Magia naturalis,* which to our

Figure 4.2 "I study myself!" said Democritus, thus declaring that he did not accept the author-ity of any teacher. Porta's collaborators in his Accademia dei segreti held "knowledge of oneself" (Belloni 1982, p. 17) in high esteem. The new explanations of the natural world offered by phi-losophy and the natural sciences raised fundamental questions about the individual's identity and self-image. This illustration from the 1607 Nürnberg edition of *Magia* II shows the author in the theater of mirrors that was his laboratory, fencing with himself; in the background, the sun heats a distillation apparatus and mixes the elements within.

ears sounds rather esoteric: to seek out natural phenomena (for Porta this includes inorganic matter, artifacts, and technical devices) whose effects we experience but cannot explain, investigate them thoroughly, describe and explain them, and test their effects in experiments. When the experiment is successful, the object of study is divested of its mystery. Only phenomena whose causes are not fully understood deserve to be called secrets.[25] This method demonstrates how highly Porta regards an approach to the natural world that is based on the evidence of the senses. Only by practical analysis of experiencable things, however small, is it possible, at best, to gain access to greater things, writes Porta in the preface to *Magia* II. It is of far greater utility to write the truth about small things than to write falsely about large. In principle, however, the infinite diversity of things is inaccessible and certainly so for any one researcher.[26]

Porta makes a strong distinction between his concept of magic and that of others who seek rather to increase the mystery of natural phenomena, the most extreme example being so-called black magic, which he condemns. Particularly in his early work, however, he treads a fine line between early scientific experiments and practices that belong to the tradition of classical and medieval alchemy and hermeticism. The latter term takes its name from Hermes Trismegistus, ancient Egyptian magus and demiurge, thrice-great Hermes, who became the messenger of the Greek gods—Mercury in Roman mythology.[27] The ideas of Marsilio Ficino, head of the Platonic Academy in Florence, coexist in Porta's work alongside those of Johannes Trithemius, legendary alchemist and abbot of Sponheim; of Cornelius Agrippa, the great hermeticist and alchemist from Cologne who lectured in Italy in the 1510s; of Albertus Magnus, particularly from his thirteenth-century book on vegetation and plants; and of Porta's compatriot Girolamo Cardano, a writer on natural philosophy whom he had met in Naples. In his extraordinary eight-volume work *A History of Magic and Experimental Science,* Lynn Thorndike makes an interesting distinction between thinkers of the sixteenth and seventeenth centuries: those who concentrated primarily on physical science (including astronomy), such as Galileo, Descartes, and Newton, tended toward a skeptical and enlightened rationalism, whereas those who focused mainly on biology, organic chemistry, or medicine (broadly, what today goes under the name of the life sciences) persisted far longer and to a greater extent in their adherence to older occult and magic views when faced with the beginning of the modern approach to science.[28]

To understand the world as a mechanism or as an organism: Porta did not opt for just one of these alternatives. Even today, these views continue to influence scientific debate as opposing poles, although now, typically, a reversal has taken place. At the dawn of the modern era, mechanics became the model for life, whereas from the beginning of contemporary culture, which is founded on mechanical principles, it is the organic that has become the model and leading metaphor of machines and programs. Today, the language used in the networks of connected machines and programs is replete with organisms, genetic processes, oceans, rivers, and streams.[29] Porta was not a specialist. He was equally interested in mathematics, arithmetic and geometry, mechanical phenomena and physical science,[30] as well as the plant and animal kingdoms. In *Magia* I, he describes pneumatic and hydraulic experiments, a section that he expands to be the entire book 19 of *Magia* II. In 1601, he published a separate treatise on the laws of levers and propulsion, their calculation, and applications. The three books that comprise his *Pneumaticorum*[31] are also a wonderful reminiscence of Heron of Alexandria and his mechanical theater of special effects machines driven by fire, water, and steam. The same year, Porta published a geometrical treatise on curved lines (*Elementorum curvilineorum*) with a discourse on squaring the circle. 1601 was also the year that he produced a study on meteorology (*De aeris transmutationibus*); however, it was not released by the censor for publication until 1610. This work is considered the most advanced of the period on the subjects of geology, weather, and marine research.

For Porta, his many studies on the wonders of life served as a springboard to the study of natural philosophy. He returns to this subject again in his attempts to discover structural commonalities between the diverse phenomena of organic nature, yet without robbing them of their individuality. In this understanding of natural magic, he follows an idea of Ficino that, in turn, owes much to Empedocles: all things are connected by sympathy because they have a deep-seated similarity to each other.[32] In a long chain of associations, the eight books of *Phytognomia* (1583) lay out with fervent enthusiasm the relationships between the forms of everything that exists under the sun: analogies between plant rhizomes and crowns in human hair, flower petals and fine eyes, fruit pips and embryos, foliage and reptiles. Porta's study of human physiognomy, *De humana physiognomia,* which appeared three years later, continues his inquiry into the relationship between character and physical traits. In this work, Porta goes a step further and links mental and physical characteristics in such a way that one appears as a reflection of the other. Again, his intention is not to reduce or make the phenomena uniform in any way. On the contrary: using a wealth of examples, Porta is at pains to demonstrate that "body and soul sustain each other and mutually modify each other"[33] while at the same time being connected in infinitely different ways. The rather monstrous analogies between the facial features and cranial shapes of humans and beasts dramatically illustrated in the book's plates[34] made this work easy prey for superficial esoteric interpretations and for the biologically inclined criminal anthropology of the nineteenth century. In 1917, when the Gabinetto-Scuola di Antropologia Criminale was founded at the University of Naples, a commemorative plaque was put up in honour of Porta, which still adorns the wall of the building that formerly housed this institute.[35]

To read the book of nature as a vast collection of signs was habitual among sixteenth-century natural philosophers and also among artists of the period. As one example, the Italian painter Guiseppe Arcimboldo delighted the courts of Europe, including Rudolf II in Prague, with his pictures of combinations of heterogeneous elements taken from nature. In an essay on Arcimboldo, Roland Barthes interprets this fascination with the monstrous thus: "The essence of what was 'wondrous,' that is, 'monstrous,' consists in crossing the line of demarcation between the species, in the mixing of animal and vegetable, of animal and human. It is *extravagance*, which changes the properties of things that God has given a name to. It is *metamorphosis*, which allows one order to pass over into another; in short, the *transmigration of souls*."[36]

The Inquisition

Paradoxically, Porta's *Phytognomia* on similarities in nature, which was part of a long tradition of interpreting external physiognomy as an expression of the emotions within,[37] should be interpreted as his attempt to react to the increasing pressure of censorship and investigation by the ecclesiastical authorities. The book's radical thesis—that human character traits impress themselves like signatures on the physical body and vice versa—appears at first glance to contradict his earlier position, in which the metaphysical is a calculable effect of the movements of the stars and planets. Porta does not reconcile the two positions until much later, when in 1603 he returns to this theme in *Coelestis physiognomonia* [Celestial Physiognomy]: he assumes that both realms of living things, the mental and the physical, are grounded essentially on astrological factors.

The reason the Neapolitan aroused the suspicions of the Inquisition is not known for certain; possibly it was connected with the many experiments that he and his fellow members in the Academy of Secrets performed. Like Bruno, Porta believed that only through operating on, and thus changing, nature could

 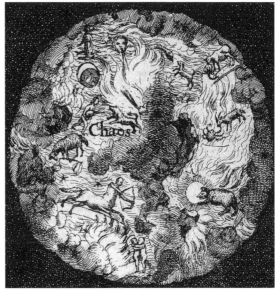

Figure 4.3 Two details from the frontispiece of the English translation of *Magia* II, depicting the themes of Book 20, chaos and nature. Nature is contrasted with art in a manneristic portrayal of a woman with six breasts (Porta 1658/1958). *Facing page, top:* Illustration from *Phytognomia* (Porta 1583, p. 143). *Bottom:* Frontispiece of the first translation into English of Euclide's "The Elements of Geometrie" from 1570 with the famous Preface by John Dee (from Werner Nekes' private collection).

the divine powers at work there be developed fully.[38] Probably less problematic were the metamorphoses he proposed for the vegetable kingdom, many of which he patented, such as methods for speeding up or slowing down the growth of grape vines or cultivating fruit without seeds. However, Porta's basic convictions and adventurous mind led him into several areas that were tabooed by the Catholic Church. He not only provides recipes for aphrodisiacs, hallucinogens, and other drugs and describes their effects, but in *Magia naturalis* he discusses how to make natural contraceptives for women (abortions at that time were horrendous tortures often ending in death), describes compounds for manipulating the gender of unborn children, and gives instructions on how to cultivate extralarge fruits. Yet what ultimately led to Porta being investigated in the 1570s and later hauled before the Inquisition in Rome to answer charges were his pronouncements on a subject where mathematics and magic were closely interwoven, *astrologia giudiziara*.[39] Judicial astrology, in contrast to natural astrology, was concerned directly with the influence of celestial bodies on the actions of individual people and involved the making of "judgments" by astrologers. For example, current political constellations were interpreted and future ones predicted by observing and plotting the movements of the planets, which were then assigned as determinants to the parties involved. The papal authorities did not tolerate any incursions into what they regarded as their exclusive province—heavenly power—and banned summarily all publications resulting from the "deluded science that clings to the stars" (Jakob Burckhardt).

Notwithstanding the attitude of the church, many powerful secular rulers were extremely keen to have their fortunes cast astrologically by great mathematicians because such charts were seen as especially authoritative. Elizabeth I of England, despised by Rome, appointed John Dee, an excellent mathematician, as her court astrologer. A specialist in geometry, Dee was an old friend of the Flemish cartographer Gérard Mercator and, in 1570, wrote a famous introduction to the English translation of Euclid's *Elements,* which played an important role in popularizing mathematics and geometry in England. His *Monas hieroglyphica* was the first work to phrase its arguments in terms of mathematics, geometry, and symbols, and put forth the concept of a smallest, ultimate, and indivisible unit that is contained in all things and from whence all develops. Even when Dee's brand of natural philosophy led him to drift more and more into the esoteric world of angels and spirits, Elizabeth continued to extend her favor and protection to the man on whom, it is thought, Shakespeare modeled Prospero in *The Tempest*.[40] Another ruler with a passionate interest in astrology and alchemy was Rudolf II of Prague. He invited artists, such as Arcimboldo, and scientists, such as Tycho Brahe and Johannes Kepler, and also—for a short while—Dee and his erstwhile partner Edward Kelley, to live and work at his court. Rudolf II was also impressed by Porta. A later edition of Porta's treatise on the interpretation of chirophysiognomy (palm-reading), reprints a letter from Rudolf, dated June 20, 1604, to his "revered, scholarly, and truly esteemed friend" in whose "great science of nature and technology" he takes great pleasure "whenever the weighty affairs of state permit."[41]

Porta did not experience the full brutal force of the Inquisition like Bruno, who suffered horrific tortures according to the methods of the Spanish Inquisition before being burned publicly as a heretic on February 17, 1600 in Rome's Campo dei Fiori; or Tommaso Campanella, who was arrested in 1599 and incarcerated for twenty-seven years, during which time he wrote his utopia of a "city of the Sun."[42] However, for at least twenty-five years, Porta lived and worked with the dangerous threat of the Inquisitors hanging over his head. The official investigations began in the mid 1570s, and in 1578 his Academy of Secrets was disbanded. By papal order, Porta was expressly forbidden to engage in any activity related to the *arte illecite,* the forbidden (divinatory) arts. He was urged strongly to give up all scientific activities and concentrate on works of literature instead. In the years that followed, Porta did in fact write many plays, particularly comedies,[43] but he ignored the tribunal's recommendation to give up research. In April 1592, shortly before Bruno was arrested in Venice, Porta received the order of the Venetian Inquisition forbidding publication of his work on human physiognomy and anything else he had written "that had not received the sanction of the Roman tribunal."[44] This situation continued until 1598, but even afterwards Porta had to fight the censors for the publication of each one of his works. He was not always successful. An intriguing late work, *Taumatologia* [On Marvels], which Porta conceived as a grand summary of all his studies and as a deeper investigation into the power of numbers (*virtù dei numeri*), remained unfinished because when he submitted the book's index to the tribunal, it sufficed for the tribunal to refuse a license to print it.[45]

Gabriella Belloni, the greatest expert on the life of Porta, writes that the Neapolitan scholar was deeply affected by the arrests of Bruno and Campanella, but at the same time, he had to avoid all mention of their names. Porta had certainly met Campanella in Naples; ironically, it was in the same room of the monastery of St. Domenico Maggiore where Thomas Aquinas had taught that, in 1590, Porta and Campanella held a public discussion on magic.[46] He

Figure 4.4 Contemporary portrait of Porta. (Original in *Magia naturalis* 1589; taken here from Mach 1921)

had probably encountered Bruno in Venice while on a longer visit to find one of its renowned glass-blowers to help in his experiments with mirrors. In Porta's book on the art of memory (*Ars reminiscendi*) of 1602, he reports encountering a person in La Serenissima who had such a phenomenal memory that he could recite up to one thousand verses without making a mistake. Giordano Bruno, who both published on and taught *ars memoria,* was famous in intellectual circles of the Italian Renaissance for his amazing powers of recollection.

Secret Writing and Ciphers

The gradual separation of the message from the body of the messenger carrying it is a process that can be traced from ancient ways of sending communications in ancient China, Asia Minor, and classical antiquity.[47] Efforts were directed not only toward speeding up delivery of messages, but also at excluding the messenger from all knowledge of the message. As a rule, messengers were slaves, with their bodies to undertake the journey, their minds to understand the message, and mouths to repeat it accurately to the recipient. In our world of networked machines and programs, the problem of keeping communications secret has still not been solved. In anthropomorphic metaphors that refer to those ancient slaves' bodies, we still refer to the *header* and *body* of a message. The header (or subject), however brief or cryptic, must remain open and publicly accessible. The supervising *postmaster* of a server requires access to the headers of messages, if only for the purpose of resolving technical transmission problems. What the body of the message conveys is supposed to remain a secret, although in principle the postmaster or higher instances of control are able to access it. Because the system is not secure, courier services were reintroduced in the latter years of the twentieth century—from messengers operating locally on foot, bicycle, motorbike, or car, to worldwide operators using aircraft. The only efficacious method, at least for a limited period of time, for keeping messages secret that are sent through the language realm to which computers belong is encryption, the art of cryptology. Thus it is hardly surprising that the Internet, a medium most admirably suited to conspiratorial theories and practices of all kinds, has innumerable sites and projects on the study of the origins of secret languages.[48]

The passion for encrypting and deciphering texts runs through the sciences like a subhistory, conspicuously so since the thirteenth century. It was a hidden component of the scholastic approach to the world, which was defined by the predominance of letters and the trivium of grammar, rhetoric, and logic.[49] Encryption was essential to the survival of the alchemists: in addition to their habitual hermetic way of writing, they communicated their discoveries about mixtures of forbidden substances, including alcohol, in the form of cryptograms.[50] Undoubtedly, one reason for Porta's intense preoccupation with the art of "criptologia" (this would have been the title of one of his last books, but its publication was not sanctioned) was the ever-present threat of the censors and the Inquisition. As late as 1612, three years before his death, his patron and founder of the Roman Accademia dei Lincei (Academy of the Lynxes), Federico Cesi, wrote in a letter that all communications to Porta should be sent via a go-between, "for if one writes to Porta, the letters are not very safe."[51]

After Porta completed *Magia naturalis,* his next major work was a four-book treatise on secret ciphers, *De furtivis literarum notis vulgò de zifferis* [On Concealed

Figure 4.5 In *Buch von der Weltpost* [Book of the World's Post], 1885.

Characters in Writing] (1563). In the preface, Porta defines what he means by the title's concepts:

What are secret characters? In the higher branches of learning, secret characters are used for writing, executed with art and ingenuity, which can only be interpreted by the person to whom it is addressed. This description would seem . . . to correspond exactly to the type of writing that is referred to as *zifera* in the vernacular of this country. . . . After we have taken brief stock of the achievements of our predecessors, we shall henceforth name only such characters a cipher by which means we may communicate with the initiated about those matters of which they must properly be informed in a secret or abbreviated form. Ciphers (*notae*) we shall name them, because they denote (*notare*) letters, syllables, and statements . . . characters that have been agreed upon beforehand and call forth these meanings for the readers, which is the reason why the persons who write down these [ciphers] are called notaries (*notarii*). If we consider their various employments, we shall conclude that they are only required in those matters such as we meet with in sacred and occult learning. Namely, in order that they will not be profaned by outsiders and such others, to whom the requisite initiation has not been vouchsafed.[52]

Porta then proceeds to take great pains to explain, for the benefit of the censor, that his work is written in the interests of the powers that be: "For so often it is necessary that we advise kings, when their deputies are absent or privy to a plot, or others in other matters, with our secret knowledge, in order that any message, were it to be intercepted by bandits, spies, or governors, who serve in far-off places (for long is the arm of kings and princes), should not yield up its secret counsel, not even if a great deal of time is lavished upon it . . . it is then that we take advantage of them [ciphers] for our own protection."[53]

However, in the revised version of this treatise, which was published thirty years later in a handy, almost paperback-size format that could easily be slipped into a pocket and carried around, Porta reveals in the title what he really means by the long arm of the rulers: "On Secret Ciphers or: On the Art of Conveying One's Own Opinion (*animi sensa*) by Other Means in a Secret Way or Finding Out the Meanings in Other Things and Deciphering Them."[54] As early as in *Magia* II, Porta describes a wide range of procedures for sending messages to friends without third parties being able to detect their existence.

These, then, are the two lineages in the the history of telematics, which occasionally converge but, from the viewpoint of technique and knowledge, are entirely disparate: on the one side are strategic focusing and acceleration of communication to serve the interests of established institutions, such as the church, the state, the military establishment, or private corporations, and on the other are the development of tactics and a culture for friends to communicate with each other, where it suffices for them to agree formally upon a code. The latter requires mutual sensitivity and respect: the willingness to engage intensely with the other. In a letter to Rudolf II, Porta proposes a bizarre telegraphic procedure, which is a fine illustration of this approach, precisely because of its impracticability. He describes the technique in connection with the power of magnetism to work over long distances. In *Magia naturalis,* Porta had described how the needles of two compasses that are far apart can influence each other and be used to send messages to a friend who is far away or even in prison. In his example for the emperor in Prague, Porta describes telecommunication that is based on blood-brotherhood. I shall not go into his meticulous recipe for the *sympathicum,* a special ointment that is essential to this experiment, but simply cite the mode of this communication over distance:

[take] two new knives and smear the salve from the point to the handle. . . . The friends must have wounds on the same part of the body, for example, on the lower part of the arm. The wounds must be kept fresh and bloody . . . above the wound, two circles must be drawn, a greater and a smaller, proportionate to the size of the wound. Around this, the letters of the alphabet are written in exactly the same order and manner, size and scale. If you desire to speak with your friend, you must hold the knife over the circle and the pierce the selected letter with its point . . . your friend will feel the same piercing pain on his wound. . . . I prick the V and he feels it, then I prick the A and he feels it, and so forth, with each separate letter. However, the knives must be smeared each with the blood of the other, mine with his and his with my blood. . . . Now after all the letters have been assembled, he will know the thoughts of your mind.[55]

This is a concept of mutual exchange that is wholly in the spirit of Empedocles, for it is generated by the binding power of sympathy—the notion of complete compatibility between the bodies of transmitter and receiver and the transmission of their autonomous, local energies. In the above example, possibility is not the mere shadow of reality, but rather a challenge to it. Separation, held to be "the alpha and omega of the spectacle"[56] of telecommunications, is thus called into question.

The techniques proposed and analyzed by Porta in his treatises on cryptography focus mainly on secret writing, that is, the transcription of texts, although he does include some simple steganographic devices, or hidden writing, where the existence of the message stays concealed during the period of its transport. A particularly perfidious ancient example of steganography was the practice of scoring messages on the scalp of a slave messenger; the hair acted as a natural means of concealment. Porta also describes the use of invisible inks, which the recipient can render visible by treatment with the appropriate chemicals, and the methods of transcribing texts rhetorically or poetically, which have been practiced since ancient Greek and Roman times: messages concealed within ambiguities, metonyms, metaphors, or allegories. Further, he discovers the potential of the newly invented printing press for adding to the arsenal of methods of concealment by using different typographies or colors of ink. Numerous

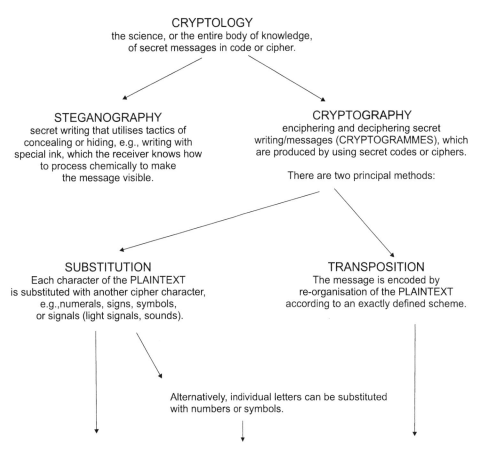

CRYPTOLOGY
the science, or the entire body of knowledge,
of secret messages in code or cipher.

STEGANOGRAPHY
secret writing that utilises tactics of
concealing or hiding, e.g., writing with
special ink, which the receiver knows how
to process chemically to make
the message visible.

CRYPTOGRAPHY
enciphering and deciphering secret
writing/messages (CRYPTOGRAMMES), which
are produced by using secret codes or ciphers.

There are two principal methods:

SUBSTITUTION
Each character of the PLAINTEXT
is substituted with another cipher character,
e.g.,numerals, signs, symbols,
or signals (light signals, sounds).

TRANSPOSITION
The message is encoded by
re-organisation of the PLAINTEXT
according to an exactly defined scheme.

Alternatively, individual letters can be substituted
with numbers or symbols.

For decryption,it is necessary to know the rules of the code or cipher being used.

Figure 4.6   Theory and praxis of secret languages

examples are cited of ways to encipher by drastically reducing the text body and combining letters, numbers, and invented characters.

The initiated were already familiar with the simple substitution method, which goes back at least as far as Julius and Augustus Caesar and is still referred to as the Caesar cipher. In this method, the encrypted messages are written as cryptograms where the position of the letters of the plaintext are shifted one or more places. The complete alphabet stands in the first line of the so-called tableau in the usual order; underneath, in the second line, the ciphertext alphabet is written according to the number of places shifted. When there is a shift of three places, for example, the second line begins with the letter *D* and ends with *C;* thus *A* is enciphered as *D,* and *Z* as *C.* The only key that the correspondents must agree upon is the number of places to shift the alphabet. In monasteries in the late Middle Ages, a great many variations of this cipher were in use.[57]

A century before Porta's book appeared, Leon Battista Alberti wrote a treatise on secret writing that was based on a philological analysis of the Latin language. Alberti describes the cryptographic game of substituting vowels and consonants with other, changing symbols. In 1499, Trithemius, alchemist and abbot of Sponheim, later of St. Jacob's monastery in Würzburg, wrote his monumental treatise *Steganographia* on how to conceal and encipher texts where even the rules for performing these operations are encrypted—theologically. At first the work circulated only in manuscript form; it was not published as a book until 1606, when it landed immediately on the church's index of censored works. In 1518, Trithemius's *Polygraphia* [Multi-alphabets] appeared, in which he develops rudiments of a *lingua universalis* (universal language). It also contains his invention of a polyalphabetic cipher with twenty-four different alphabets, an idea taken up by Athanasius Kircher around 150 years later. The abbot of Sponheim was inclined to dramatic gestures. In the preface to *Steganographia*—"steganography" was often used as a synonym for cryptography in this period—he also provides the ecclesiastical authorities with good ammunition for rejecting outright what he is describing: "Henceforth it may come to pass [if there is wide access to the secrets of steganography] that conjugal fidelity will no longer exist, for any wife could, without the slightest knowledge of Latin but educated through holy and chaste teachings in any other language, gain knowledge of the despicable and unchaste inclinations of her lover, whereby the husband might even act as the messenger and praise the contents [of the hidden message]. In this very same way, not needful of concern, the woman could send back her desires in eloquent words."[58]

In *De furtivis literarum notis vulgò de zifferis,* Porta assembles all that was known in his time about secret writing, knowledge that was spread out over centuries and not easily accessible, to produce a proper manual. He obviously received excellent assistance from his publisher and printer, for his special symbols did not exist as type and either had to be entered in each copy by hand in writing or with specially made woodcuts. Particularly striking are the pictograms, probably designed by Porta himself, which stand for letters, words, or agreed-upon combinations of words and are reminiscent of ancient Egyptian hieroglyphs. As signs located somewhere between abstraction, mystery, and representation, Egyptian hieroglyphs exerted a fascination on men of learning, from Renaissance scholars to text artists of the baroque. Even today, the arsenal of simple cryptography includes the method of concealing a short, secret message within a longer, seemingly innocuous one (for example, a religious tract); the message is revealed when a specially made template, or grille, is laid over the text and the words of the message appear in the holes of the grille. More important is a system of substitution first suggested by Alberti that has been used widely throughout the history of diplomacy and espionage, which are very closely related. In Porta's more sophisticated version, thirteen alphabets are listed one above the other in a square tableau, whereby the last thirteen characters are arranged at random. Each alphabet is assigned a pair of letters (from *AB* to *YZ*). The two parties communicating agree upon a password that indicates which alphabet is to be used to decipher the message. Then it is simply a matter of assigning the letters given in the cryptogram and deciphering them using the appropriate alphabet.

From the perspective of media archaeology, two systems described by Porta are especially interesting. In the first, he presents a system for encryption that is based on two discrete elements. Two horizontal and two vertical lines are drawn, which cross each other at right angles (as in a game of tic-tac-toe). In the nine spaces of this framework, the alphabet, which has been reduced to twenty-one letters,[59] is entered according to a scheme agreed upon by the correspondents. The three spaces at the top each contain three letters, and the other six fields have two letters apiece. A cryptogram produced by this system is written not as text, but as symbols. The exact arrangement of two, three, or four rectilinear lines containing the selected letter is given, and the letter's position is designated by another geometric form—a dot. As each space can contain up to three different letters, one dot denotes the first, two dots the second, and three dots the third. Thus the code consists simply of combinations of dots and dashes, like the Morse code developed and used centuries later by telegraphy. The only difference is the way in which the two codes are written: the Morse alphabet is written as a continuous sequence of dots and dashes, whereas in Porta's system the two elements are noted in groups. Reading this code very quickly becomes an exercise in fast and precise pattern recognition.

Porta's second original suggestion concerns a concept for generating and interpreting texts that has fascinated cryptologists from Trithemius and Alberti to Bruno and Kircher: Ramon Llull's *Ars generalis ultima,* also used by Werner

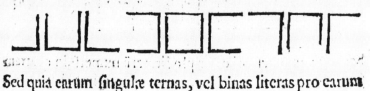

do | e p | f q

g r | h s | i t

Tali propofito fchœmate ; eius areæ fingulæ iuxta faciem, quam gerent depingentur, vt exemplum indicabit.

Sed quia earum fingulæ ternas, vel binas literas pro earum diftributione complectuntur, carū vnaqueq; fuo charactere

punctis diftincta pro ordine, quemadmodum collocata fit, defignabitur,vt fi in charactere prima collocetur,vnico tantum puncto difcernatur, fi fecunda duplici, fi tertia triplici. Res exemplo clarior fiet.

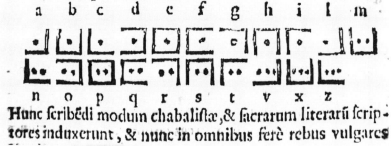

a   b   c   d   e   f   g   h   i   l   m

n   o   p   q   r   s   t   v   x   z

Hunc fcribēdi modum chabaliftæ,& facrarum literarū fcriptores induxerunt, & nunc in omnibus ferè rebus vulgares

Figure 4.7 An example of a code that uses the substitution method from Porta's treatise on secret writing. The tableau containing the letters of the alphabet generates cryptograms, which use only two discrete elements: a line and a dot. (Porta 1563)

Künzel and Heiko Cornelius as the title of their pilot study on the Majorcan scholar.[60] This system amalgamates the arts of combination and interpretation, of cabalistic and astrological readings, in an attempt to reveal a global, intelligible scheme for interpreting the complex holy Christian Scriptures. The Majorcan scholar's most important basic assumption is that the three great monotheistic world religions, which are founded on words and texts—Islam on the Koran, Judaism on the Talmud, Christianity on the Bible—are similar in essentials and can be linked with each other. Llull reduces the entire knowledge of the Bible to nine axiomatic concepts (such as goodness, greatness, eternity) to which he assigns nine letters of the alphabet (from *B* to *K,* without *J*). Five different modes (proportions, questions, subjects, virtues, and bonds), which are again subdivided each into nine terms, differentiate the nine axiomatic concepts further by assigning groups of meanings, which can then be used to construct manifold internal combinations with the nine-letter alphabet. The idea was to provide scholars well versed in theology with a system for using the Bible as an apparatus and reading the texts like data sets.[61] However, such a system can only function if the basic precept of any mechanical system is given; namely, that it is possible to formalize whatever the system is designed to process. Llull recognized this quite clearly: "The subject of this art is to answer all questions, provided that whatever it is possible to know can be formulated as a concept."[62] Apart from the ingeniousness of this design for a late-medieval expertise system, even more fascinating is the fact that Llull translated his system into actual artifacts. Each consisted of two rings and a disk on which he wrote the nine letters for the axiomatic terms, the hidden meanings, and possible combinations with terms from the other classes. The rings and disk, later named a *volvelle,* could be rotated in either direction around a central pivot. Thus with the aid of a sort of toy, the entire categorized knowledge of the Bible was transformed into a work of variable combinations. In the long and rich history of *ars combinatoria,* however, Llull's invention was not without earlier models. His volvelle also bears a strong resemblance to the astrolabes and devices constructed by Arab astronomers before A.D. 1000 to calculate the movements and positions of stars and planets or to establish connections between astronomical and geological data.[63]

The design highlight of Porta's encyclopedic work on cryptography is his presentation of encryption devices that operate with what he called circular writing. He also says that the writing is arranged in the form of a "rota, that is, like a wheel,"[64] which again conjures up associations with the circular "wheel-like maps" of Arab and medieval cartographers in which the Earth is depicted as a disk with the inscriptions arranged correspondingly.[65] Even today, circular ciphers have proved among the most effective in cryptography. Like Llull's model, Porta's also consists of two graduated concentric circles with a movable disk in the middle, which can be rotated to the position of choice. However,

Figure 4.8 Two of Porta's decorative deciphering and enciphering volvelles. The central disk can be lifted and turned; it is affixed to the page by a gold-colored thread through the middle where the hand of God, pointing, rests upon a cloud. *Top:* Cryptography as drama: The smiling figure of a woman on the left side has a sad-looking counterpart on the right. (Porta 1563, p. 73)

Porta's interest in this device was not to encode biblical knowledge for the purposes of answering questions about the Scriptures. The circles of his instrument contained the letters of the alphabet and Roman numerals, and the disk was inscribed with pictograms of his own invention. According to what is decided upon by the parties communicating, any meaning can be assigned to the three components, which are then written down in a glossary. If the rotating disk has the letters of the alphabet on it, it becomes an instrument for the substitution system described above that makes encryption and decryption of the ciphertext easy.

Porta took Llull's hermetic philosophical and theological expertise system and transformed it into an easy-to-use cipher system—potentially, for a wide range of people. His printer's execution of these text generators in the first edition of *De furtivis* is really beautiful. There are two examples of them: the rotating disk is fixed to the page with a fine gold-colored thread, which acts as a pivot and allows the disk to be raised and turned.[66] (Porta attached great importance to practical experience, which in this case must have been very expensive for the printer.) The final part of the volume consists of an extensive index of words and various possibilities for substituting them with numbers, letters, or pictograms, as he demonstrates in many examples using the methods described. Porta had stressed an essential feature of the art of cryptology in the introduction: if it is to be at all practical, it makes enormous demands on memory and exactness. Fittingly, *De furtivis* was published again in 1566 in one volume with the Italian translation of Porta's treatise on the art of memory, *L'arte del ricordare*.

Of Glasses and Refraction

Glasses as prostheses for the human eye were produced in Europe since the thirteenth century, probably in Venice, the contemporary center of glass blowing. Centuries before, however, glasses of a special kind existed in China. These glasses did not allow the wearer to see better, but instead prevented his eyes from being seen. Judges at the imperial Chinese courts had such glasses made, with lenses of cloudy gray quartz, so that the counsels for the prosecution and the defense could not make any deductions from the judge's reaction to anything said during the course of the trial. Thus, long before dark glasses were used to shield the eyes from bright light, they served to hide that feature of the human face which reveals the most about the soul within. In John Cassavetes' film *Faces*, for example, or Jean-Luc Godard's early films, sunglasses identify the existentialist characters who set themselves apart from surrounding reality with their eyes hidden behind their shades.[67]

Why and how certain technical artifacts originated—what interplay of idea, blueprint, exact description, and construction led to their development—are especially difficult to reconstruct when a great number of researchers, from a variety of countries, disciplines, and epochs, have investigated different aspects. Optics is one such example. For over 2,500 years, it has been the subject of physical, biological, and philosphical inquiry. Even if we take only the main concepts, we are dealing with literally dozens of investigators from the ancient cultures of China, Greece, Rome, Arab countries, and modern Europe, who all engaged more or less rigorously with their predecessors in the field and, at best, achieved some small advance in knowledge. Standard reference works, such as encyclopedias or histories of science, have enormous difficulties in offering an overview. In fact, to my knowledge, none was even attempted in the twentieth century.[68] I suspect this lack exists for two reasons, which at first glance seem to be contradictory. First, ever since Artistotle the faculty of vision has been privileged over all the other human senses with which we perceive the world. The language of science brims over with metaphors related to vision and the visible because, obviously, science depends crucially on visual experience and observation—an aspect that engaged Michel Foucault intensely in his various archaeologies of power. Second, although little is known with relative certainty about thought processes and their mechanisms, neurobiologists assume that around 60 percent of all information that reaches the brain is of visual provenance and that the brain uses a considerable proportion of its capacity (about 30 percent) to process this information. Further, the physiological basis of vision is by no means fully understood, especially since vision is regarded now as a complex neurophysiological process and no longer primarily as an optomechanical one (a view precipitated by George Berkeley's theory of vision, formulated in the early seventeenth century). Progress is slow; much remains unknown about the technical devices for producing visuals and their psychological dimensions. Research on processes of perception has not advanced much farther than the findings of Gestalt psychology, which dates from the early twentieth century. On the technical side, the situation is even more astonishing: all optical systems in cameras, for still or film photography, are still based on the geometrical laws of central perspective, which are over five hundred years old.[69]

It will help us to locate Porta within this history of investigating vision in the fifteenth and sixteenth centuries if we follow a classification proposed in 1675 by Zacharias Traber, a Jesuit mathematician whose terminology refers to Euclid. Traber's treatise on the "nervus opticus" is divided into three books: optics, catoptrics, and dioptrics. The first concept covers the entire doctrine of sight and light, which, from a scientific point of view, is subdivided further into biological and physical phenomena. Since classical antiquity, dioptrics has concerned the refraction of light in transparent bodies, later including the geometry of lenses. Catoptrics deals with reflections produced by planar surfaces, although it was taught and described together with dioptrics under the name of catadioptrics. In these two subfields of optics, one can pinpoint different foci of researchers' interest, which can be characterized from a media-archaeological viewpoint as follows: the "diopricians,"—which include the great scientists Kepler, Galilei, Descartes, and Newton whose work promoted a "physics of the visible" in the seventeenth century[70]—were interested primarily in problems of "looking through," whereas the "catoptricians" were fascinated by problems of "looking at." This juxtaposing of the two views, in both senses of the word, continues to have implications and consequences for image technologies today.

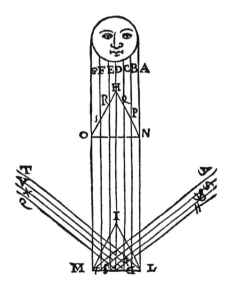

Figure 4.9 Stylized representation of the sun and the refraction of its rays from Porta's treatise on optics, *De refractione* (1593, p. 124).

Electronic visual display systems, whether they operate with the Braun tube or liquid crystal, belong to the "looking through" category. All media requiring projection, including cinematography, are techniques of "looking at." The former, dioptrics, are indebted to the idea of *perspicere,* of seeing through something in the sense of insight or understanding; the latter, catoptrics, are more oriented toward the illusionizing potential of projection, the production of artificial reality.

Porta's main interest was catoptrics and its strange balancing act between truth and falsehood, image and reality.[71] This focus was intimately bound up with Porta's fundamental relationship to nature. He believed that nature could only reveal and develop its hidden powers through the intervention of the researcher, and this belief is reflected in the way he deals with optical phenomena. He was not particularly interested in the possibilities offered by media devices as prostheses. What fascinated him most of all were transformations, metamorphoses, and production of visual spectacles, which we cannot see with the naked eye under normal conditions. "I shall deem my comprehensive work to be accomplished when I have described a number of catoptric experiments,"[72] writes Porta in the preface to book 4 of *Magia* I, after discoursing in detail in book 3 on alchemy, which defines projection as the highest stage in the transformation of the base to the noble. Porta makes it perfectly clear that he does not intend to discuss what is already known or what other authors before him, such as the Greeks Euclid and Ptolemy or Witelo of Wrocław, the greatest opthalmologist of the thirteenth century, have already written about; moreover, he does not even mention contemporary researchers in this field, such as Cardano. Forget the simple flat glasses, he tells the reader, if you want to see more than what already exists.

Porta's tone and language become increasingly ecstatic in the course of the book, and he admits to having dropped his pen more than once through sheer agitation at the thought of the improbable phenomena in the experiments he describes. In fact, the first edition of 1558 contains the seeds of most of what the physicists of the visible and engineers of technical vision will probe in detail over the centuries to come. In *Magia* II, published thirty years later, the topics are formulated in more detail; *De refractione* has a stronger mathematical and geometrical bias in conjunction with an added treatise on the eye. However, the very fine book 4 of the original *Magia naturalis* is like a nucleus: it contains the entire micro-universe of modern illusionizing by means of technical apparatus.

Using as an example the camera obscura, a crucial device in the history of optical media, I shall suggest how the process of its invention may have proceeded and indicate Porta's position within that process.

The effect that the camera obscura utilizes is well known to us. Lying in bed in the morning, if we didn't draw the curtains completely, the sun shines through the narrow opening and throws image fragments of the world outside onto the opposite wall. Probably everyone has observed this phenomenon—and enjoys it—without taking much notice or bothering to write it down. It is simply pleasant and relaxing to watch these black-and-white minifilms of a morning while we are still halfway between the there of sleep and the here of being awake. From a European perspective, it was allegedly Aristotle who, in the fourth century B.C., turned this everyday experience into a scientific observation. He described the phenomenon elaborately in the case of light rays passing through foliage and made use of it for studying eclipses of the sun. However, Joseph Needham, the eminent historian of Chinese science and civilization, points out with some emphasis that there are descriptions of this effect from China dating from the fifth century B.C., in particular the writings of the natural philosopher Mo Ti, who calls the room in which the shadow images are projected the "locked treasure room."[73] In the following centuries, Archimedes, Ptolemy, Heron of Alexandria, and others investigated further using mathematics and geometry. The effect was formulated most precisely at the turn of the first millennium by the outstanding, Persian-born natural scientist Ibn al-Haytham, whose wide-ranging work on optics included the translation of texts by the Greek authors and his own brilliant additions to this knowledge.[74] Shortly after Ibn al-Haytham's death, the Chinese astronomer Shen Kua described in his treatise *Mêng Chhi Pi Than* (1086) his discovery of the focus (or focal point), the exact center midway between the object and projection surface, and described its function for seeing via optical instruments with impressive examples of flying birds and moving clouds. In this, he allegedly made use of the the specialist knowledge of the Mohists from the third century B.C.[75]

In the thirteenth century, the Polish mathematician Witelo and the Englishman Roger Bacon wrote their important works on vision and light. Witelo cited extensively from the work of the ancient Greek authors and Ibn al-Haytham, and thus reintroduced them to Europeans. The Franciscan natural philosopher and mathematician Bacon, who studied and taught in Paris and Oxford, concentrated on defining more precisely a number of optical phenomena. In addition to the work of the ancient Greeks, Bacon also drew upon the nineth-century treatise by Ya'qûb al-Kindî on mirrors and their laws of refraction (*De aspectibus*), which in turn has strong parallels to Ptolemy's work on optics.[76] Bacon studied not only the laws governing the diffusion of sunlight when it passes through a narrrow aperture, but also the properties of eyeglass lenses (which were just becoming available), the possibilities of telescopic vision, the position of the focus in concave mirrors, and the varying focal lengths of convex mirrors. In his "opus magnus" of 1267–1268, Bacon assigned pride of place to mathematics in the ranking of the sciences and, with deep conviction, privileged experiment and observation over the speculative approach. He spent the last ten years of his life in prison, for prowess in mathematics, not yet an established discipline sanctioned by the church, was regarded as a gift of the devil. Among Bacon's legendary technological visions was a machine "with which men can rise like birds" and fly.[77]

The works of the ancient Greeks were being read again because of the discoveries of Arab scholars and this trend resulted in a growing body of literature on optics, including on the camera obscura as an instrument for astronomical studies. Villeneuve of France deviated from this tradition somewhat by investigating its potential for entertainment. In the fifteenth century, Leon Battista Alberti and Leonardo da Vinci profited from this accumulated knowledge and, in darkened rooms with small apertures, made sketches and detailed studies of projections of objects outside as inverted images. Truly spectacular is da Vinci's sketch with two openings, side by side, to produce slightly shifted double views of objects. In the sixteenth century, one finds vague suggestions in the writings of the Milan mathematician Girolamo Cardano for a possible use of lenses to improve projection, which the Venetians Daniele Barbaro and Giovan Battista Benedetti formulated more precisely as biconvex lenses. They also proposed the use of deflecting mirrors, which were actually put into effect when the camera obscura became a drawing aid. Extraordinary new ways of calculating the geometry of light rays are said to have been invented by Francesco Maurolico in an isolated monastery near Messina in Sicily. His studies on optics, the first of which was completed in 1523, anticipated in many essential points the *Supplement to Witelo* of 1604 in which Johannes Kepler presented an exact description of the pupil's function as a flexible opening and used geometry to calculate the light cone through the eye.[78]

Whether Porta was acquainted with all the work of his predecessors and contemporaries is not decisive; it must be remembered that access to literature was still very difficult in his day. Still, it was Porta's description of the "dark chamber" that transformed it into a sensation. He pried it out of its narrow context of application in astronomy or, in Alberti and Leonardo's case, in architecture, and opened up a wide range of new uses for the apparatus.

The second chapter of book 4 of the 1558 edition of *Magia naturalis* is devoted to the question "How one can see in a dark chamber that which is illuminated by the Sun outside, even its colours."[79] Porta calls his projection room a *cubiculum obscurum.* Remarkable technical details illustrate what an extraordinary technical imagination he possessed. Of the objects from the outside world that are to be projected, Porta begins by naming only moving objects and suggests hanging a white sheet or paper on the wall where they will be projected. He takes into account the distance of the projection "screen" from the aperture (thus reflecting upon the sharpness of the image), the size of the projected image, as well as the sluggishness of human perception (the eyes need time to become accustomed to the dark after the brightness outside). The technical sensation of what he is proposing he describes in rather awkward language but quite clearly enough for others to repeat the experiment. By using a lens that reduces the divergence of the light rays entering through the opening, the image (*idolum*) can be seen in its natural colors and the right way up if the lens is positioned correctly between the sun and the objects. Porta admits that he is not able to work this out mathematically and will leave this task to future researchers; but he cannot resist saying that, so far, all those who have claimed to be able to do the calculations have produced nothing but "fables." Porta stresses that it is essential to use a "mirror" (the word "lens" for ground, transparent glasses was unknown at this time, and authors used the term *speculum* [mirror,] for both the flat and curved variety), which does not diffuse the rays of light (i.e., is not biconcave), but focuses them (i.e., is biconvex): "speculum . . . non quod disgregando dissipet, sed colligendo uniat."[80] There follows one of Porta's typical mental leaps: the description of how one can use the *cubiculum obscurum* as an aid for drawing leads him to make two further suggestions, which have far-reaching ramifications. Instead of using the natural light of the sun, one could just as well use artificial light and, instead of using existing objects, one could make some especially for the purpose of projecting them. Here, associations with the masks and manmade objects in Plato's cave metaphor spring immediately to mind. In *Magia* I, these ideas are touched upon only briefly, but in Book 17 of *Magia* II, a wealth of details are added to the description. This amplification is probably due to the fact that in the intervening decades Porta concentrated on his playwriting—under the orders of the Inquisition—and was obviously very familiar with stage techniques. Porta suggests that opposite the projection wall, outside of the viewing room, landscapes or architectonic settings should be built, peopled with actors, and illuminated with strong light. Then it would be possible to view hunting scenes, battles, or any kind of play in the dark chamber, and it could be arranged that the sounds of trumpets or the clash of weapons be heard. After this exemplary discourse on media praxis, Porta releases us from his dark room, but not before drawing our attention to a problem that continues to be a central concern for media theory up to the present day: the reality test. Porta says that he has performed his dark-room experiment with friends, many times. However, they obstinately clung to the impression of having experienced natural reality, even after he had explained to them the "illusion"—he actually uses this word—and the laws of optics involved.[81]

Porta's main interest, however, continued to be the lenses themselves and their possible effects. In the book on refraction and its forms (*De catoptricis imaginibus*), which had grown to twenty-three chapters with many subdivisions, Porta assembles everything he can glean from the literature as well as what he can imagine. A great deal that has been attributed to later authors can, in fact, be found here. For example, the controversy over who should be given pride of place in the discovery of the telescope fills whole bookshelves.[82] Should it be Galilei, who used it for the first time in 1609 for astronomical observations? Or Johann Lippershey, a Dutch grinder of lenses who submitted the first patent for a telescope in 1608 in the General States of Holland? Or Kepler, who in 1611 described it exactly as an instrument for astronomy? Or perhaps the Jesuit astronomer Christoph Scheiner, who claimed to have discovered sunspots with the aid of telescopes before Galilei, and then got the chance to punish his powerful and more famous rival at the Inquisition's tribunal by using his good connections with the Vatican to influence the verdict against Galilei?[83] There are many contenders for the honor. Galilei—who, shortly after the publication of his *Sidereus nuncius* [Starry Messenger], had finally been appointed in 1610 to the position of court mathematician and philosopher to Cosimo de Medici, Grand Duke of Tuscany,[84] a post he had long coveted—had a very strong position in the historiography of science. For a long time this role went uncontested, even though Kepler had explicitly acknowledged in his work on dioptrics, which contains the treatise on the astronomical telescope, that he was indebted to the Neapolitan Porta for suggestions of considerable importance.

In the case of the telescope, I think there are two important points: first, practice existed before theory. Galilei developed his first device intuitively and by following hearsay, not according to exact calculations that had already been worked out in detail. He established the use of the instrument in astronomy, just as Porta had at first installed the convex lens in his camera obscura according to the principle of trial and error. Yet, fifty years earlier, Porta had sensed the possibility that vision could overcome great distances by using specially ground lenses, an experiment he describes in book 17 of *Magia* II; he only goes into more detail in *De refractione*. Second, Porta was not primarily interested in contributing to improved visibility. It was the invisible that he wanted to hunt down, and he was fascinated by access to things that would remain unseen and intangible without the use of aids. Thus, his description of a glass with which one "can see further than one can even imagine" becomes an idea that oscillates between a scientific instrument and a model of a medium. The "perspective," as he calls it in the text, conjures up associations with that important medium of the twentieth century, television, which during its technical model phase in the nineteenth century was actually known in Germany as *Perspektiv*. Paul Nipkow's master patent of 1884 for mechanical-electrical television was essentially a combination of telescopy—breaking down images or objects into points for consecutive transmission—and electricity, as the precise accelerator of the scanned and transmitted dots and reconstituter of the image.[85] In Kepler's commentary on Witelo, he makes it quite clear that it was this aspect of Porta's preparatory work, the inspired vision of possibilities, that played a decise role in his own. As a convinced supporter of the ideas of the atomists, Kepler proceeded—as had Ibn al-Haytham and Maurolico before him—from the assumption that entities consist of an aggregation of a great number of nondivisible elements. From these particles, in the form of light dots, the light rays move out infinitely far in all directions and in straight lines forming narrow cones until they encounter resistance. In the perceiving eye, they pass through the focusing convex lens of the pupil (which Kepler also called the "window"), are then refracted by the cornea and crystalline interior of the eye, and finally are projected as a cone on the retina, which Kepler calculated precisely.[86]

To the end of his days, Porta made no secret of his annoyance with Galilei's ignorant and arrogant attitude toward Porta's preliminary work. In 1610, when he was over seventy, Porta became a member of an academy for the second time. The academy, Federico Cesi's Accademia dei Lincei in Rome, was named after

Figure 4.10   The Jesuit Father Christoph Scheiner suffered from the fact that his work in astronomy was little regarded, being overshadowed by that of Kepler and Galilei. In this portrait (detail from Braunmühl 1891), he stakes his claim to co-discovery of the telescope. On the right is part of a drawing showing the sunspots Scheiner observed.

Facing page: At the beginning of Scheiner's major work, *Rosa ursina sive sol* (1626–1630), he presents the entire array of his observation instruments. At the bottom of the page is his helioscopic telescope of which he built examples measuring up to 22 meters. Using this device, sunspots could be projected onto a sheet of paper and then traced. Scheiner sits in the background, making calculations and giving instructions.

an animal that had been fabled for its sharp sight since ancient times, to which Porta had contributed his own observations in *Magia* I. In fact, Cesi took the drawing of the lynx from the title page of *Magia* II for the society's official emblem.[87] After the four founding members, Porta was the first to be invited to join, and the next, in the following year, was Galilei. Not the least due to the ongoing dispute between the two new members with their very different temperaments, the academy divided its responsibilities for recruiting new members regionally. Galilei was responsible for Tuscany, and Porta for the Kindom of Naples in the south. Bertolt Brecht devoted the second scene of his play *The Life of Galileo* to the affair of the telescope, which opens with the polemical four lines:

All is not great what a great man does
And Galilei was fond of eating well
Now listen and don't be grim about it:
The truth about the telescope.[88]

Seeking the originator(s) of other achievements connected with catoptric theater—the dramatic presentation of effects using mirrors—quickly becomes a journey through a maze. To the present day, Athanasius Kircher is credited with inventing the device that Gustav René Hocke named the "metaphor machine." Yet in *Magia* I, Porta describes a contrivance that utilizes the same effect and later became a standard technique for manufacturing illusions in the cinema:

In the following manner a mirror can be set up such that when you look into it, you shall not see your own face but some other form that is not to be seen anywhere round about: fix a flat mirror on a wall perpendicularly above [another] wall. Incline it from the top at a specific angle. In the wall opposite [the flat mirror] cut a hole the size of a painting or a statue, which one places in the corresponding size in front of the hole in order to conceal it in this way from the person who looks [in the mirror]. This will make the matter even more wonderful. . . . From its fixed position, the mirror must capture the image in such a way that the gaze [of the spectator] and the visible object meet in the looking-glass. . . . Thus, when a spectator comes here, he will never see his own face, nor anything else besides. And when he stands opposite the mirror and reaches the intended spot, he shall see the painting, or some other thing, which he cannot see anywhere else.[89]

The description of such an arrangement, together with exact calculations of the functional angles of the adjustable mirrors, appears approximately four centuries earlier in *De speculis,* also known as *Pseudo-Euclid,* a work that actually proved to be a compilation of fragments by a number of classical authors (Euclid, Heron, Ptolemy, and probably Archimedes). The very first section describes an experiment where "a mirror is positioned so that an observer may view the image of an object in it but not his own." Before proceeding with the technical details, the commentators give us an indication of how difficult the question of the original inventor is: "This problem is the same as No. 18 in the work on catoptrics by Heron [of Alexandria]. In addition, it appears . . . in Valentin Rose: *Anecdota Graeca et Graecolatina,* and also in Witelo V, 56. The Risner edition of 1572 also cites Ptolemy 9 th. 2 catopt. Alb. Magnus mentions this problem with reference to Euclid in *Prospectiva.*"[90] In summary, when the knowledge that is in an invention has been developed over centuries, the question of who actually invented it first becomes rather pointless.

The *Pseudo-Euclid* describes many of the most cunning effects that it is possible to produce with flat, convex, and concave mirrors and which Porta claims are his inventions. However, two of his descriptions do contain very original additions to the arrangements which I have not discovered anywhere else. The first is described in the original edition of *Magia naturalis:*

So may a man secretly see, and without suspicion, what is done afar off, and in other places, which otherwise he would not be able to do. But you must be careful in setting your glasses. Let there be a place appointed in a house or elsewhere, where you wish to see something, and set a glass against your window, or hole, that is toward your face. Let it be set up securely, or fastened to the wall, if needs be. By moving it here and there and inclining it in all directions and looking on it, and coming toward it you shall attain in making it reflect the right place. And if it be difficult, you cannot mistake, if you use a [diopter] or some such instrument. Let this be set perpendicular upon a line, that cuts the angle of reflection, and incidence of the lines. In this way you shall see what is done in that place, very clearly.[91]

Porta goes on to describe variations on this construction that implement several mirrors due to the particular layout of the locality. Strangely enough, this rather bizarre arrangement is almost exactly the same as one installed by the psychoanalyst Sigmund Freud at Berggasse 19 in Vienna's ninth district in order to observe secretly his wife Martha and her sister Minna. Their bedroom was in the farthest corner of the apartment. The small adjustable mirror fixed to the window frame of his study afforded Freud a privileged and covert look into the private sphere of the two women.[92]

TRACTATUS [PSEUDO-] EUCLIDIS DE SPECULIS.

1.

Praeparatio speculi, in quo uideas alterius imaginem et non tuam.

Sit *ab* paries supra superficiem *bg* orthogonaliter erecta, et *bd* sit 5 speculum, quod inclinetur secundum quantitatem tertiae anguli *abg* recti sitque speculum quadratum. Deinde protrahatur linea *be*, donec angulus *abd* sit tertia recti. Deinde producatur a linea *edb*, quae est cum superficie speculi, linea una, quae sit linea *eg*, orthogonaliter; angulus ergo *beg* est rectus. Sitque locus uisus punctum *g*, a quo ad punctum *d* protraham lineam. A puncto quoque *d* producam lineam cadentem supra superficiem *bg*, donec angulus *zdg* angulo *edg* aequalis. Et protraham *zh* perpendicularem supra superficiem *bg*. Et producam lineam *it* lineae *db* aequidistantem, quae est speculum. Et ponam *zi* aequalem *db*. Et depingam in linea *it*, quae est tabula, *zt* quamcunque uoluero formam, et ponam eam in loco *zi*, scilicet lineam totam.

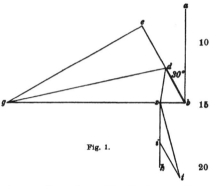

Fig. 1.

Cum ergo considerauerimus a loco *g*, uidebimus formam in speculo, 25 nostram uero formam non uidebimus. Et haec est huius forma.

Figure 4.11 First page of the *Pseudo-Euclid* describing an arrangement of mirrors, which Athanasius Kircher also used for his "allegory machine." (Björnbo and Vogl 1912, p. 97)

The second originality is inspired by a purpose that is quite the opposite of surveillance; it recalls Porta's passion for secret writing and friendship as the motivation for his inventions. In his discussion of parabolic mirrors, he describes how to write messages in inverted script on such a mirror and project this text at night into a dark prison cell where a friend is incarcerated. In the writings of the Neapolitan scholar there often surface reminders, such as this one, of the ever-present threat with which he lived. The fact that this experiment was reproduced prominently in treatises on optical instruments well into the seventeenth century serves to underline that the threat to unorthodox thinkers from the Inquisition and their secular officers continued for a long time.

Vilém Flusser came from Prague, a city where alchemists have their own streets named after them. When the Nazis forced him to flee his native city, he went first to England and then to Brazil, before finally returning to Europe. Flusser had no doubt whatsoever that magical thinkers, with their recklessly experimental approach to the phenomena that interested them, are among the founders of modern science.[93] In his lectures, Flusser often jumped back and forth between the reality of facticity and fecund speculation, or sketched the identity of a thought that operates within the strong tension of *curiositas* and *necessitas* (curiosity and necessity), as Porta defined the two most important motivations for the work of the researcher.[94] Flusser charismatically embodied such an identity. He fired European debates on the media in the 1980s with enthusiasm, when, after structuralism, Marxism, and Lacanism, people were thirsty for new impulses. The great abstract works bored artists and others, who wanted to change the world using the latest media, for they were unable to discover in these texts any relationship to their own work of transformation. By contrast, Flusser succeeded in arousing passionate motivation to try out the possible shift "from subject to project,"[95] both in theoretical and practical media work with all its contradictions and paradoxes. For established academe, his thinking, characterized by its mental leaps between the disciplines, is unacceptable even today.

Porta was far more committed to the magic tradition of thinking than to the emergent European rationalism with its rigid divisions between the subjects that understand and the objects that are understood by them. The senses and the mind represented for Porta the kind of vicissitudinous unity that Erwin Schroedinger appreciated so much in Democritus's thought and which he found exemplified in a dialogue from the Fragments: "the intellect says: 'Ostensibly there is colour, ostensibly sweetness, ostensibly bitterness, actually only atoms and the void,' to which the senses retort: 'Poor intellect, do you hope to defeat us while from us you borrow your evidence? Your victory is your defeat.'"[96]

Marveling and with passionate interest, Porta opened up the world around him for himself and others. First and foremost he explored the world here on Earth, with its absurdities, tensions, and turbulences, not the celestial world of the church nor the conceptual world of the mind. For this, Porta came in for much harsh criticism, not least from many who came after him who regarded themselves as the guardians of pure (natural) science. Of course, this disdain did not deter such people from plundering the Neapolitan's rich treasure trove whenever the opportunity presented itself. For example, in *Physikalische Optik*

Figure 4.12 *Center:* The first letter, *M,* of Porta's *Magia naturalis* (1558) against a ground with two entwined bodies. Alchemistic imagery symbolizing the union of unlike elements reoccurs throughout his entire oeuvre.

"When bodies melt, they finally come to their senses," wrote the physicist Johann Wilhelm Ritter in his *Fragments* (1810/1984, p. 77), "only then do they understand one another. So, too, it is with us: the 'warmer' we are, the more we grasp and understand — we thaw." The illustrations show distillation apparatus (top) and heating with a burning glass (bottom) from Porta's treatise on alchemy, *De destillatione* (1608, p. 40, 30).

(1921), the Austrian physicist Ernst Mach acknowledges Porta in several places for his original ideas and, at the same time, attests to his "significant excess" of "foolishness"; attributes him with an "unscientific, uncritical way of thinking," and reproaches him severely for merely entitling book 20 of *Magia* II "Chaos": "the entire book deserves this title."[97] For Mach, science must be formulated precisely, like the flight of a missile, and it must be pure. Mach has no notion of the secret as the other side of what is evident, no conception of the interaction between chaos and order as found, for example, in the most fascinating of all drama worlds from Porta's era: "There are more things in heaven and Earth, Horatio, than are dreamt of in your philosophy," says Shakespeare's Hamlet in the play written at the turn of the sixteenth century. And Michel Foucault formulates succinctly in two questions all that needs to be said about the arrogant stance of people who have the privilege of earning a living from knowledge, research, and communication of their subjects: "which kinds of knowledge do you seek to disqualify when you ask, 'Is it a science?' Which human subjects, who speak and carry on a discourse, which subjects of experience and knowledge do you seek to 'minorize' when you say 'I, who hold this discourse, hold a scientific discourse, I am a scientist'?"[98]

Porta was not a free rider of factualism. He was a juggler of the possible, which for him included risky games with the impossible. Johann Wolfgang von Goethe, who praises his "blithe and diverse knowledge" and in the same breath condemns "a decided propensity to folly, to the bizarre and unobtainable" as well as Porta's refractoriness in declining to reduce variety to a common denominator, nevertheless concludes with fair words for the Neapolitan: "Although one cannot see in him an intellect that would have been capable of summoning the sciences to unity in any sense whatsoever, one is compelled to recognize him as an alert, ingenious collector. With assiduous, restless activity, he explores thoroughly the field of experience; his notice reaches out everywhere, his collector's passion never returns unsatisfied." And a few lines before this, Goethe writes, "With reluctance, we now take our leave of this man of whom a great deal more remains to be said."[99]

Light and Shadow — Consonance and Dissonance: Athanasius Kircher

It would be a very bad dissimilarity of character, if any one musical tonality had the prerogative of being more perfect or imperfect than the others.

—ERNST FLORENZ FRIEDRICH CHLADNI, *ENTDECKUNGEN ÜBER DIE THEORIE DES KLANGES*

Tuning

The few existing portraits of Giovan Battista della Porta depict him with strangely heavy eyelids. Theatrical effects with mirrors and distorting, magnifying, or duplicating lenses play an important role in his work, but his studies of nature exhibit an equally strong relationship to attractions that appeal to the senses of smell, taste, and touch. With his experimental and magical approach to the natural world, Porta alternated between alchemy, the study of all living organisms, and the nascent physics of the visible. The world of sounds, tones, harmonies, and rhythms, however, did not occupy much of his attention—it only crops up in minor facets in the oeuvre of this tone-deaf ex-pupil of the Pythagoras School in Naples. In his *Magia naturalis,* for example, there are various descriptions of pipes that act as amplifiers for the human voice as well as straight and spiral-shaped instruments that prolong tones. He experimented with the way sound travels in circular architectural structures and presented a "whispering gallery," as can be experienced today under the dome of St. Paul's Cathedral in London. He also devoted some attention to the Aeolian harp, an instrument on which the wind produces varying harmonics. In his book on physiognomy, he discusses the voice as an expression of character. In his study on magnetism, he examines the phenomenon of tarantism, a dancing mania in medieval Europe accompanied by hallucinatory delirium, which was popularly believed to be caused by the bite of the tarantula. For Ficino, Agrippa, and Campanella, just as for Porta, tarantism was a "primary example of musical magic"[1] where the interaction of mind and body was particularly evident; Athanasius Kircher even wrote several polyphonic compositions for it. Notwithstanding these examples, it is obvious that the world of sounds did not interest Porta greatly. In his body of work, the passages on hearing, sound, and music are marginal.[2]

The English scholar Robert Fludd, who began publishing his weighty tomes shortly after Porta's death, had a very different set of priorities regarding sensory perception. To date, historians of media whose allegiance is to the paradigms of the visible and the image have not adequately addressed Fludd's work, which is rashly dismissed because his worldview is regarded as overly mystical and "reactionary."[3] However, from a media-archaeological point of view, his work is not exclusively oriented toward the past. Rather, it represents a pivot between the runaway heterology of Porta and Kircher's attempt to organize existing knowledge about the world's phenomena into a consistent universal system of signs, artifacts, and their relationships to each other. In a monumental hermetic undertaking, Fludd sought cohesion for the strands of natural philosophy, which were beginning to drift apart, in a single idea that was not overtly articulated in the things themselves but constituted their hidden structure and driving force. In this he followed the Neoplatonic ideas of Ficino, who understood the "reflective force" of the "world-soul" as "the direct cause of the order, or harmony, of the world," which creates and organizes "the mathematically conceived analogy . . . and joins the individual to a whole."[4] For Fludd, the most important art of all was music. In this, he was in agreement with many of his contemporaries who sought a model for theory and praxis that would be capable of expressing everything in a direct way and in a single form. Musical tones were understood as indicators of reality.[5]

By training, Fludd was a doctor. In 1598, he completed his master of arts at Oxford and then traveled for several years in Spain, France, Germany, and Italy, where he became acquainted with the works of Albertus Magnus, Ficino, Cardano, Campanella, Paracelsus, and probably Porta. Many of these writers' ideas appear in his books. From 1605 to 1609, he was again in Oxford, where he completed his doctor of medicine. This degree took several attempts because

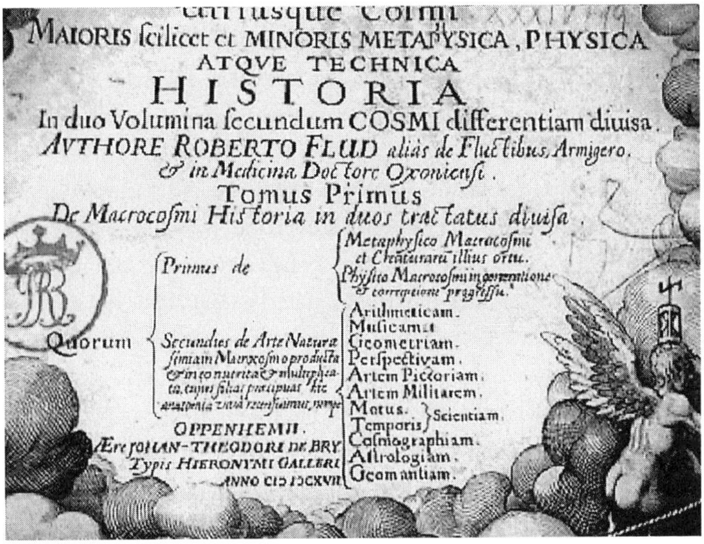

Figure 5.1 *Top:* The title page of the second volume published in 1619, which gives the Jesuit College in Cologne as the former owner. *Bottom:* Detail from the title page of the first volume of Robert Fludd's history of the macrocosm and microcosm with a brief table of contents (Fludd 1617).

Light and Shadow—Consonance and Dissonance

his strong adherence to the ideas of the Rosicrucians, and particularly of Paracelsus, did not recommend him to his examiners. During the struggles between Reformation and Counter Reformation, Fludd found himself between the fronts, as Catholics and Protestants alike rejected him. To his roots in medicine, both as theory and practice for the healing of humankind, he remained committed all his life. Many technical inventions, can be traced back to Fludd, such as the adaptation of the thermometer to measure body temperature.[6]

Between 1617 and 1619, Fludd's gigantic *opus magnum* on the history of the macrocosm and microcosm appeared. In the first volume, he unfolds its structure in the form of a widely branching "master plan."[7] Physics and metaphysics of the macrocosm are the subject of the first *tractatus*. Under the general heading of "arte naturae," the second deals with individual fields of natural philosophy and their methods, from arithmetic to mechanics and geometry plus their various applications. This magnificent folio alone has almost one thousand pages of small print. The second volume focuses on the microcosm, which for Fludd means the individual human. In two further tracts, he deals with human physical and metaphysical anatomy, the connections to the macrocosm as well as to his own, very different, scientific fields of research, which range from theology and metaphysics to music, applied engineering, and meteorology. A third, planned tract was never realized, as the project was too ambitious. Instead, Fludd drew up a master plan for a new grand-scale intellectual venture, the "Medicina Catholica," which he also failed to realize. Like the additional sections on the macrocosm and microcosm, these fragments were published as separate, independent studies.[8]

To me, the worldview of this doctor of medicine and natural philosopher, with his close relationship to Paracelsus's doctrine, is rather remote and inaccessible. More than seventy years after the Polish astronomer and jurist Nicolas Copernicus had revolutionized astronomy, and at the same time as work by Kepler and Galilei, as well as Francis Bacon's cool conception of science with its watchword of "knowledge is power," Fludd's outlook seems in many ways anachronistic. In the labyrinthine windings of his encyclopedic work, however, there are many chambers that are well worth a visit. Deleuzians, for example, would be delighted with the chapters on "vulgar arithmetic or, the algorithm," on "arte militari" and its explanations of war machines of all kinds, or the special set theory for military applications, which he calls cohort theory. Fludd's discourse on *ars memoria,* which has strong parallels with that of Porta but is presented more impressively, has already been discussed in depth by Frances

Yates. However, the chapters on hydraulics, kinetics, and pneumatics, or the section on horology and the devices for measuring time, add very little to the work of his Neapolitan predecessor.

The real attraction of Fludd's work lies in one instrument and his interpretation of it. I am referring to the monochord, an instrument that was used in ancient Egypt and Greece to produce notes of the harmonic series and to measure the mathematical relations of musical tones. Fludd uses this one-string instrument to describe the world. In principle, he follows the discovery, attributed to Pythagoras, that "the division of a single string in ratios of small whole numbers (e.g., 1:2, 2:3, 3:4) produces musical intervals, which, composed in the harmony of a song, can move us to tears and, as it were, touch our very soul."[9] This discovery became the basis for the analogy of reality to numbers, as determined by Pythagorean doctrine. Through numbers and their proportions, anything that could be measured spatially became a principle of harmony, that is, something metaphysical. Fludd's starting point is, to begin with, strictly geometrical. Two uniform triangles pushed together, which in some illustrations take the form of three-dimensional pyramids or cones with elliptical or round intersections, constitute his basic model. One triangle's base rests in the bright celestial sphere, sharing its base line almost equally with the equilateral triangle that represents the Holy Trinity, and its apex points toward the ground. The base of the other triangle rests in the dark matter of the Earth's interior and its apex reaches the divine sphere. Both of the triangles' centers are intersected by the sphere of equivalence, which holds the entire construction in balance. At the center is the sun as *anima mundi*. As God's mediator for the animation of the world, the sun is responsible for forming the particular from the formless sphere of matter. This concept is not so much a reference to the Copernican heliocentric worldview as to a fundamental idea in alchemy. Through the various mixings and separatings during the alchemical process, the driving force inherent in matter is gradually liberated. Fludd's two triangles should be understood as being in a reciprocal dynamic relationship: earthly matter is in a state of striving to ascend toward the Divine, and the divine principle constantly works downwards toward the Earth. In the in-between develops the rich realm of different things, the phenomena of what can be perceived, and also of what can be imagined. For, the upper sphere beyond the sun is filled with nonvisible attributes of the celestial or divine.

Analogous to this construction, Fludd designs his instrument, combining geometry with arithmetic.[10] The string of the monochord spans the entire

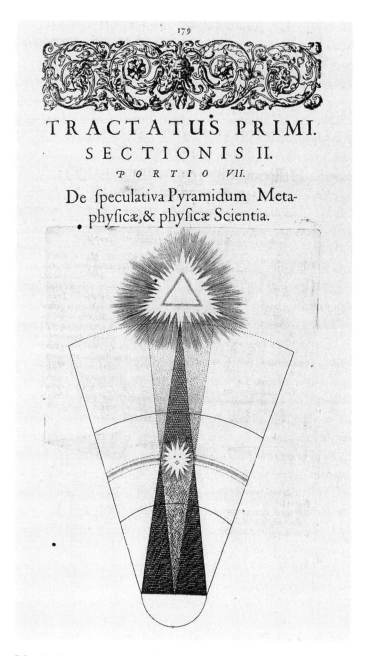

TRACTATUS PRIMI.
SECTIONIS II.
PORTIO VII.

De fpeculativa Pyramidum Meta-
phyficæ,& phyficæ Scientia.

Figure 5.2 Fludd's two dynamic pyramids/triangles with the sun as *anima mundi* at the center, that is, at the point where the dynamics from the metaphysical to the physical and vice versa are ideally congruent.

vertical hierarchy from extreme darkness to bright luminosity. Since Pythagoras, the scale—that is, the series of intervals—had been determined by the length of the string, not yet by vibration frequency. The string extends over two octaves of a tone; at its exact midpoint—at the point where Fludd's *anima mundi* resides—lies the division between the two octaves. Two-thirds of the string gives the interval of the fifth, three-quarters give the fourth. With the two octaves (1:4) and the three simple ratios of 1:2, 2:3, and 3:4, which even those with no musical training can recognize as the musical intervals that make up the basic system of consonance, Fludd conceives his world as an instrument: a harmonious construction with multiple variations and sections. The two octaves mirror the triangles of the Divine and the material. The intervals relate to the different spheres of Earth and the heavens in a graded system of correspondences. Fludd does not make further divisions, as calculated and made by Arab scientists around the turn of the first millennium, parallel to their major work on optics, and formulated theoretically in Johannes Kepler's *Harmonice mundis,* such as the minor third (5:6) or the sixths (5:8 and 3:5). Fludd remains within the Pythagorean system where the furthest division is into fifths and lower fifths. God assumes the function of He who tunes the instrument most perfectly due to His omniscience; Fludd refers to God several times as *pulsator monochordii.*

Fludd's monochord is like a media artifact with which he attempts to encompass the great variety of relations of the world in a simple and symbolic form. In one of his replies to Kepler, he states: "What he [Kepler] has expressed in many words and at great length, I have condensed and explained with hieroglyphic, deeply significant figures, not because I am enamoured of images . . . but because I . . . have decided to bring together the many in the few, to collect the extracted essence, to discard the sedimented essence."[11]

Fludd's design expresses vividly a basic problem of musical pitch. From his perspective, he solves it elegantly, yet it remains a matter of dispute even in our age of the electronic reproducibility of tones. Two approaches exist for determining the intervals between notes: first, the mathematical one, which is based on numbers and their relations to each other and views the exact determination of pitch as a perfect quantity; second, the approach that is oriented on the physiological aspect of tones, their production and perception, which, instead of considering the theoretically infinite number of possible notes, only operates with those that can be heard as distinguishable consonances. Even in classical antiquity, music theorists were split into these two camps.[12] The Pythagoreans, particularly their working group of *mathematikoi,* declared numbers and their

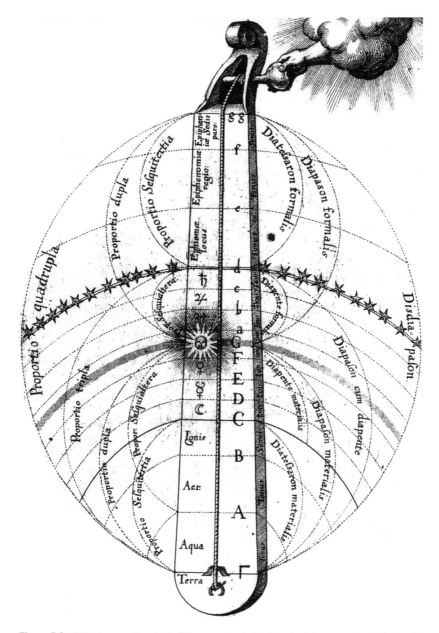

Figure 5.3 "Musica mundana": Fludd's monochord with the principles governing the intervals and the hand of the Great Pulsator, who is tuning the instrument (Fludd 1617, p. 90).

relations to be the starting point for correct pitch, which Plato raised to a philosophical doctrine and rounded off metaphysically in his *Timaios.* The school of Aristoxenos, which followed Aristotle and, in the 4th century B.C., produced the first great treatise on the "elements of harmony," rejected the notion of numbers as the determining factor. Only as a secondary step are numbers able to formulate the harmony that is produced solely by experience, by what is hidden, by hearing and musical intuition: "to use calculation to construct intervals that cannot be produced by either voice or instrument and that the ear cannot discern"[13] would make no sense at all.

In the praxis of rendering sound digital, and in general in the debate on the capabilities of computers for artistic production, this ancient dispute is again a hot topic. The drive to produce ever finer divisions in musical tones and intervals in order to process the microstructures of sounds has reached the limits of what is formalizable. To my mind, however, this problem appears to be less a musical one than an economical and technical one, which does not need to be solved by mathematics. Particularly from the perspective of even greater industrial involvement in procedures to standardize the preparation of acoustic material, there does not seem to be much point in searching for further subdividable mathematical relations beyond what the senses can perceive as differences. As for musical praxis, this quest evokes the monster-organs and harpsichords built in late Renaissance times, which were supposed to achieve an even more perfect modulation through multiple divisions of the octaves. Their function was primarily supportive. For example, the archicembalo, built around 1550 in Venice, had thirty-one notes or keys for each octave and was "designed to make it possible to accompany singers and other instruments in any key of any pitch desired, without compromising the major thirds of meantone tuning."[14] With the aid of these monster keyboards, it was possible to produce new and unusual modulations. Initially, in computer-based electronic music, the search for ever-new divisions resulted in an exciting phase of modulated diversity. In the meantime, however, it appears to have arrived at the point where only aficionados of formalized music and computer programmers get excited about it. One of the duties of the Pythagoreans, it is claimed, was to smooth their bed sheets after rising in the morning so that all impressions of their bodies disappeared.[15]

Fludd took the side of Aristoxenos of Tarentum. He declared God to be the highest and ultimate authority for the correct tuning, inexpressible in numbers. In his model of the world as monochord, the Divine principle has to take the dynamic path through dark matter so that diversity of form can arise. Applied to music, this process can be conceived of as the experience of the listener. It is through this process that the Great Pulsator, mediated by the sun as *anima mundi,* also provides for the correct temperament. Without discussing the question in mathematical terms, Fludd argues in favor of distributing the "imperfections," which arise of necessity from the intervals of fifths, among all the intervals.[16]

The model of harmony developed by Fludd was fiercely criticized by his French contemporary Mersenne, who had published his own fifteen-hundred-page *Harmonie universelle* (1636–1637), and by another great theorist of cosmic harmony, the mathematician and astronomer Johannes Kepler. Between Kepler and Fludd a dialogue developed that lasted for many years, much of which has been published, on the ideal way to achieve perfect harmony. In the appendix to his *Harmonice mundis* of 1619, Kepler accuses his "English friend" of having developed his universal harmony purely imaginatively and, ultimately, through the nonpermissible comparison of what cannot be compared—namely, light and shadow. By contrast, he, Kepler, had developed his doctrine of world harmony through analyzing the movements of the planets.[17] Fludd replied with an assertion that, more precisely formulated and four hundred years later, became the focus of a fundamental debate in mathematics and was still controversial at the beginning of the twentieth century:[18] one must make a fundamental distinction between things natural and things mathematical. Opinion and abstraction have very different ontologies with regard to knowledge. It is the "business of ordinary mathematicians to concern themselves with the shadows of quantities; however, the alchemists and hermetists grasp the true marrow of natural bodies." Kepler's reply was equally drastic: "*I am holding the tail,* but I hold it in my hand. *You could be clasping the head* with your mind, if you are not dreaming."[19] Eva Wertenschläger-Birkhäuser interprets the escalation of this dispute between the analyst and the dreamer with the apt remark that each was fighting his own shadow in the other. In essence, both Fludd and Kepler take archetypal images as their starting point. For Kepler, this is the sphere or circle, from which he derives the basic geometrical forms that are necessary for producing harmony. For Fludd, it is the triangle, the symbol of the Holy Trinity and Pythagorean basic figure; however, he integrates the form of the ellipse into it, into the base of the pyramids.[20] Both men use geometry as a prestabilized harmony, a concept that, in the same century, was to become pivotal for Leibniz. The essential difference between the two protagonists is that Kepler, standing on the threshold to modern science, assigns a higher priority to quantification, whereas for Fludd the highest principle is a metaphysical quality that, ultimately, is not quantifiable. There results for art—in this case, music—a decisive difference: inherent in Kepler's symbol of the sphere is the idea that not only is it possible to return to the starting point, it is a law. Despite setting intervals, the notes of an octave can be identical. In Fludd's pyramid, the form of the spiral is implicit. In a spiral, the beginning and end points approach and move away from each other in a dynamic relationship.

At the time, it seemed as though the English physician and philosopher was fighting a lost battle. His approach, strongly influenced by hermeticism and alchemy, in which he attempted to unite the diametrical opposites of light and dark, mind and matter, good and evil, masculine and feminine, did not stand a chance. The thinkers of the Enlightenment, fixated on light and concepts, marginalized researchers like Fludd, relegated them to the periphery of scientific discourse, or ignored them entirely. This state of affairs prevailed until relativity theory and quantum mechanics demanded a radical rethinking of the old opposites of calculation and imagination, dimensions and boundlessness, materiality and intellect. This is bound up with the epistemological status of their disciplines. "Molecules, atoms, electrons, quarks, or strings are . . . not the building blocks of nature, they have not been discovered but invented," as the physicist Hans Primas from Zurich says in his essay on "dark aspects of natural science," where he argues strongly for the unconscious as a productive force to be included in the scientific explanation of the world and its phenomena.[21]

The opening of his first tract on *The Structure of the Macrocosm and on the Origins of its Creatures* illustrates that Fludd's thinking was not only turned toward the past, but his imagination also touched upon constellations of problems that pointed toward the future. In section five, Fludd discusses the relationship of darkness and light as an intricate theological problem relating to the relationship between darkness and privation. For this, he found an appropriate and bold image. So much ink was used to print it that the page of the original is wavy. The first illustration of the book, on page 26, is a pitch-black square, a symbolic image of unformed matter. Measuring 144 by 146 mm, the copperplate engraving is not quite symmetrical; however, there can be no doubt that Fludd intended it to be a square. All four sides have the same legend indicating that matter should be understood as being in a state of infinite expansion: "Et sic in infinitum." In the accompanying text, Fludd not only risks criticizing a Catholic institution in the shape of St. Augustine, he discusses the problem as

Figure 5.4 Fludd 1617, p. 26.

essentially an aesthetic one, as a problem of the genesis of forms, albeit with an exciting psychological twist. For Fludd, privation is a category in the relation of darkness to the experience of lack, the dialectic of presence and absence:

In St. Augustine's writings against the Manicheans, he says that privation [*privatio*] is nothing other than darkness, which is defined by the absence of light. However, when one looks more closely at the meaning of darkness [*tenebrae*, which also means "shadow"], one sees that it means more than "privation." For according to Moses, darkness was upon the deep waters before light was created and the Earth took on form. One can only speak of privation, however, with reference to a particular point of reference, that is, when something is absent that was present before. In this case—and I concur with St. Augustine—all privation is darkness, for a light-giving form is absent, but the opposite is not the case. Thus, it is clear that since the beginning of the world darkness or privation can only arise because of transience. And such is the nature of the basic elements and elements of the sphere below.[22]

In other words: darkness should not simply be understood as the absence of light, or evil as the nonpresence of good. Both must be understood as complementary. They interlock with each other like the two dynamic pyramids in the image representing Fludd's worldview.

A Roman College as a Control Center

Athanasius Kircher's universe of knowledge and faith appears to be a good deal more complex than Robert Fludd's. It is spectacularly impressive by its volume alone. In an article on Kircher, the physicist and writer Georg Christoph Lichtenberg remarked that whenever Kircher picked up a pen, a folio was the result.[23] On over sixteen thousand pages of thirty-two published works,[24] Kircher sets out a staggering cornucopia of phenomena and their possible connections. Aptly characterized as a polymath (by Godwin), Kircher was one of the last of this type of scholar who set out to combine and unify everything that can be formulated as text in a single body of work. One of the commandments that Ignatius of Loyola, founder of the Jesuit order, gave to his brothers was that each one of them, as a microcosm, had to embody the macrocosm and, particularly, knowledge about God and nature in its entirety. Here, theology and science were still together. In the worldview of the Jesuits in principle there was no epistemological separation between religion and philosophy or knowledge of nature. The one must imbue and enrich the other. This view did not apply, however, when dogmas of the Catholic Church were involved. Neither Giordano Bruno nor Galileo Galilei found in Kircher an advocate of their ideas, and his works contain many passages distancing himself from the magician Porta, the Rosicrucian Fludd, and the hermetic mathematician John Dee, while at the same time being brimful of their ideas.

Kircher lived through the reigns of ten different popes. His imposing network of clients and patrons—to which it seems that all representatives of secular and clerical power in the Catholic parts of Europe belonged—enabled the texts to appear in expensive deluxe editions with opulent copperplate engravings and many symbol characters that had to be cast specially. Some of the works, for example *Musurgia universalis* [Universal Art of Music], with its first edition of fifteen hundred copies, were real bestsellers, carefully edited and designed by the author, his publisher, and engravers for an international readership. His books were published in Naples, Cologne, Augsburg, Rome, Leipzig, Avignon, and above all in Amsterdam.[25] The inscription "Athanasius Kircher S. J. (Societas Jesu)" on the title pages of the folios functioned like a brand name in the early European book market.

Kircher was obviously an extremely industrious and gifted communicator. However, what many commentators present in the secondary literature as the incredible achievement of one man was to a considerable extent the results of an organization that one may justifiably term an excellently appointed and strategically operating media concern. Since the official recognition of the Jesuit order in the papal bull "Regimini militantis" in 1540, the Societas Jesu of St. Ignatius had developed into an individual elite order to preserve and propagate Catholic doctrine and faith on a global scale. It was headed by a "general" who was relatively independent of the Pope, and its followers understood themselves as "courageous warriors."[26] In the founding years of the order, its spiritual father from Spain and his followers wore sackcloth and ashes so that they might mix with the poorest of the poor—beggars, prostitutes, lepers, and cripples— whom they desired to convert first of all and to liberate from the hell of their earthly existence. The Jesuits lived according to strict rituals of self-castigation as laid down by Loyola in the *Spiritual Exercises*. In his history of the Jesuits, René Fülöp-Miller tells a bizarre story about this period: after excessive self-torture and extreme fasting, Loyola became so ill that the doctors gave up hope, and a few devout women begged for his clothes as relics. The mistress of the house where he had been taken in, "wanted to grant their wishes and, in order to take out the clothes of the supposedly dying man, she opened Inigo's wardrobe, whereupon she started in horror. In the cupboard, hanging neatly in a row, were the most dreadful instruments for the mortification of the flesh: a penitential belt made of woven wire, heavy chains, nails in the shape of the cross, and underclothes with iron thorns woven into them."[27]

However, even at that time it was clear to Loyola that the image of a mendicant order and charitable social organization, whose members delighted in their own martyrdom, was not exactly well suited to the plan of preserving the traditional Catholic worldview for the future, keeping it strong and influential, and saving it from the reformers Luther and Calvin and the heads of state who sympathized with them. Moreover, this work had to be accomplished in a period of the rise of world markets, constant armed conflicts, the spread of international

Figure 5.5 Portrait of Athanasius Kircher. In *Mundus subterraneus,* 1665, vol. 1.

transport systems, and the nascent systematic sciences of nature. Fülöp-Miller writes, "When the Jesuits in Cologne continued to invest a lot of time and effort into missionizing rural areas, Ignatius rebuked them for this and wrote that such activities were only commendable as a beginning. There was nothing worse than pursuing such small successes and losing sight of the greater tasks: the Jesuits were not merely striving for the conversion of the peasant masses but for far greater ends."[28] Loyola began work on a profile of his organization where the members saw themselves as the avant-garde, not only in the fields of theology and philosophy, but also in astronomy, mathematics, physics, painting and sculpture, architecture, music, theater, and literature. All means justified the end of maintaining and promoting the Catholic world. Highest priority was given to an education system that was both advanced and open to the new and unusual. Willingness to embrace asceticism mutated into the duty to exercise discipline, which included the punctilious education of the intellect and promotion of physical health. The order sent its missionaries to the farthest corners of the globe to establish the Catholic faith—to Mexico, Salvador Bahia and the Amazon region of Brazil, Africa, India, Japan, and China. Their modus operandi in such lands followed rather modern guidelines: with sensitivity and understanding, they must adapt to the existing cultural circumstances, learn to listen and pick things up. They were not simply to attempt to remold foreign cultures to their own worldview but rather were to integrate their world-view into these cultures. Thus, in the last decades of the sixteenth and the first of the seventeenth century, the Jesuits built up a worldwide network of missionary work, education, and art, with the Vatican as its ultimate political and supervisory authority.

When, in 1633, Kircher began to teach mathematics as a professor at the Collegium Romanum in Rome, the Jesuit information and communication system was already in place and functioning well. From all over the world the missionaries, like correspondents, sent in their finds, reports, observations, and interpretations of cultural particulars and constellations to their spiritual headquarters, where they were collected, archived, evaluated, and utilized in teaching and publications. The fathers received their instructions for their missionary work from the Collegium Romanum, the academic seat of the Societas Jesu. Built over the remains of an ancient temple of Isis, even today much of it looks more like a fortress than an academy.[29]

Thus, Kircher was situated at the very heart of the power center of knowledge, and he made masterly use of its network, to which, for example, his former

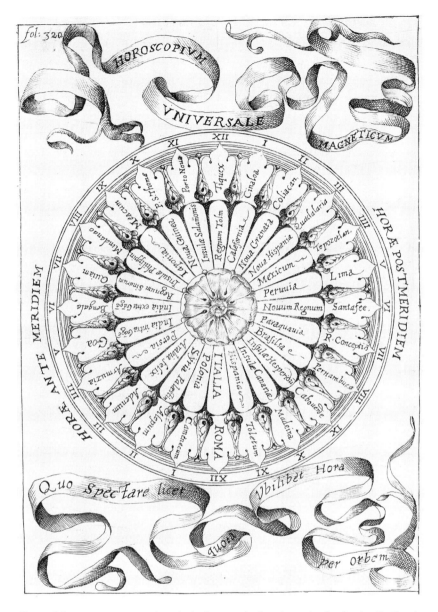

Figure 5.6 The magnetic clock, with the "rosa ursina" at the center, for showing the time in various parts of the world that featured significantly in the Jesuit network. It illustrates the entirely different political and cultural geography of the seventeenth century. Mean time, naturally, is the time in Rome. In *Magnes sive de arte magnetica* (1641), Kircher's first book publication.

student Gaspar Schott and the astronomer Christoph Scheiner belonged. His books give the impression that their author is a cosmopolitan globetrotter; however, with the exception of one long trip to Malta, Sicily, and Naples, Kircher did not travel much beyond the immediate environs of Rome. Information on foreign animals, like the chameleon, or the exotic percussion instruments and documents of all varieties that appear in his texts was collected by other researchers and members of the order. *China illustrata* (1667), his popular work on Chinese civilization, was based totally on the travel reports written by other Jesuits, namely, Albert d'Orville and Johannes Gruber—for Kircher never visited the Far East. Yet his treatment of the reports demonstrates such care and insight into the subject matter that the Bibliotheca Himalayica in Katmandu, Nepal, produced a reprint edition in 1979 respectfully acknowledging the work, in spite of many misunderstandings of details, as the "first comprehensive collection of material on China, India, and the adjoining regions, including Tibet, Nepal, and Mongolia."[30]

Combine and Analogize

The operating method of the Societas Jesu in the seventeenth century can be described from a media-archaeological perspective as governed by two principles, which were also of decisive importance for Kircher's own work. These principles were the international network of a thoroughly hierarchical and centralistically structured system of religious faith, knowledge, and politics, combined with the development of advanced strategies for the mise-en-scène of their messages, including the invention and construction of the requisite devices and apparatus.

Kircher's "concept of a closed world order, entirely permeated by interconnected and relational structures"[31] corresponded to the Jesuits' external network of knowledge and the elaborate connections between its complex branches. In his book *Ars magna sciendi sive combinatoria* [*The Great Art of Knowledge or Combinatories*] (1669), Kircher treats this view in detail. Taking Llull's art of combination as a basis, he develops his concept of the world as a boundless collection of different phenomena, which need to be dismantled into combinable and calculable units that can then be re-assembled into harmonious sequences and ensembles. The inner nature of things, in Kircher's view, is not accessible either empirically or experimentally. His world can be comprehended and played out using the signs and symbols through which the principles of its construction are organized and reconstructed. For Kircher, numbers are pivotal—"regula et norma omnium," as he says in his book on the universal art of music.[32] They have the

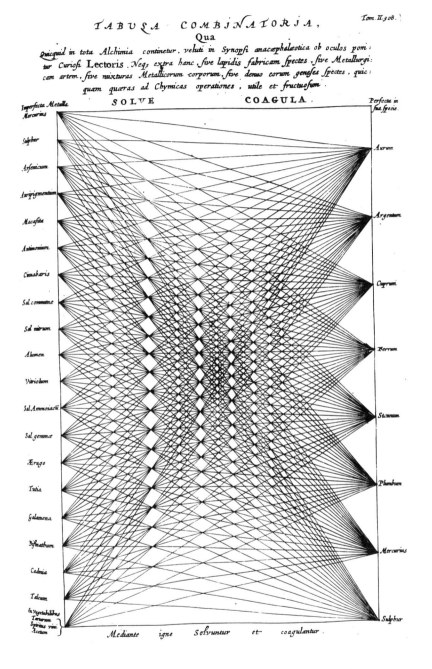

Figure 5.7   One example of the many combinatorial diagrams appearing in Kircher's *Ars magna sciendi sive combinatoria* (1669, p. 308).

unique power to link the singular with the compound, to develop plurality from unity, and vice versa. Three years before *Ars magna sciendi* was published, another scholar, who was just twenty years old, wrote his first work on the combinatorial art with the title *Dissertatio de arte combinatori.* Gottfried Wilhelm Leibniz was a great admirer of Kircher as well as of the late-medieval father of this art, Ramon Llull.[33]

The second essential operation by which transformations are effected is, according to the *Ars magna sciendi,* analogization. Time and again we find in Kircher operative principles from the tradition of natural magic. Steered by the central art of combination, the diversity of individual entities are brought into a system of similar differences and different similarities. Everything that can be seen, heard, or felt is integrated into a true and correct structure that is maintained in harmonious motion by the Great Pulsator. This can work only if the symbols and the things are of the same nature; if language, as well as music, are understood as the expression of nature.[34] As in the works of the natural philosophers of the sixteenth century, numbers, language, and images share the same space as the natural world. The division into extended and thinking entities, which was introduced by his contemporary, Descartes, was unknown to Kircher. His work stands, a monument made of splendid folios of paper, linen, and leather, like the very symbolization of Michel Foucault's thesis in his *Archaeology of the Human Sciences* that "such a linkage of language and things within the same space . . . is only conceivable if the written word is privileged absolutely"[35]—whereupon the Vatican set very great store.

Universal Art of Music

Although Kircher's universe is diverse, his thinking is striking in its strict bipolarity, a trait that reveals his general affinity to the "Baroque antithesis"[36] linked to this critical juncture in history. The Catholic Church was engaged in a bitter struggle against the reform attempts by Lutherans and Calvinists, which it waged as a struggle of good against evil, the divine against the demonic. Kircher's "universal order of things"[37] is determined by the confrontation of consonance and dissonance, which corresponds to the pair of opposites of light and shadow in the realm of visible things. In between the two lies the diversity of individual phenomena. The task of the scholar, and the artist, is to explain the development of diversity from the One, the Divine, and to integrate it, in all its unwieldy dissonance, in such a way that harmonious unity results once again. The transformation of base into precious matter, the cancellation of polarities

Figure 5.8 Between the eye and the ear: Title page for *Ars magna sciendi sive combinatoria,* one of Kircher's magnificently elaborate engravings. The goddess of wisdom floats above the four Empedoclean elements. With an eyeglass on a long handle she is pointing at a tablet on which are inscribed the twenty-seven categories from which "the entirety of human knowledge" can be combined, as the text below states. The left column lists the prime concepts of Raimundus Lullus's combinatorics and at the top, written on the droplets, are the scientific disciplines.

through the process of mixing and remixing of their substances, are also fundamental notions in alchemy.

Like many of his Catholic predecessors and contemporaries, Kircher learned
to sing in Latin long before he actually understood the language. Music occupies an important place in his view of the world. In *Musurgia universalis,* he defines his idea of music's power to transform in an elegant play on words: music
is "monophonic polyphony and polyphonic monophony" [discors concordia
vel concors discordia].[38] Its mode of operating is arithmetic. In the strict sense
of the Pythagorean doctrine of proportion, which was actually borrowed from
geometry,[39] he understands music as a "scienta subordinata," a discipline subordinate to mathematics. A good eighty years earlier, this same point had been
made, forcefully and in best Shakespearean English, by John Dee in his introduction to the English translation of Euclid's *Elements:* "Musicke I here call that
Science, which of the Greeks is called Harmonie . . . Musicke is a Mathematical Science, which teacheth, by sense and reason, perfectly to judge, and order
the diversities of soundes hye and low."[40]

In book 10 of *Musurgia universalis,*[41] Kircher constructs his model of harmony
as an arithmetical structure and makes God the ultimate musical principle. On
the book's title plate, which is richly ornamented and adorned with angels, musical praxis is given the lowest position. Harmony is realized through disciplined adherence to the doctrine of Pythagoras—who is depicted pointing to
the hammering blacksmiths from whom he allegedly heard intervals for the first
time—and through musical genius, in which the divine principle expresses
itself. Mastery, however, can be achieved only by the *musicus* through a unity of
theoretical studies and equally disciplined praxis. (Johann Sebastian Bach was
a great admirer of Kircher; the art of the fugue, with its fleeting movement from
subject to exposition to countersubject, can also be seen as realizing a process of
transformation from unity through exciting plurality to unity. Its contingency
principle is that, as they flee from or chase each other, the fugue's voices, whether
vocal or instrumental, must avoid entering into an inner contradiction that
would tear them apart.[42] Bach's musical exercises constructed on arithmetical
examples are legendary, as are his contributions to tempering and chromati-
cism.[43] Magic, musical genius, and mathematics complemented each other
suberbly.)

Although Kircher took great pains to distance himself from Robert Fludd,
the fact remains that the basic ideas in their notions of world harmony are very
similar. Further, they do not differ substantially from the concepts in *Harmonie*

universelle published 1636 in Paris by Marin Mersenne, who also received his education from the Jesuits but subsequently joined the religious order of the Minims.[44] Compared to Kircher's *Musurgia universalis,* Mersenne's book is more rigorous, applying classical antiquity's idea of harmony to mathematics. Kircher, however, is the only one of the three to go further and adapt the concept for the structure of the state. According to his idea, it should be possible to bring peace to the warring factions of the political order within a "harmonia politica."[45]

One of the most effective and base tactics for conquering people's souls is the Catholic notion of purgatory, that strange place located between heaven and hell, which fascinated Dante Alighieri, as evidenced by his *Divine Comedy,* and Sandro Botticelli, whose cycle of artworks graphically illustrate the *Comedy.* Purgatory is where the imaginary orgies of martyrdom take place, souls purged step by step of their sins through suffering and penance, or—if they are not steadfast enough and fail these trials—they are sent to eternal damnation. Purgatory is an experience of the utmost limits, as purifying as numbers that are the gateway either to order or chaos. A highly dramatic place located between Earth, heaven, and hell, it is the most important and controversial locality in the Catholic faith. It is also eminently suitable for mise en scène, and Ignatius of Loyola's *Spiritual Exercises* contain an abundance of stage directions. They are a doctrine of the emotions. Translated into media techniques, they read like the shooting script of a film:

Fifth Exercise: It is a meditation on Hell. It contains, after the Preparatory Prayer and two Preludes, five Points and one Colloquy. . . . The first Prelude is the composition, which is here to see with the sight of the imagination the length, breadth, and depth of Hell. . . . The second, to ask for what I want: it will be here to ask for interior sense of the pain which the damned suffer. . . . The first Point will be to see with the sight of the imagination the great fires, and the souls as in bodies of fire. Second Point. The second, to hear with the ears wailings, howlings, cries. . . . The third, to smell with the smell smoke, sulphur, dregs and putrid things.[46]

Kircher's complex world of sound and music strives, on the one hand, to present conclusive evidence for the existence of God with the aid of numbers and their logic. Mathematical-physical processes and acts of divine manifestation are all to be as one.[47] However, as soon as Kircher leaves the level of mathematical calculation and turns to giving meaning to the realm of sounds, the theory

In fine illorum inferi, tenebræ & poenæ.

Figure 5.9   Plate from one of the first illustrated editions of Ignatius of Loyola's *Spiritual Exercises*. In the exercise on hell, one of the additions to penance is "to chastise the flesh, that is, giving it sensible pain, which is given by wearing haircloth or cords or iron chains next the flesh, by scourging or wounding oneself, and by other kinds of austerity." (Translation by Father Mullan, S. J., 1914, p. 29)

of the affects becomes central, tied closely to theological arguments. Perfectly harmonious and, in this sense, beautiful music can have powerful healing effects, but only for mental illnesses, not for physical ailments. Disharmonious sounds have the opposite effect and plunge the soul into turbulence. Kircher's idea of "musica pathetica,"[48] a music of pathos that would move and carry away anyone who heard it, was very much in line with the prevailing contemporary attitude toward music in Italy. The "ultimate purpose of music" was seen as "the production of a joyous emotion."[49] Further, Kircher is also in agreement with the *Poetics* of Aristotle, for whom music is the most important force for organizing the emotions.

The acoustic mechanical theater, which Kircher developed elaborately and had built, belongs to the category of staging spectacular effects. It contributed considerably to the fascination exerted by the Museo Kircherianum, which the Jesuit established in the Collegium Romanum. Sought out by learned visitors to Rome from all over the world, the museum was one of the city's most popular attractions in the second half of the seventeenth century. Set up after the manner of a *wunderkammer* (cabinet of wonders), the museum exhibited curios from faraway places, which had been sent to Kircher or brought back by his correspondents, and his own constructed artifacts: fossils, books, maps, mathematical and astronomical instruments, mechanical and hydraulic clocks, stuffed alligators, skeletons, skulls, distillation vessels, and reproductions of Egyptian obelisks, whose hieroglyphs Kircher claims to decipher in *Oedipus aegyptiacus* (1652–1654).[50]

The museum was also full of marvelous optical and acoustic devices. The concept of technology that Kircher elaborated and presented here was, on a complex level, entirely characteristic of natural magic. Technology stood for the spectrum of artificial constructions where "the operative force or agent was not obvious to the eye."[51] Many of the devices were not original but reconstructions or copies of Heron of Alexandria's theater of illusions, where figures operated by hydraulic or pneumatic power performed a variety of movements. Kircher and his assistants built organs modeled on mechanical glockenspiels that functioned like audiovisual automata. The mechanism was driven by water power and turned a cylinder on which music programs were stamped on tin foil. These controlled metal pins that opened and closed the organ pipes. Automata depicting miniature scenes, which were also driven by the mechanism that turned the cylinder, moved in time to the music like film sequences.

Kircheriana Domus naturæ artisq theatrum
Par cui vix alibi cernere poſſe datur.

AMSTELODAMI.
Ex officina Janſsonio – Waeſbergiana Anno MDCLXXVIII.

Figure 5.10 Title page of Giorgio de Sepibus's book (1678) on Kircher's museum in Rome, showing the entrance hall with Kircher greeting visitors. The rows of talking busts can be seen on both sides of the passage leading off.

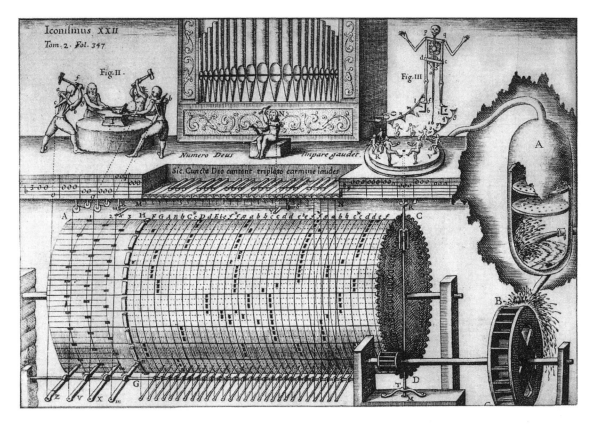

Figure 5.11 Organ from Kircher's *Musurgia universalis* (1650). Among the most popular figures featured by early music automata were representations of purgatory and the hammering smiths, who were supposed to have inspired Pythagoras to his theory of musical intervals. Pythagorean teaching on harmony, with its preponderance of numerical expression, obviously invited transformation into automata. Speaking of composers of computer music as a whole, Xenakis says they are all basically Pythagoreans (Xenakis 1966).

One of the exhibits that impressed visitors the most was a gallery of metal heads, which stood along the walls at the museum's main entrance. In a perfect mise en scène of God as omnipresent watcher and prompter, the heads would begin to speak whenever anyone passed by, and no one knew where the voices came from. (It is claimed that the idea of mysterious talking sculptures originally came from Albertus Magnus.) Book 9 of Kircher's *Musurgia universalis* contains many designs and descriptions of devices for eavesdropping, which, used the other way round, can amplify sound and stage miraculous events. These include long conduits linking rooms that are far apart, huge funnels whose vast openings cover entire courtyards, and intricately branching systems for listening in which are acoustic companion pieces to Jeremy Bentham's Panopticon of 1790—his design for a completely transparent prison.[52] Foucault's claim that the faculty of sight is distinguished by its misuse as an organ of control does not appear to hold good in light of these sophisticated acoustic constructions for spying.[53] However, the hundred-eyed Argus of classical mythology does not have any acoustic counterpart, such as a god with ears sprouting all over his head.

Kircher devotes much attention and effort to the conduits for carrying sound, which he calls "channels." His *Phonurgia nova* is like a textbook on the contemporary state of the art of acoustics and its laws; however, it was severely criticized when it was published because it contained much that was already out of date.[54] Like Porta and others, Kircher assumes that sound travels in straight lines like light, only much slower: "The sound or echo is an imitator or follower of light."[55] One of his core theses concerns the directly proportional relation of volume and velocity. Because Kircher assumes that sound will be reflected, as light is on mirrors, when it hits a smooth solid surface and will even be amplified by reflection, he favors spiral-shaped channels with a polished interior surface for effective transmission. His architectonic constructions for eavesdropping and for amplifying speech have huge snail-shell-like structures built into them which also lend them an anthropomorphic character. However, he does not intend these facilities to be used exclusively for cloak-and-dagger purposes. In *Phonurgia nova,* the first technical sensation is the sketch of a house where a quartet is playing in a closed room. Above the musicians is a huge funnel built into the ceiling with the narrow end passing through the wall to the outside. In this way, the music produced in the closed room can be heard two or three miles away by people far from its origin who do not know from whence it comes.[56]

Magical Image Machines

Natural scientists in general are far better at presenting their ideas orally and visually than scholars in the arts and humanities. There are various reasons for this, including, in the last decades, a certain trend toward the Americanization of academic institutions. A deeper reason for the pressure to communicate in a way that is intelligible to all is to impress the disbursers of public and private funds and to legitimate for the taxpayer the enormous sums required for ambitious

Figure 5.12   Panacousticon: Kircher's design for a surveillance system of courtyards and public spaces where every word can be overheard. Kircher believed that the spiral-shaped conduits would act as amplifiers. "Fig. I" (right) shows one of his "talking heads," which in this arrangement functions as an eavesdropper. (Kircher 1650)

projects. Furthermore, academics in the arts and humanities have come to rely much more exclusively on the written text, which they regard as their original and privileged medium. For scientists, on the other hand, it is a matter of course to argue their case using graphic presentations and images.[57] So far, no one has produced a history of the media innovations that natural scientists and engineers have dreamed up to captivate their audiences in lecture theaters, though such a study would be well worth the effort. John Dee, for example, impressed Oxford initially not with his mathematical genius but with a spectacular *coup de théâtre*. At a performance in Trinity College in the late 1540s, he used pulleys, pneumatic apparatus, and mirrors to make one of the play's protagonists fly up into the stage's heavens, riding on a big metal scarabaeus. He also presented other kinetic effects with this mechanical monster. The audience of students and professors was enchanted, but at the same time they suspected Dee of practicing demonic magic. (Later, Shakespeare is purported to have made use of Dee's theatrical tricks.)[58]

After finishing his studies in philosophy in 1623, before he became a professor of mathematics, philosophy, and oriental languages in 1628 in Würzburg, Kircher taught Greek for a short time at the Jesuit school in Heiligenstadt. There he arranged a stage performance with many mechanical theatrical wonders. The audience was captivated to such a degree that the Elector of Mainz came to hear of Kircher and gave him a lucrative commission for maps. Kircher was able to refute successfully the charges of magic by demonstrating that his special effects were "merely the product of his knowledge of mathematics and physics."[59]

Baroque mechanical theater was a highly developed media world of special effects. It even led to the establishment of a new profession, as many theaters began to engage a "capomaestro delle teatri, who was responsible for inventing the many and diverse devices that allowed the most complicated, surprising, bizarre, magnificent appearances and events to take shape."[60] A large proportion of these inventions originated from the work of the theatrical specialists in the Jesuit order. Theatrics that would cause the audience's hearts and souls to tremble constituted an acknowledged method in their strategies aimed at conversion: "Many Jesuit theatres had traps for scenes in which apparitions appeared, or disappearances were staged; further, flying machines and others for producing clouds . . . enabled the Jesuit directors . . . to let gods appear in the clouds, ghosts to materialise, and eagles fly into the sky; the impact of these stage effects was enhanced by wind and thunder machines. They even found ways and means to stage the parting of the Red Sea for the Jews, floods, storms at sea and other difficult scenes with a high level of technical perfection."[61]

Allegedly, Kircher's two major works on hearing and seeing were also written as a reaction to certain critics who, after his first studies on Egyptian hieroglyphics, had accused him of being merely speculative. With these two works, which appeared within a few years of each other, Kircher wanted to prove his secure grounding in mathematics. *Musurgia universalis* was devoted to arithmetic, and *Ars magna lucis et umbrae* [The Great Art of Light and Shadow] to geometry as a subject that served the foremost discipline of mathematics. For Kircher, music was applied arithmetic and optics applied geometry. In fact, in the first

Figure 5.13   Frontispiece of "The Bear-Rose or the Sun" (Scheiner 1626–1630)

few chapters of *Ars magna lucis et umbrae,* he attempts to give an overview of the contemporary state of knowledge concerning optics, which was already quite a formidable task at this time. However, not only Kepler's seminal works had been published in the meantime. In addition to studies on the telescope and comparative work on the eye and optical lenses, the mathematician and astronomer Christoph Scheiner had also published geometrical calculations of specific problems, such as nearsightedness and farsightedness, following in the tradition of the Arab eye specialists of the turn of the first millennium. The first treatises on microscopy appeared in various European countries. Mersenne, who built up a network of scientist friends in Paris that could compete with the one in Rome, published studies of the most important laws of optics before his works on music and combinatorics appeared. In 1638, Jean-François Niceron published his first studies, followed in 1646 by his magnificent book *Thaumaturgus opticus,* on perspective and its cunning applications, from the camera

Figure 5.14   Title page of Kircher's *Ars magna lucis et umbrae,* engraved by Pierre Miotte, from the Amsterdam edition of 1671. The basic design and details of the picture are a variation on the frontispieces of Christoph Scheiner's works that had appeared several years earlier: *Oculus hoc est fundamentum opticum* (1619) and *Rosa ursina sive sol* (1626–1630). Kircher takes Scheiner's ideas and both develops and condenses them. Four elements comprise the polymath's frame of reference—ecclesiastical authority, reason, secular authority, and the human senses. Shining down upon this quartet is the Tetragrammaton, the name of God in four Hebrew letters: Yahweh, the Unutterable. Within this framework is one of Kircher's characteristic dualisms, depicted as an allegory. The male figure on the left is Apollo, the sun, bringer of light and day; his skin is incrusted with symbols of scientific and alchemical approaches to reality and at his feet is the principal symbolic figure of the transmutation process, the two-headed black bird. At the tip of his scepter shines the same eye, the monocular gaze, that writes the Book of Reason (top right). Opposite Apollo sits Diana, the feminine principle, shrouded in shadow and only dimly illuminated by the moon, richly adorned with glittering stars, with her feet upon the wondrously colorful two-headed peacock, which in alchemy the black eagle, or crow, turns into. Diana's scepter is topped by Minerva's owl, which is very close to the Eye of Reason. In her right hand, Diana/Night holds a parabolic mirror that deflects the beam from the Perceiving One to the temporal sphere and, vice versa, the rays emitted by the world reach his eye. The second beam from the eye of the Perceiving One is deflected via a mirror in a cave to the realm of the senses—an allusion to both Plato's ray theory of vision in *Timaios* and the cave metaphor in *Politeia.* In the picture the telescope, symbol of the baroque age's investigation of the visible world, functions as a projector, like Scheiner's helioscope. Epistemological, theological, scientific, magical, and mythological energy fields are interwoven here, resulting in a construction that admits multiple interpretations. At the center of the lower half of the picture is a portrait of the patron who contributed significantly to the book's production. The dedication to Archduke Ferdinand is also mirrored in the central glass at the top, which contains the book's title.

Light and Shadow—Consonance and Dissonance

obscura to anamorphosis. What is more, ten years before the first edition of *Ars magna lucis et umbrae*, *Discours de la méthode* was published by Descartes, Mersenne's long-time friend and fellow student at the Jesuit college of La Flèche, followed by an appendix on "Dioptrics as a geometrical theory of the behavior of light in transparent media," in which he provides the law of refraction.[62] With such advanced experts in the field of vision and the visible, the polymath Kircher, whose focus was collecting, combining, and analogizing, could not seriously compete. At best, he could present excerpts with his commentaries. What he did brilliantly, however, was to analyze and evaluate these published findings to produce his comprehensive treatise on applying the laws of optics in draftsmanship and apparatus design, and describe the effects that could be produced. For most historians of media and art, subjects that are heavily oriented toward images and pictorial objects, *Ars magna lucis et umbrae* founded the legendary status of the Jesuit from Geisa in Germany, who, due to the force of historical circumstances, made a successful career for himself in Rome as an outstanding innovator of technical envisioning during the transition from the Renaissance to baroque.

Already in the first edition of this work (1646),[63] on almost one thousand pages Kircher lays out a universe of ideas, drafts, models, sketches, and building instructions that has no parallel on this scale. In the Amsterdam edition, which appeared a quarter of a century later (1671), the illustrations are even more lavish and the compendium is supplemented by the addition of numerous innovations that had appeared in the meantime. In 1680, Kircher's assistant Johann Stephan Kestler collected Kircher's contributions to the field of applied optics in a special volume, *Physiologia Kircheriana experimentalis.*[64] In all these mammoth works, only a few details are technically original; Kircher mostly revisits and processes many classical texts on optics, often without acknowledging his sources. The entire riches of Porta's theater of mirrors are found here in a reworked version as well as the mathematical and philosophical treatise by Mario Bettino, published 1645 in Bologna, which deals with perspective, anamorphosis, burning mirrors, and the projection of secret messages using parabolic mirrors.

What makes Kircher's work stand apart is immediately comprehensible on a superficial level: nobody before him had presented the material using knowledge of optics in such an imaginative and truly impressive way. Excellent craftsmen produced the highly ornate engravings and xylographs according to his sketches and instructions.[65] In addition, there is an inner vibration pervading his entire oeuvre that is not easy to pin down. It was already perceptible in Porta's work, and in Kircher it is even more pronounced. The phenomena and apparatus that he presents convey very clearly contemporary knowledge about the respective discipline or even fall short of it. At the same time, they undergo a strange shift in the direction of grand-scale, generalizable models where fact and fiction, calculation and imagination, the gravity of the laws of geometry and mechanics, and the individual wayward imagination mingle. Godwin writes that in Kircher's conception of technology, there is still room for the dreamer. To this characterization I should like to add that there is still room for the ardent enthusiast. This shift is also a kind of intensification, as Kircher is convinced and enthusiastic about each and every detail that he presents to the reader. His delight in the ability to present the world as a mise en scène, to effect metamorphoses of its symbolic representations, is apparent in every fragment of text and every illustration.

For example, the *laterna magica,* the original device for projecting bright images in dark rooms. More than two centuries before *Ars magna lucis et umbrae* was published, there were attempts to project figures. The Venetian Giovanni da Fontana, rector of the Padua art academy and an enthusiastic pyrotechnist, was the artist of an outrageous sketch, ca. 1420, of a laterna projecting an obviously female devil, complete with pubic hair, onto a wall. Porta also drew on predecessors' work in describing his *cubiculum obscurum.* In the first edition of *Ars magna lucis et umbrae,* Kircher is essentially reporting on the state of the art. Then competitors began to appear upon the scene, who quickly grasped the media potential of the magic lantern. As a by-product of his physiological studies, Christian Huygens started painting skeletons in 1659 after the manner of Holbein's *Dance of Death* on glass and projecting them onto the wall with the aid of biconvex lenses. Through simple animation, he made them dance or appear without heads. Beginning in 1664, the Dane Thomas Wallgenstein traveled around various European countries, including Italy, with a portable projection apparatus. He was so successful with his magic lantern that he was able to sell several of them.[66] Kircher was extremely annoyed about this exploitation of what he considered to be "his" invention. In the appendix to the 1646 edition of *Ars magna lucis et umbrae,* under the title of "Kryptologia nova," he had proposed the lantern as a device for projecting secret messages with the aid of a concave mirror—thus taking over Porta's idea. Now, however, someone had beaten him to it and turned the idea into a marketable media apparatus for projecting artificially produced images. In the 1671 edition, Kircher launched a counterattack.

He accused Wallgenstein of plagiarism and presented engravings of two different scenarios for using magic lanterns, which were not only the first complex graphic representations of this apparatus but also the most impressive for many years to come. Technically, they are incorrect. Kircher places the transparent strips of glass with the images in front of the lens instead of between the light source and the lens; moreover, when two convex lenses are used, as he describes in the text, the projected images will be upside down. These errors are most likely the fault of the engravers, but, with Kircher's enormous output, it was impossible for him to check all the proofs thoroughly. He used the magic lantern himself both for lectures and in theatrical productions, and, in both editions of *Ars magna lucis et umbrae,* he discusses with great technical competence many uses for the projection of images in dark rooms. Perhaps the most decisive aspect is that in his description of the "thaumaturgic construction"[67] of the lantern, one can feel the tremendous power of such an apparatus for staging illusions. Magic lantern and catoptric theater are welded together as a medium that is highly suitable for presenting "satirical scenes" and "plays that are tragedies."[68] The dark room becomes a screening room, and the projection equipment is in a cubicle, invisible to the spectators. With the two subjects of his projected images in the engravings, the frightening Grim Reaper and the female figure in the crackling flames of *purgatorio,* one can imagine just how powerful an instrument for the projection of imaginary signifiers is being described here. After this, it was only a question of perfecting technical and dramaturgical details and then marketing the medium.

In a similar fashion Kircher transformed another arrangement of mirrors, which is described coolly in the *Pseudo-Euclid* as "positioning of a mirror such that an observer may view the image of an object but not his own reflection," and which Porta had metamorphosed into a mysterious chamber that played with the visible and the invisible. Besides introducing magic into a technical device, Kircher expands its possibilities and defines it philosophically and aesthetically by treating it within the category of experiments with metamophoses.[69] When the mirror is tilted, the object that the observer sees in place of his or her own head is no longer a hidden sculpture or other material object; it is now an image. Below the mirror, Kircher installs an octagonal drum on which are painted pictures of the sun and of seven different animals' heads. On entering the chamber, the observer sees first the sun, the mediator of all that exists on Earth, and then his or her own allegorical transformation into an ass,

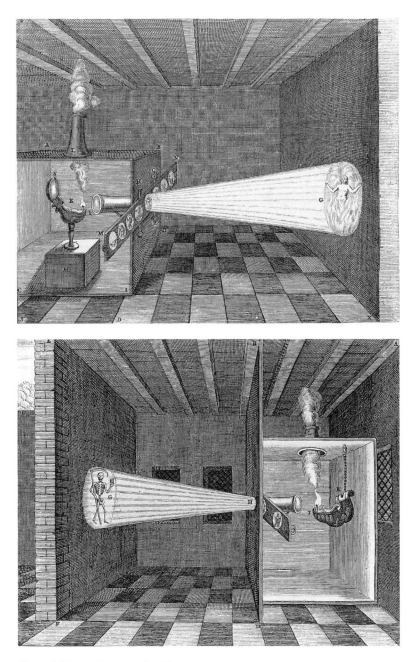

Figure 5.15   Kircher 1671, Book X.

a lion, or other creature. The images are painted with great skill in close-up so that they substitute exactly for the head of the observer when he or she takes the position indicated by marks on the floor. This artifact also conceals the technical process being used, an aspect that Gaspar Schott emphasizes in his more detailed description.[70] Who or what is at work should not be recognizable. The drum with the painted images is concealed in a casing, which is only open at the top, underneath the mirror. Neither the mechanism nor the mechanic who turns the crank should be seen, although theoretically the observer who enters the chamber (whom Schott refers to as the "in-looker") could also operate the crank and tilt the mirror. Kircher considered the dramatic effect of this "metaphormachine," as Hocke calls the apparatus,[71] to be so powerful that he adds a second observer, who can watch the first observer's interaction with the illusion-machine or its images. Observer B does not have a body, but is only an eye floating in front of the opening left for iit. This voyeuristic variant allegedly functioned especially well when the chamber was dark and an artificial light source was used for the projection.

With such technical artifacts and their specific arrangements, Kircher established a tradition of visual apparatus that, in the following centuries, was both highly effective and the dominant model. Based on the concept of purifying the soul through catharsis, media machines were designed and built in such a way that their functioning mechanisms remained a mystery to the audience: the projected world must not be recognizable as an artificial construct. Above all else, the intention was that the effects should take the onlookers by surprise, captivate them, and prevent them from giving free rein to their imagination and reason. For Kircher's age, the concept was advanced in its construction details, but already antiquated in terms of perception and aesthetics. It follows Aristotle's dramaturgy of catharsis in *Poetics:* "with the aid of pity and fear a purification of . . . the emotions can be contrived."[72]

Yet Kircher's optical world was not a unified or closed system; his imagination often breaks the bonds of what is feasible. In one of his metamorphoses, for example, he proposes a modified version of the allegory machine, which is both impressive and technically impossible. Using a cylindrical mirror, the machine could be constructed in such a way that the projected images of the figures appear to hang in the air, a kind of nonpresent presence. Even in the second edition of *Ars magna lucis et umbrae,* he fails to explain exactly how the mirror-cylinder can be integrated into the apparatus.[73] A particular product of his fertile imagination regarding things technical is the "polymontrale" catoptric

Figure 5.16 Kircher's metamorphosis apparatus for the allegorical transformation of an observer. On the upper side of the case containing the drum with images, one can see that a rectangle has been cut out so that the mirror arrangement can project the images. On the left, outside the chamber, is the disembodied eye of the voyeuristic observer *B*. The figure at the top of the page shows glass cylinders for Kircher's planned projection of figures that appear to float in the air. (Kircher 1671, p. 783)

Light and Shadow—Consonance and Dissonance

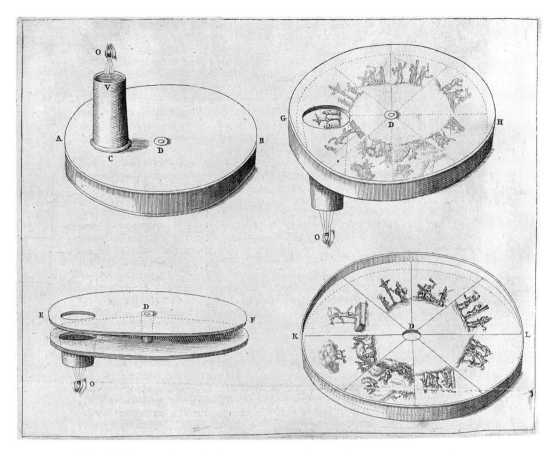

Figure 5.17 Toward the end of the chapter on the theater of mirrors in *Ars magna lucis et umbrae,* almost hidden away, there is a design for a device called a "smicroscopin." Through an ocular, the observer can view Christ's passion, depicted in eight scenes, at optional speed. The glass disk with the eight image segments is exchangeable and rotates between two fixed metal disks. (Kircher 1671, p. 770)

theater.[74] This piece of media furniture was on show in his museum in the Collegium Romanum. When open, it was a cabinet where the top functioned as a stage. Movable screens flanked it on all sides and appeared as windows giving onto an infinite world of images. This effect was produced by various types of mirrors—the illustration shows over sixty mirrors—fitted on the inside of the screens, which were fixed by hinges. Kircher is reported to have made many improvements to this apparatus and added even more mirrors. The closed part of the cabinet held the objects that "performed" on the stage, for example, an artificial tree, flowers, a book, kinetic models of human figures, and even living animals. A mechanism with levers raised the objects onto the stage. With the aid of a handle at the side of the cabinet, the operator could make the inert objects move; kinetic objects were also used, like marionettes or Kircher's movable hydraulic sculptures. Depending on the positioning of the mirrors, the objects appeared infinitely multiplied, upside down, elongated, or changed in many other ways. The optical effects produced by the mirrors, however, were not enough for Kircher. He suggested that the action on the stage of this theatrical cabinet should be accompanied by specially created sound effects or music. An accompanying marginal sketch depicts an architectonic realization of the catoptric theater in baroque splendor, showing a room in a house that is completely lined with mirrors. This project was actually realized in one of the privately owned palaces in Rome, where one could try out all the possibilities of an existence that was in a constant state of metamorphosis.[75]

Combinatorial Boxes for Personal Use

Designs for artifacts that would allow those not privy to specialist knowledge to share in the aura of the great arts of music, science, or cryptography occupy a prominent place in Kircher's work. It must have been a source of great pleasure to him to develop these tools, for he made some of them himself, taking great care in their construction and doing the lettering himself. These fanciful objects are all of a similar design: small boxes of wood or cartons containing specific organizing systems that use thin, sliding slats. The slats are positioned vertically one behind the other, and the units of information inscribed on them are arranged so that, following certain rules, these can also be linked horizontally. In the Jesuit education system, these artifacts all fulfilled a didactic purpose as learning aids.

The nucleus for such artifacts for combining and calculating in Kircher's theater of devices is his *cassetta matematica* (mathematical box), an example of which is now in the Institute and Museum of the History of Science in Florence, surrounded by a wonderful collection of other early instruments for arithmetic and computing.[76] On opening the box, the first thing to attract notice is a horizontal strip inscribed in black on a white ground with a menu of nine different branches and applications of mathematics: arithmetic, geometry, *fortificatoria* (dealing with calculations for military fortifications), *chronologia* (measuring time by regular divisions, in this case, the cycles of the moon and movements of

Figure 5.18 Kircher's *cassetta matematica* (mathematical box), or *organum mathematicum* (mathematical organ), of 1661, from the Institute and Museum of History of Science, Florence. The triangular tips of the slats have different colors, the slats themselves are of warm reddish brown wood, and the box is black with brown intarsia.

the planets), *horologia* (science of constructing sundials), astronomy, astrology, steganography, and music. Assigned to each of these headings are twenty-four wooden slats, one behind the other, which, according to each of the nine mathematical areas, are of different colors and marked with the letters of the alphabet *A* through *I*. Each slat has up to twenty-four spaces, which contain arithmetical operations from the various fields. These have been selected and arranged in such a way that they can be combined with the spaces on the other slats, and with that of arithmetic as the main slat. This carries examples of basic operations of dividing, multiplying, and square and cubic roots. On the lid and the front of the box are two rotatable discs, which symbolize the universal character of the instrument: one represents an astrolabe, and the other shows the time of day around the (known) world in geographically important countries and their capitals. The *cassetta matematica* is a handy size (44.5 × 31 × 25 cm), which fits comfortably on any writing desk. The rotating discs are not the only reference to the combinatorics of Ramon Llull; the division of mathematics into nine areas is not imperative—it matches the Majorcan mystic's division of biblical wisdom into nine guiding concepts. By dividing each slat into twenty-four spaces—which correspond to the number of letters in the Latin alphabet (including the seldom-used letter *K*)—Kircher also demonstrates clearly on a formal level what the *ars combinatoria* was essentially for him: the elaborate and artistic linking of linguistic elements.

When Mersenne claimed in his *Harmonie universelle* that music is only algebra translated into sound and that with the help of the sequential method, any lay person can become a composer within the space of an hour or less, he was heavily criticized.[77] In principle, Kircher shared Mersenne's view of music as a discipline subordinate to mathematics, and he underlines this view in book 8 of *Musurgia universalis,* which focuses on the mechanical arts in music,[78] by proposing an apparatus for composing music. The construction is similar to his *cassetta matematica.* On the front side are clefs and at the back, arranged in the form of tables, are keys that can be used in the composition. Inside the box are slats: each is marked on the front with sequences of four-note chords and on the other side are rhythm variations that can be combined with them. This *arca musarithmica,* as Kircher named his musical treasure chest, was intended to accompany his *Musurgia universalis,* which notes the chord sequences as rows of numbers and explains the method of composing music with this device.

Because of Kircher's extremely vague descriptions and very confusing terminology in book eight of *Musurgia universalis* ("Musurgia mirifica," p. 185ff), the precise functioning of the composing box and the quality of the music it produces are very controversial among music historians, and contradictory interpretations abound. Accounts are so divergent that one often has the impression the authors are describing completely different devices.[79] Briefly, we shall take a closer look at this device.

To begin with, the opulent illustration of the "new invention of a box for composers of music" (*arcae musurgicae novum inventum*) is misleading. From the text description, it is quite clear that it is a small box, "which is as high as it is deep and both [measurements] are a hand's breadth" (p. 185). Its width Kircher gives as only "half a hand's breadth." Forming an impression is made more difficult by the fact that Kircher describes the box as divided into three equal compartments—which the illustration does not show—but he assigns various composing functions to these compartments: (1) rhythmic combinations of any song phrases consisting of "specific polysyllabic segments" (the example "Cantate domino . . . ," p. 186f. refers to this); (2) compositions with several strophes and their various combinable metric structures; (3) compositions of "artistic and embellished song phrases" as instructed by a "rhetorical art of music" (p. 189f.). Taking an example from the second compartment, "poetic music," which is further subdivided into six smaller compartments each with any possible number of columns, demonstrates how difficult it is to ascertain the concrete use value of this device and how Kircher contributes to the lack of clarity. After explaining how one makes a "selected song phrase into an Anacreontic sextuplet," he continues:

This is the manner in which all other monomial metres with several strophes are put together. However, when the text theme has several strophes, i.e., it has been put together from several songs, then the columns must be taken out of the compartments, which are inscribed with the appropriate metres. An example will serve to clarify this matter [this list of metres is inserted in the text]:

1. Eleven-syllabic [metre].
2. Anacreontic metre.
3. Archilochian iamb.
4. Euripedian iamb.
5. Alkmanian iamb.
6. Adonisian metre.

Therefore, when someone composes something with several strophes, where the first strophe is a phaleucic eleven-syllabic, the second Anacreontic, the third Archilochian, the fourth Euripedian, the fifth Alkmanian, and the sixth Adonisian, one must proceed as follows: Of all the closed compartments, one opens the lid of the one inscribed "for eleven-syllabic," then the one inscribed "for Anacreontic metres," as a third the lid for Archilochic iambs, the fourth for Euripedian iambs, the fifth lid for Alkmanian iambs, and, finally, the sixth for Adonisian metres. Then, in the order that one opened the lids, one takes out a single column from each of the compartments and lays them carefully in the same order next to each other. The columns arranged in this way one can combine at will with each other and select any horizontal row for the desired song phrase. When one has decided on the beats, one can begin with the composition of the song phrase in the exact way that was described in the previous section and one will arrive at the desired result.[80]

For media archaeology, interest centers on the essence of Kircher's design: from the device's collection of pregiven musical, poetic, and rhetorical patterns or their formal representations, it is possible to put together variable, harmonic compositions. It is only necessary that the individual patterns are capable of being combined with each other. This is the principle of a sequencer in electronic music, which stores sound sequences and delivers them to other instruments, such as a synthesizer or a computer music program, for further processing. At the end of chapter 3 ("On using the *arca musurgica*"), Kircher effusively formulates his conclusion and demonstrates that he is well aware of the far-reaching consequences of his invention: "It is apparent from that which is put forward here the infinite number of possible combinations, which are given by the different ordering of the five columns. Assuredly there are so many that had an angel begun with the combinations at the dawning of the world, it would not be finished today" (p. 188).

However, Kircher thought that it was also possible to use the *arca musarith-mica* without any explanation or instructions. Rather than intending the instrument to replace musical genius in any way, he sought to place it at the disposal of nonmusicians, so that they could produce their own simple compositions. The musical sequences contained in the box are selected in such a way that any combination produces consonant constructions. In this, too, Kircher follows Mersenne exactly, but mentions him only briefly. On the whole, Kircher's description of the sequential composing apparatus is rather vague. He makes a point of saying that he developed the box for friends, to whom he will personally explain its correct use. Probably the only existing example, which Kircher built himself, is the one he presented to Duke August

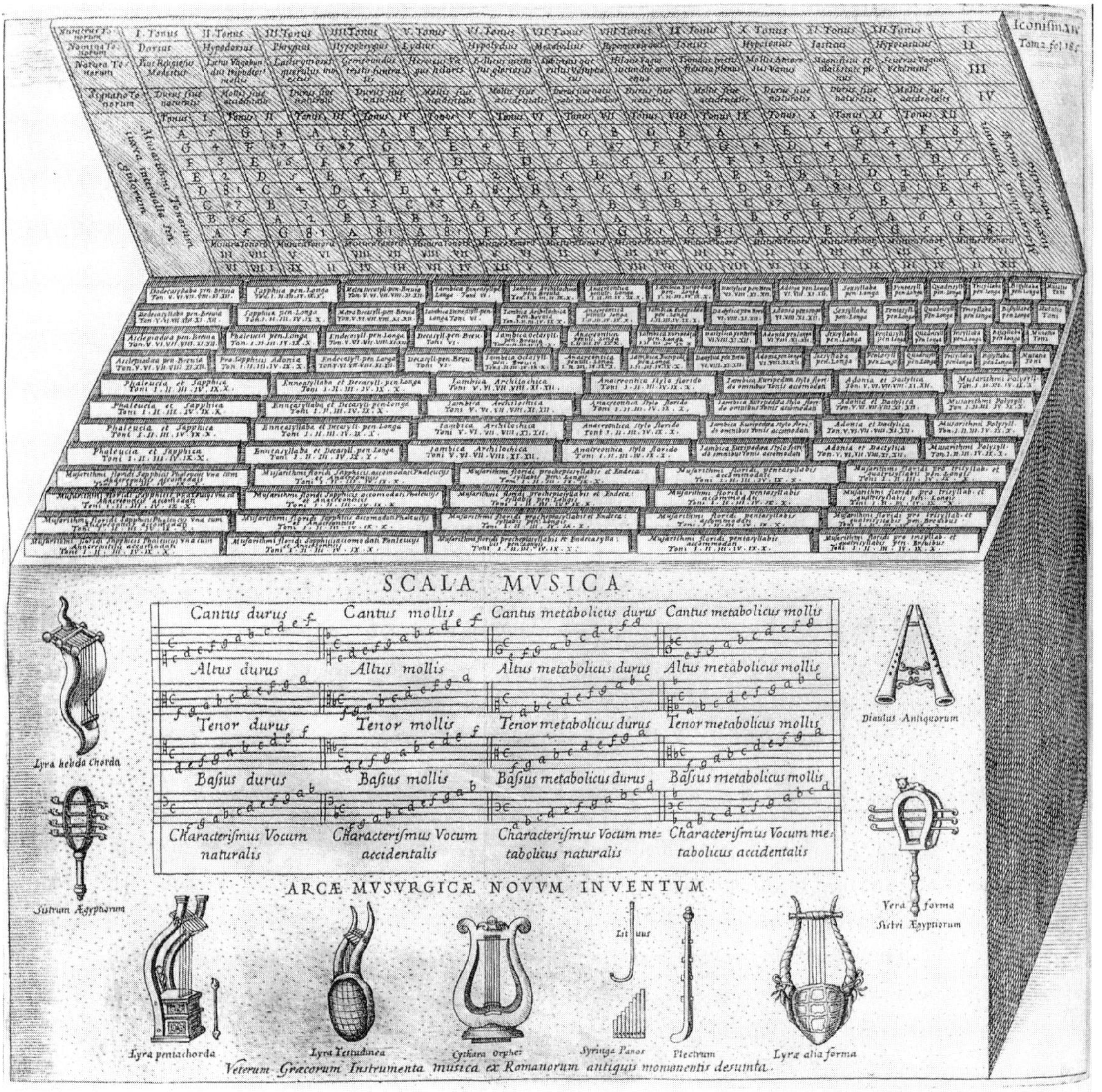

Figure 5.19 The *arca musarithmica,* (box for rhythmic sequences of notes) the device Kircher intended as an aid for amateurs to compose their own simple music pieces from pregiven musical phrases. This combinatorial box is one of the highlights of *Musica mechanica,* Kircher's book on the universal musical arts (1650, vol. 2, plate XIV).

von Wolffenbüttel, a passionate collector of cryptological ciphers. Today, it is still in the library that bears the Duke's name.

While Leibniz was working there as a librarian, he apparently experimented quite often with the smart box.[81] This other important German polymath shared with Kircher and other contemporaries, such as Wilhelm Schickard and Blaise Pascal, a delight in combining, constructing devices for calculating, and devising methods for preserving. Not only did he design a machine in 1672 that was capable of all four basic calculating operations,[82] but, in the course of his forty years' service to the dukes and electors of Hanover, he also developed ways of preserving food. In 1714, shortly before his death, Leibniz described in a manuscript (which was treated like a secret military document) "the means by which troops may keep up their strength during long marches or other strenuous exercises"; he recommends for this purpose that they "partake of *Kraft-Compositiones,*" that is, preserves, particularly "meat extracts, the composition of which is known to me."[83]

Kircher was a linguistic genius. He taught Latin, Greek, the biblical languages Hebrew and Syrian and other oriental languages, and was also able to make himself understood to his visitors from all over the world. At a banquet given in Rome on Boxing Day, 1655, to celebrate the conversion of Christine of Sweden to Catholicism, he is reported to have welcomed the Queen in thirty-four languages.[84] Above all else, the *ars combinatoria* were for Kircher the artistic linking of different linguistic systems. He was convinced, for example, that many of the characters used in the Chinese languages were related to ancient Egyptian hieroglyphics; they only needed to be interpreted, as he proceeded to do in *China illustrata.* At least at this level—the reality of symbolic material— he could attempt to keep things together that elsewhere were disintegrating into diverging interests and positions. He devoted a number of manuscripts and books to this thematic complex, including the major works *Oedipus aegyptiacus*[85] (1652–1654), considerable sections of *Ars magna sciendi* (1669) and *Musurgia universalis* (1650), which also contains examples of ciphers in the form of musical notations, as well as his late work on the confused tangle of languages in ancient Babylon, *Turris Babel* (1679), in which his obsession with symbols makes bizarre reading.

Polygraphia nova et universalis (1663) is based on a longer manuscript dating from 1559, *Novum inventum linguarum,* and Kircher collaborated closely on it with Schott, who discusses some of the same major topics in his collection of technical curiosities.[86] Obviously, as the titles suggest, these works are about inventing a new universal language, which Kircher and Schott immediately associate with concepts for secret languages. This combination proved to be problematic. The two ideas are diametrically opposite with regard to usage: a universal language seeks to simplify communication, whereas cryptography renders the immediate intentions of communication temporarily indecipherable, that is, it makes it more complicated.

The project of a *lingua universalis* was commissioned by Emperor Ferdinand III to try out an easily understood lingua franca for his multilingual Hapsburg empire. However, the emperor's special initiative was probably superfluous, because historical circumstances brought this idea to the fore when the Thirty Years' War caused destruction and divisions on an unprecedented scale. Scientists and scholars in the affected countries also suffered as a result. The creation of a language that would be understood by all, that would correspond to the diversity of things in reality, was a notion that occupied the best minds in many places during this period, rather like a sophisticated parlor game with a serious political intent. Such a universal language would also neutralize, at least symbolically, the divisions in thinking, religion, and politics. Mersenne and Descartes corresponded intensely on the subject; the Spanish Jesuit Pedro Bermudo, the Pole Jan Amos Comenius [Komensky], Leibniz, the Englishman John Wilkins, and the Dutchman Francis Mercury van Helmont all wrote extensive treatises on the subject; as a result, "In the seventeenth century, more books were written about language than in any other period before."[87] Motives and intentions were, of course, as many and different as the authors. These ranged from the development of a means of communication for the deaf to attempts at reducing complex Latin to a simpler language. After all, for the vast majority of people, the language of scientists was already cryptic. For Descartes and Leibniz, the close relationship between linguistic and philosophical systems was uppermost in their considerations. The Frenchman held that a universal language would only be of value if it were capable of expressing a universal, generally recognized philosophical theory. Such a theory, however, could only arise from a perfect, paradisiacal world. As he could imagine such a theory and such a world being possible only in dreams, Descartes regarded all the efforts being invested in a universal language as a waste of energy, and he relegated the project to the realm of poetry. Leibniz, however, was more tenacious. The logical outcome of his deliberations was arithmetical, calculations: the most general and effective form of making oneself understood.[88] The most radical semiological approach was taken by Francis Bacon, one-time lord chancellor of England, secret agent, theorist of science, and Rosicrucian. In 1605, he proposed a method of encryption, where the code probably consisted of only two elements. Each letter of the alphabet was expressed by five-letter combinations of *a* or *b;* for example, the letter *A* was represented as *aaaaa* and *Z* by *babbb.*[89]

Kircher's *Polygraphia nova,* like so many of his treatises on other subjects, attempts to combine various approaches. The universal language that he proposes in the first *syntagma* of the book has a "hybrid" character.[90] It consists of letters, words, numbers, and rules of inflexion and case for which he invented a special graphic notation. Essentially, his method of multilingual writing is an aid for producing texts of a particular kind, which exhibit similar structures irrespective of the language they are written in. Kircher concentrates on the media form of the letter, developing an elaborate strategy for how these may be written and read in five different languages using a single code. There are three prerequisites: reduction of the languages' lexical range to the minimum necessary for communication, including more complex issues; assignment of binding grammatical and syntactical rules; and a key, to be agreed upon, for writing and reading the communications. Once such a system is thoroughly formalized, it is then possible to translate it into a mechanical device or apparatus.

The five languages Kircher selected were those he considered the most important in the Europe of his day: Latin, Italian, French, Spanish, and German. He said it was possible to include other languages without any difficulty, and in fact in his preparatory study he worked with eight languages, including Polish. The first step was to reduce the language to 1,226 words. At first sight, this number appears rather small; yet considering that all the plays by Shakespeare contain but five thousand different words, this number does allow significant complexity. The selected words are then divided into groups of words; in Kircher's handwritten manuscript, there are fifty-four groups, but in the published volume these have been reduced to thirty-two. The various groups are formed according to categories in which meaning and grammar are mixed, for example, philosophical and religious concepts, technical instruments, emotions and actions, reptiles and fish, but also verbs denoting perception and general verbs, prepositions, and adverbs. Here, Kircher assumes that the semantics and grammar of the languages he uses are roughly identical. In two thick lexicons with pages divided into five columns, all the words are listed. The first volume is for writing the communications. Words are arranged in alphabetical order and each is assigned one number in Roman and one in Arabic numerals. The Roman numerals stand for the group where the word can be found and the Arabic for the position of the word under the particular letter of the alphabet. The leading alphabet is, naturally, Latin. *Amor,* for example, carries the numbers II.6, which are the same numbers assigned to *amore, amour,* and *liebe* (the German word is written in the lower-case in Kircher's glossary, as it can stand for both the noun and the verb). This is a simple example; for other words, it is not easy to find corresponding words in the other languages. Words that the writer needs, such as proper names, which are not in the glossary are simply written in the cryptogram in the appropriate language. The companion volume, designed for readers of the communications, whom Kircher aptly calls "interpreters,"[91] is sorted according to the Roman numerals of the word groups. Under the heading "II.," one finds at position 6 in the first column (Latin) the word *amor,* and so on. The grammatical rules also follow Latin. Kircher reduces them to the declension of active and passive verbs and to the cases in singular and plural. Sentence construction also follows Latin, which was the contemporary Esperanto of science and learning and, moreover, the language of the Catholic Church.

It is apparent from the text that this first *syntagma* of *Polygraphia nova* did not fire its author with much enthusiasm. It was the execution of a commission. This changes, however, in the second and third *syntagmas,* which are devoted to cryptology in the more strict sense. Here it becomes clear what the long title refers to: the plan is a new version of Trithemius's *Polygraphia.* In the second *syntagma,* Kircher runs through the method for encrypting a short secret message within a longer, apparently innocuous text. The procedure is actually quite simple, but it requires extensive glossaries: each letter of a cryptogram is represented by set combinations of words given in a glossary[92] consisting of forty lists. Each of these lists of five columns (one column per language) is headed by a Roman numeral and contains twenty-two entries; one for each letter of the Latin alphabet, shortened by removing *J, U, W,* and *X.* The listed combinations of words mainly derive from standardized phrases used in letters. List I begins with "Ich hab bchommen" (I have received) and stands for the letter *A* in the cryptogram; "dein Buch" (your book) is the seventh entry in list II and thus stands for the letter *G,* and so on. Kircher accepts the differences in spelling or word order and leaves it to the user to rectify them. As an encryption method for finding and assigning the right word combinations to letters, the procedure is time-consuming, and the same applies to its use for decryption. The method literally cries out for simplification in the form of a device, and Kircher obliges. In the *arca glottotactica,* (which Kircher understood as a device for meaningful use of language), the lists of words are inscribed on slats that hang in a wooden box, an arrangement with which we are already familiar. In this case, the menu bar contains the Roman numerals for the word combinations. There are five slats under each numeral, one for each language. Thus, the box contains 200 text slats, each with 22 entries, which makes a total of 2,200 information elements that can be combined with each other.

In the third part of *Polygraphia nova,* which he calls the technological *syntagma,* Kircher appears to arrive at the real purpose of his elaborate venture: the "re-coining of universal linguistic ideas in a cryptographic table."[93] On the basis of the so-called multiplication cipher of Trithemius, Kircher develops his own system, which consists of only single letters and numbers. Trithemius had proposed a substitution matrix as a system for writing cryptograms. The top, horizontal line of his table was the twenty-four-letter alphabet and underneath this, the alphabet was listed twenty-three more times where each line was shifted one place. This scheme represented a considerable advance on the ancient Caesar cipher with its two alphabets, written one above the other, for it offered many more possibilities of encryption. Kircher developed this system into mathematical combinatorics by assigning numbers to letters. These do not follow the strict, but easily decipherable logic of substitution ciphers, where letters are moved one, two, or three places but create the impression of a random series. Written as a matrix, this system fits onto a single page. To the lay person it looks rather baffling, in spite of the diagonal arrangement of certain rows of letters, which is the result of multiplication. Using the cipher was, however, greatly facilitated by the *arca steganographica,* which Kircher not only described but actually built. In this device, the columns of the table are mounted on wooden slats and positioned under the letters of the alphabet. The parties communicating must agree upon which slats, or alphabets, they will use to correspond.

The simplest method of encryption using this system is to select just one slat and, instead of noting down the letters, to substitute them with numbers and incorporate these, for example, within the text of a letter. In more complex forms, several of the alphabets would be used. Using the steganographic ark to read and write the encoded texts simplifies the process considerably because the slats being used can be placed side by side in the appropriate shifted positions.[94] That his system worked, and was also a source of pleasure to the privileged few who had access to it, is attested to in Kircher's correspondence, for he received several letters wholly or partly encoded according to his system. However, Kircher himself allegedly composed only one letter using his invention.[95]

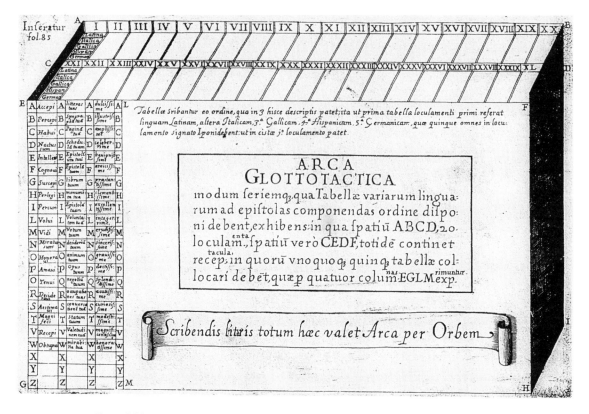

Figure 5.20 *Top*, text in small print: "The registers are arranged after the manner of the three shown here, namely, the first register in the first compartment is Latin, the second Italian, the third French, the fourth Spanish, and fifth German. All five must be placed in the compartment marked *I*, as can be seen in the illustration of the first compartment." The text on the banderole in the illustration says: "This box is useful when writing in any part of the world." (Kircher 1663, p. 85)

The Vatican in Rome had an officially appointed secretary who dealt exclusively with matters concerning cryptography. Notwithstanding, the works of Llull, Alberti, Bruno, Dee, and Trithemius on the art of combination and secret ciphers were regarded by the Catholic Church as the devil's work; thus, they were circulated clandestinely as manuscripts or, if published as books, were immediately placed on the index. What Kircher did was to translate their ideas and concepts into manageable praxis, tone down their exaggerated claims of use-value, render them accessible and, by putting his name to them, make them

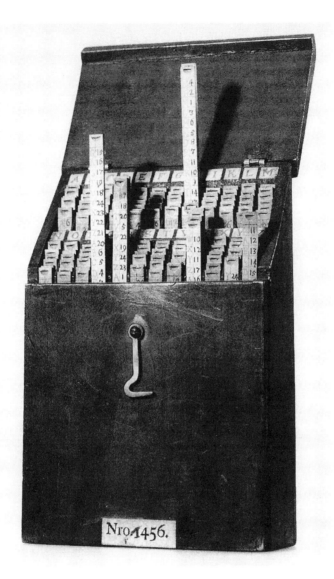

Figure 5.21 The *arca steganographica,* made by Kircher himself, for encrypting and decrypting letters, preserved in the Herzog Anton Ulrich Museum in Braunschweig (Strasser 1988, p. 173). Kircher's explanation: "According to the series of letters and numbers, the registers are written alphabetically according to a combinatorial table. Six registers at a time begin with the same letter of the alphabet and are placed in the compartments intended for them, as can be seen here. . . . With the aid of this box, you will learn of the connections between them. Whatsoever you desire to write, it will provide this for you in several languages." (Kircher 1663, p. 130)

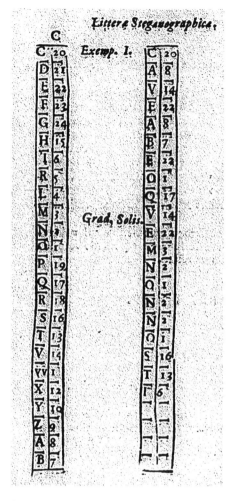

Figure 5.22 "To . . . keep a secret especially safe, one can preface it with any sort of heading that refers to numbers, as demonstrated in this example. By selecting [the heading] 'Positions of the Sun' *(Gradus Solis)* from which one may quickly construe [the content of] the letter, the secret is so well concealed that no one will have the idea that there is anything more behind it. The numbers are deliberately arranged in a vertical column following the conventions of astronomers in order that no one's suspicions will be aroused as to the hidden meaning, as might occur if it were arranged horizontally due to the commas and full stops. This is the first and simplest of all the possibilities." (Kircher 1663, p. 132)

truly socially acceptable. This he accomplished at the expense of his predecessors from the band of magicians,[96] whom he frequently criticized severely in his works (Bruno he had to ignore completely) in order to establish and maintain a fitting distance. Surprisingly, he chose to attack them on rather inconsequential issues.

One such vituperative polemic is found under the heading "Apologetica" and occupies almost the entire appendix to the *Polygraphia nova.* The target is a proposal for a "silenographic mirror,"[97] which Agrippa, Paracelsus, and others are reputed to have constructed. The idea was to use a strong mirror to project encoded messages over long distances such that even the moon could be used as a screen. All these dreamers allegedly traced the idea back to vague hints made by Trithemius, and Kircher acts as though he is amazed "that even the renowned Giovanni Battista Porta appeared to place his trust in this thoroughly tasteless piece of machinery."[98] Kircher condemns the idea as a delusion of alchemist megalomania, which merely serves to increase unnecessarily the mysteriousness of the world: "Who does not see that all these devices are not conceived for the sake of any useful purpose but, instead, to conceal boastfully unfathomed secrets with a mysterious apparatus?" At the end, he modestly states that his own proposals for steganography are in stark contrast to such arrogance:

In order that such a taint of vanity may not be attached to me, it is my belief that I should only include in this present book on polygraphy that which I have promised in all my texts on the achievements of the art of secret writing so that the reader may see that what I have promised I have kept, with the help of Almighty God. In those cases where success has been denied me with respect to the completeness I had hoped for, may the reader at least credit me for making the attempt. Perhaps one day another, with greater gifts than I, will come and, following in my footsteps, will again take up Ariadne's thread to penetrate the innermost sanctum of the steganographic labyrinth. Until that time, all manner and varieties of writing, of whatsoever kind they may be, should be comprehended as having been wrought for the greater glory of God, on Whom alone our thoughts should dwell.[99]

Be that as it may, Kircher could not forget the idea of using the moon as a screen. In the second edition of *Ars magna lucis et umbrae,* he returns to it again and discusses "how on the moon's disk any manner of story" might be presented, and does not rule out the probability of such an invention at some future point in time.[100]

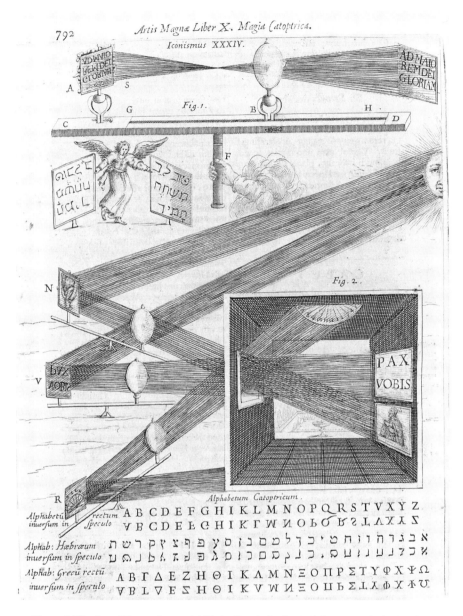

Figure 5.23 One of the most powerful illustrations in book 10 of *Ars magna lucis et umbrae* (2nd ed., 1671, p. 792). At the top is depicted the steganographic method of projecting inverted text using an adjustable biconvex lens. At the bottom, the text to be projected is demonstrated using the examples of the Latin, Hebrew, and Greek alphabets. The center shows the "dark chamber," here as a multimedia theater for simultaneous projection of images, letters, and numbers.

Kircher was a good servant of the Catholic God, "one of the strong men who dealt alchemy a mortal blow from which it never recovered," wrote one of his biographers, a doctor of letters from near Kircher's birthplace Fulda, in the mid-nineteenth century.[101] Furthermore, he was a tenacious and clever fighter for his own physical and intellectual survival and continued well-being. He was also very lucky. His autobiography[102] is replete with accounts of dangerous and dramatic events. As a child, he was run down by a herd of galloping horses at an equestrian event but sustained no injuries. At the beginning of the Thirty Years' War (1628–1631), when he was still a student and fulfilling his first teaching assignments, he was constantly on the run from various advancing Protestant armies, an odyssey that took him all over Germany. During these travels he was set upon several times by bands of marauding robbers and murderers. In the winter of 1621, he fled with some friends from Paderborn, where he was studying philosophy and physics, to Cologne where he intended to finish his studies on these subjects. While he was attempting to cross the frozen Rhine at Düsseldorf one night, the ice gave way and he fell in, but miraculously survived. John Fletcher, an expert on Kircher, lists fourteen different types of such life-threatening events in his biography of the Jesuit, including some that occurred during his years in Rome.

Kircher's monumental efforts to revive a universal republic of letters in a period when the world was fragmenting along fault lines of religious, political, and economic conflicts (and when others were attempting the same task, but through mathematics), can be interpreted as a rescue attempt. In this endeavor, Kircher's belief in his own mission and that of his Societas Jesu formed the utterly unshakable imperatives for his thought and action. The world that he creates in his prodigious oeuvre is highly ordered and beautiful. It is full of harmony, effects, illusions; it is calculated, dreamy, and fantastic: an ideal media world.

6

Electrification, Tele-Writing, Seeing Close Up: Johann Wilhelm Ritter, Joseph Chudy, and Jan Evangelista Purkyně

> Whoever thinks that there is nothing more to discover is wrong! He mistakes the horizon for the limits of the world!
> —JOHANN LORENZ BOECKMANN, *VERSUCH UEBER TELEGRAPHIE UND TELEGRAPHEN*

Virtus electrica

In 1767 a most bizarre book was published. In veneration of the Virgin Mary, its author took the name Josephus Marianus Parthenius. This was the pseudonym of Giuseppe Mazzolari, a graduate of the Collegium Romanum, who taught rhetoric and classics in Florence and Rome. Comprising six books and an appendix, his strange work had a succinct but entirely apt title: *Electricorum.* After a dedication to Ignatius of Loyola and a short preface to the reader, the Jesuit unleashes a veritable poetic fireworks display. On 247 pages of hexameters in classical Latin, Mazzolari unfolds his hymn to the phenomena and forms of electricity.[1] Written in the tradition of Lucretius and his magnificent poetic work on nature, *De rerum natura,* it belongs to the genre of didactic Latin poetry that the Jesuits particularly cultivated as an element of their offensive education strategy.[2] Mazzolari's book, however, does not contain only a poem: many technical terms, names, and concepts that appear in the poem are explained at length in footnotes, which represent a parallel, discursive text. The author has assembled everything about electricity that was known in his day that he could lay his hands upon, ranging from the speculations of classical authors about the magnetic properties of amber and early texts by natural philosophers on magnetism (including texts by Kircher and some of his devices in the Museo Kircheriano), to mathematical calculations by Ruggiero Giuseppe Boscovich from Ragusa[3] and the invention of the Leyden jar in 1745. This electrical capacitor realized the principle of storing and boosting electricity, discovered by Pieter van Muschenbroek and Ewald Jürgen von Kleist (an ancestor of the German poet Heinrich von Kleist). The sixth and last book of *Electricorum* begins with a tribute to the physicist, writer, and politician Benjamin Franklin of Philadelphia. In the late 1740s, Franklin had caused a sensation with his ideas for lightning rods and a bizarre proposal published in his letters collected in *Experiments and Observations on Electricity* (1751) for a banquet that featured a cock "killed by an electrical shock, roasted over a fire lit by electricity on an electrically powered spit."[4] Franklin was competent to make such imaginative projections because he had developed an artifact that was comparable to the Leyden jar, which stored electricity temporarily for specific uses. It consisted of a square plate of glass covered on both sides with silver paper and was charged by friction. Mazzolari combines his tribute to Franklin with an effusive expression of thanks to "Roger," his rather familiar form of address for Boscovich in the poem: "Thus far, the author has taken Franklin as his guide, whom he has named Anglus; now, however, when he embarks on a deeper examination of electrical power [*virtus electrica*], he freely acknowledges, as is just and fitting, that all that is set down in his book he has learned from Boscovich."[5] The encyclopedic poem is followed by an appendix in which Mazzolari discusses his views with other Roman scholars, including his brother, also in Latin hexameters. At the back of the volume are two inserts with drawings of machines: one is a finely crafted carillon driven by electric power, an adaptation of Jean Baptiste de la Borde's *clavecin électrique* (electric harpsichord) of 1759, and the other a so-called *machina electrica,* which warrants our special attention.

This electric machine is first mentioned in book 1 of *Electricorum,* both in the text and the footnotes. In the poem, Mazzolari praises a fellow brother of the Societas Jesu, who is introduced in a footnote as Josephus Bozolus. Bozolus, known as Giuseppe Bozzoli, was also at the Collegium Romanum, where he taught physics and philosophy and gave occasional demonstrations of spectacular experiments with electrical phenomena. By training a classical scholar, Bozzoli was best known for his translations of Homer's *Iliad* and *Odyssey.* Mazzolari writes of his colleague:

. . . well-versed in the subjects of Minerva, who brings forth all things,
he works diligently and untiringly with glass
until he has created a showpiece with new forms worthy of attention.

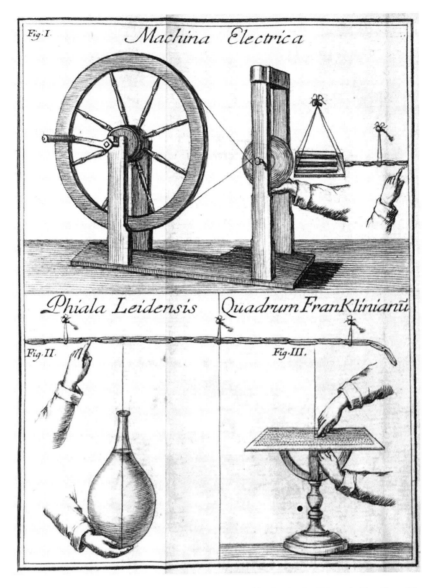

Figure 6.1 The following lines of poetry were composed by Parthenius [Mazzolari] on Bozzoli's machine for producing electrical charges by friction:

How can I explain how he moved his hand like a fan
at the end of the hanging chain
and the flame appeared again and again
and why a soft murmur could be heard in the rush of air?
Why he did first arrange the little balls connected to it in a long row
then made the flame spurt, and, finally, in masterly fashion let the fire die down in front and re-appear behind?
Of course, this also demands great skill . . . (Mazzolari 1767, p. 32 f.)

It is also a mystery how he produces most different sparks
and puts them to uses hitherto unconceived of,
and how he contacts an absent friend . . . using unusual signals.
Two steel threads unwound from a bound chain
he draws to a length that equals the distance of where his friend is.
However, to deceive the eyes of the public
and to conceal his curious invention cleverly
he buries the wires deep in the earth
yet in such way that the ends appear at the surface
where the friend, who has knowledge of the matter, waits and observes
    the secret signals.
He produces the current [*fluctus*] from the interior of a glass that turns on
    its own axis
by making the sphere tremble in the usual way
and at the place where the two steel wires are, next to each other but not
    touching
and at a fixed distance from one another,
when all is prepared, he makes as many sparks as are necessary for the
    following purpose:
they [the sparks] actually signify single elements [he explains in a footnote
    that these are the letters of the alphabet]
which, when assembled into . . . words reveal and express the mind's thoughts
    in sentences with meaning.
With the aid of these indicators and the faithfully mediating spark
    [*interprete flamma*]
the absent one speaks in words to his distant friend.[6]

From the generation of an electric charge to its storage in the Leyden jar to the reception of signals using Franklin's plate device, Mazzolari describes here an entire process of electrical transmission of messages over distances. He explains the details in footnotes; for example, that experiments with various materials showed that metal wires are the best conductors; that it is sensible to run the wires under the ground to conceal the exchange of messages from others; that it should be possible to develop a simplified language where each letter would be represented by a certain number of sparks: "Together with the friend, it would not be difficult to compose something akin to an alphabet and establish a method of speaking; the way this would be contrived and determined, as simply as possible, would be entirely at each person's discretion."[7]

Figure 6.2 The great electrification machine of the Haarlem doctor Martin van Marum, ca. 1785. At the time of its construction, it was the largest and strongest machine for experimenting with electricity produced by friction. The machine consisted of two circular glass plates with a diameter of 1.6 m separated by a distance of 20 cm. The charge produced was so strong that it attracted a strand of wool forty feet away. (Teyler Museum, Haarlem, which van Marum joined in 1784)

The literary form of Mazzolari's paean of praise (an encyclopedic poem in Latin) is extravagant, and its media-technical perspective (Bozzoli's electrical apparatus for transmitting written messages over distances)[8] is precocious, but it is by no means a rarity with regard to its subject. In the mid-eighteenth century, electricity began its ascent to being the prime focus of fascination in the applied natural sciences. Since classical antiquity, the strange phenomena connected with it—such as lightweight materials like leaves adhering to particular materials or stones, the spectacular electrical discharges that occur during thunderstorms, and the mysterious St. Elmo's fire that dances over ships' rigging—had aroused people's curiosity as well as their fear. Even for Porta, who devoted a great deal of attention to amber and the magnetization of metals in his *Magia naturalis,* these remained inexplicable natural secrets; one could only describe their effects and, in the case of amber, experiment with it and perform magical tricks for naive audiences.[9] In 1600, these phenomena received their general name when the London doctor William Gilbert, who like John Dee was in the service of Queen Elizabeth I for a time, published his book *De magnete magneticisque corporibus* [On Magnets and Magnetic Bodies]. Gilbert analyzes many materials with regard to their magnetic properties, divides them into the categories of natural and artificial, and gives them all the name *"electrica,"* which derives from the Greek word for amber, *elektron.*[10] After this, it appears to have become de rigueur for well-educated natural scientists to apply themselves to this subject. Kircher produced a folio on it; Leibniz and Newton gave it much thought. Things began to move at a practical level when, in the 1650s, Otto von Guericke, mayor of Magdeburg in Germany, began to experiment with creating a vacuum. Following Kircher's *Musurgia universalis,* he had set out to prove that sound requires a medium—air—in order to travel.[11] Then, von Guericke's interest turned in a different direction. In a text published in 1672, he describes how electricity can be generated artificially by rubbing preparations of sulphur balls with one's dry hands. He also discusses problems of conduction, its influence on nonmagnetized bodies, and the light effects of electricity. In the early eighteenth century, Francis Hauksbee, a curator of the respected Royal Society in London, built electrification machines, which generated electricity by means of friction. He also discovered the possibility of using this current to produce artificial light in glass vessels from which all air had been pumped. By 1729 Stephen Gray and Granville Wheeler had experimented in England with materials that were capable of conducting electricity. At a public demonstration Gray showed that the human body was an excellent conductor. He arranged for small, lightweight boys to hang over an electrification machine, sent electric current through their gracefully floating bodies, and let them attract pieces of metal, which stuck to their fingertips. An even more spectacular demonstration was staged by Christian Friedrich Ludolff at the newly founded Berlin Academy of Science. In the 1740s he also used a human being as a conductor in order to prove that electricity can be used to ignite fire. In this bizarre experiment, when the person touched the electrification machine with one hand, sparks flew from the fingertips of the other hand, which were hot enough to ignite preheated alcohol. The test persons used by Ludolff in his experiments were usually young women. Using their bodies, he demonstrated that fire was no longer an object external to the investigator's activities but could pass through the test subjects in the form of energy and could then be produced artificially from them. When the Leyden jar was invented, such experiments became even more bizarre and sensational. Antoine Nollet, teacher of physics at the court of Louis XV in Paris, made 180 soldiers form a chain, holding each other by the hand, and electrified them all simultaneously. It is reported that he repeated this experiment with the entire brotherhood of a monastery. For a brief moment, seven hundred monks' bodies experienced an artificially induced ecstatic state.[12]

Demonstrations such as these soon made research on electricity the foremost fashionable scientific discipline, enlightenment taken literally. Bourgeois salons and the courts of the nobility enthusiastically staged demonstrations at which the well-to-do audiences could fancy themselves way ahead of their time—time that otherwise flowed along at a sluggish pace. A similar function was fulfilled by the singing, flute-playing, writing, or allegedly chess-playing automatons, which were also in vogue in this period.[13] However, the electric charge that could be generated by rubbing glass or sulphur crystals was very weak. The physical laws of electricity were still very poorly understood, notwithstanding the fact that in 1777 Georg Christoph Lichtenberg had demonstrated the bipolar nature of discharges on a dielectricum by creating graphic patterns. He captured the effects of the negative and positive poles of electrodes on dark resin-coated plates upon which he had scattered powdered red lead and sulphur. For the first time, Lichtenberg's palpable traces produced by electricity allowed lay persons to observe the effects of electricity in a comprehensible form. In essence, however, people still believed that electricity was a property of particular materials and organisms, that is, a natural phenomenon. To prove this belief, Luigi Galvani, doctor of medicine, obstetrician, and professor of anatomy in Bologna, began to conduct his experiments in 1780. Galvani studied the

Figure 6.3   Electrical kiss, a public experiment performed by the German natural philosopher Georg Matthias Bose in the mid-eighteenth century, engraving ca. 1800. (Deutsches Museum, Munich)

electrical stimulation of nerves and muscles, for which he mainly used dissected frogs that were impaled on butchers' hooks and hung on iron railings outdoors. In 1790, he made a decisive discovery by chance. Dissatisfied with the weak reactions that could be induced in the animals in calm weather, Galvani took them into his laboratory and, with his assistant, began pressing the metal hooks in the animals' backs against metal plates. The frogs' leg muscles began to twitch as strongly indoors as had been observed outdoors only during thunderstorms.

Galvani remained convinced that electricity was a property of the animal kingdom, that is, purely organic. He believed that what he had discovered through his experiments was merely a method whereby the electricity inherent in nerves and muscles could be better brought out and its presence demonstrated. The frogs were unimportant; they merely served as conductors and instruments to register the electricity: they functioned as organic oscillographs. In practice, however, Galvani had actually invented the first battery cell, for the electricity was generated through contact between two different metals—the zinc of the butchers' hooks and the iron of the plates—connected by a damp conductor. The circuit was completed when his assistants held each other by their free hands. When Galvani published his findings in 1791, he was unable to give a complete explanation of the physical processes involved but, in the meantime, had tested numerous metals and found that contact between copper and zinc, or silver, gave the best results. The good doctor was not particularly interested in reactions between inorganic substances, which he regarded primarily as tools to prove the existence of animal electricity.[14] The precise physical explanation was provided by his countryman Alessandro Volta: electric current flows when a conducting medium—for example, cardboard soaked in a liquid, a so-called electrolyte—is pressed between two suitable metals, which are connected to each other by a conductor outside of the electrolyte. Volta constructed an apparatus on the basis of his findings, which made it possible to produce electricity artificially and store it much more efficiently than had been possible with friction apparatus or the Leyden jar. When a number of these battery cells were connected in series, it was possible to increase the strength of the current. Thus, in 1799–1800, the voltaic pile opened up a new possibility: the ability to produce potentially unlimited electric current, depending on how many piles were connected in series, as a product that was relatively independent of animate nature.

The effects of these inventions and Volta's battery were dramatic and spread rapidly, affecting many areas of life. Napoleon Bonaparte, who was appointed

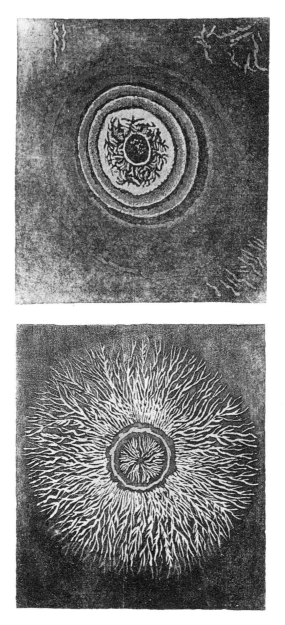

Figure 6.4 Lichtenberg figures from *Georg Christoph Lichtenberg's physikalische und mathematische Schriften,* vol. 4, Göttingen 1806. "Figures, which positive electricity brings forth, are different from those made by negative [electricity], like the Sun is to the Moon." (*Göttingscher Anzeiger* newspaper, April 9, 1778)

first consul of Paris in 1799 and crowned himself emperor of France in 1804, had this potential instrument of power demonstrated over and over again. He even offered a reward for innovations in this field. The electrical effect soon became a metaphor for the present, and also for a specific political situation: Europe, after the French Revolution, was a maelstrom of turbulence, and polarizations abounded. The state of being excited—"galvanized" or invigorated,[15]—became synonymous with electricity. Neither did it escape the attention of the Marquis de Sade, who in spite of his forced moves between prisons and medical institutions remained one of the sharpest observers of his day. His companion novels *Justine* and *Juliette,* whose protagonists can be interpreted as the positive and negative poles of moralism and depravity, were published in 1797. In her passion for crime and self-squandering decadence, Juliette is "electrified by the present."[16]

"Ritter is Ritter [Knight], and we're only his squires."

This was ideal stuff to feed the dreams and nightmares of the romantics, a school of poets and philosophers of mind and nature, who felt such a strong affinity with both science and nature.[17] At the eighteenth *fin de siècle,* a group of young intellectuals gathered in Jena, Thuringia, and declared individual subjectivity to be the decisive and final authority. The appearance of this radical group of thinkers and artists coincided with the period when the German classicists had all but left their *Sturm und Drang* phase behind them. Addicted to classical antiquity, these authors began increasingly to turn to traditional universalizations, while Goethe became more involved in the business of day-to-day politics. The romantic circle in Jena included August Wilhelm, Karoline and Friedrich Schlegel, Dorothea Veit (later the wife of Friedrich Schlegel), Ludwig Tieck, Ludwig Achim von Arnim (who was a physicist before he devoted himself entirely to writing), Clemens Brentano, and Novalis, and was supplemented by occasional guests. Following Fichte's scientific doctrine and Schelling's natural philosophy, they reembarked on a quest to seek the unity of the world and to formulate this unity poetically.

Sixteenth-century writers on magical natural philosophy had given heterologous phenomena free rein while, at the same time, respecting the individual identity of things and their designations. The universalists of the seventeenth century had attempted to unify things in the form of numbers and to formulate their relationships in general laws. In order to do this, it had been necessary to pry out of nature what they sought to formulate. The separation of mind from

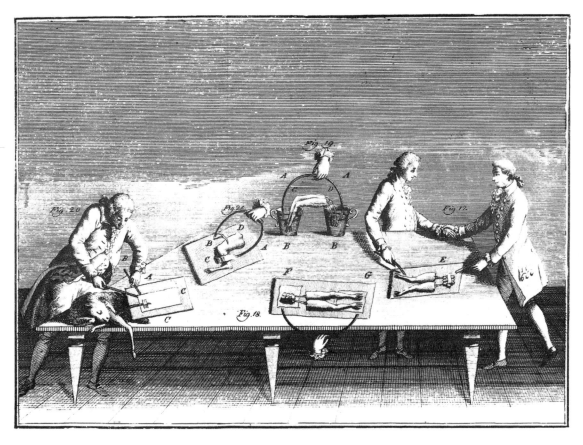

Figure 6.5 "Galvani de viribus electricitatis IV." (Ostwald 1896, p. 35)

matter, as reasoned by Descartes, appeared to have spawned unbridgeable divisions, which sensitive intellects experienced as intensely painful. Now, at last, it seemed as though a general principle had been discovered, beyond the God of the Christians, wherein the many and varied natures of things and the observer perceiving them could become united once more. The Great Clockmaker relinquished his place to a phenomenon that had not been discovered through theology but science—a phenomenon that was initially believed to be natural and, moreover, had feminine connotations. Microcosm and macrocosm could now come together in a new way. In electricity, the early romantics found confirmation that "the pulse of humanity is the rhythm of the universe"[18] and vice versa.

Like no other Johann Wilhelm Ritter, a young apothecary from the village of Samitz, Silesia, embodied this ideal of a human subject knowing and feeling with the cosmos. In 1796, when he had just turned twenty, Ritter enrolled at the University of Jena as a "foreign" student to read pharmacology (Silesia was ruled by Prussia, and Jena was located in the Duchy of Saxony-Anhalt). The son of a clergyman Ritter was an autodidact who was not interested in a normal university education. He was dying to pursue his passion for experimenting with physical and chemical processes in an academic environment. His principal interest was galvanism, its associated phenomena and effects. About a year after his arrival in Jena, Ritter had the opportunity to give a demonstration before an academic audience, when the twenty-nine-year-old Prussian mining official Alexander von Humboldt, who had already made a name for himself, invited him to deliver a critical assessment of his "experiments on excited nerve and muscle fibres."[19] Ritter wrote reams on Humboldt's treatise in a very short space of time and presented the most important highlights of his deliberations on galvanism in dead and living matter on October 29, 1797, to a most enthusiastic audience. Shortly afterwards, he published the expanded version of this lecture, which became his first book. It quickly found its way onto the desks of the seekers after truth and beauty—from Schelling and Novalis to Goethe and

JOHANN WILHELM RITTER

Figure 6.6   Portrait of Ritter. (Worbs 1971, n.p.)

Schiller—and laid the foundations of Ritter's status as a cult figure of the nascent romantic movement. Only thirteen years later, he was dead. He lived and worked to excess, without a care for his own physical well-being, and did not live to see the age of thirty-four. "Ritter [Knight] is Ritter, and we're only his squires," wrote Novalis in a letter to Karoline Schlegel on January 20, 1799,[20] expressing his profound admiration for the young scientist who had made his own body his laboratory workhorse. A deep bond of enduring friendship united Ritter and Novalis; when Novalis died in 1801, aged twenty-nine, the physicist mourned his friend deeply. After this personal loss, Ritter gradually distanced himself from the loosely knit group in Jena.

In the few short years of his working life, Ritter produced an amazing body of work. His publications amounted to some 5,500 pages of research findings, which were published in monographs, essay collections, and specialist journals. Additionally, much of his diary and correspondence can be accounted scientific treatises.[21] However, his contemporaries who were established scientists ignored Ritter and regarded him as an overexcited eccentric who did not even have a doctorate.[22] He had been dead for eighty years when first recognition came, in a speech at the first meeting of the newly founded German Electrochemical Society on October 5, 1894, in Berlin. Wilhelm Ostwald paid tribute to Ritter's exceptional work, which had led to the discovery of ultraviolet light, invisible radiation beyond the violet end of the spectrum; his invention of the storage battery; his pioneering work in physiological reactions to external stimuli and subjective perception; and, above all, his position as founder of the discipline of electrochemistry.[23]

Alessandro Volta was a physicist. Unlike Ritter and Galvani, he did not believe in the existence of organic electricity, nor was he particularly interested in the conducting liquids between the different metals of his batteries. For Volta, electricity was generated through contact between the heterogeneous inanimate matter of the metals. Ritter, by contrast, saw them all as inseparable parts of a process whereby the electrolyte connecting the metals was essential for generating electricity. Ritter proved that the chemical process produced the electrical charge[24] and thus combined galvanism with voltaic physics in electrochemistry. Before Volta, Ritter proposed various designs for apparatus that functioned with different solutions of salts as the chemical media. However, Ritter had not taken the time or the trouble to describe these designs exactly because his interest in electricity was more far-ranging than just an effective source of energy. Only after the Italian scientist caused a sensation with

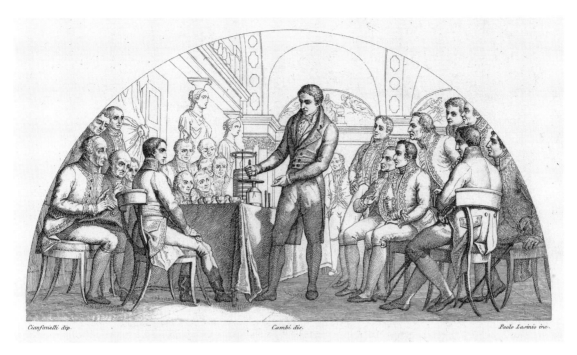

Cianfanelli dip.   Cambi dir.   Paolo Lasinio inc.

Figure 6.7 In 1801, Alessandro Volta demonstrated his experiments with electricity to Napoleon Bonaparte and members of the physics and mathematics section of the Institut de France. On the table is the pile named after him. The illustration appeared in a Florentine commemorative volume for Galilei, published in 1841.

his voltaic pile did Ritter begin to devote more attention to the construction of instruments. This he did with characteristic excessiveness. In 1802 he designed an experiment that utilized six hundred plates of copper and of zinc, each measuring 30 cm². Two years later, he increased the number of plates to two thousand of each metal and connected them in units of one hundred cells. This was the period when Ritter both discovered that electric current flowing continuously produces chemical corrosion and proved the principle of the storage battery by experiment: "A pile of copper plates in an electrolytic liquid was connected by a circuit to a voltaic pile and charged; afterwards, it was itself capable of generating current."[25]

Ritter did not have a regular income, and his experiments were enormously expensive. He spent every penny he earned on them and was always heavily in debt to his friends. He was often forced to sell books from his library, existed on a very meager diet, and is reported to have worn the same shirt for six weeks at

Figure 6.8 The accumulator, after Johann Wilhelm Ritter, was the first device to store energy on an electrochemical basis. (Deutsches Museum, Munich)

a time without washing it because it was the only one he had. In the early years, Goethe greatly admired the young man and invited him to demonstrate his batteries at the Weimar court so that the impecunious scientist might earn a little money. Because Goethe was unable to attend, Ritter wrote to him the next day expressing his feelings about such demonstrations in general and also revealing something of what drove him, the passionate experimenter: ". . . experimented yesterday in front of the Duke and the entire court. Whether it was well received I could not tell. I did not discern that there was anyone among the society gathering who had any knowledge of the art and this, of course, is a depressing feeling for one who can only appeal to such persons."[26] For Ritter physics was not just a scientific discipline. He lived his experimental praxis, it was a worldview that governed his life, and in this sense, it was art; moreover, it was only good when it succeeded in penetrating the deepest regions of unexplained phenomena.

"All energy . . . originates from polarity," wrote Ritter in *Posthumous Fragments from the Unpublished Work of a Young Physicist,*[27] which was published in 1810, the year of his death. Of all the "heterodoxies," as he called them, that he loved to study, galvanism and electrical phenomena were his obsession. In his first book, the expanded lecture, he had asserted that electricity not only flows through all organic bodies but, as a law of motion, also determines nonliving organic matter. Ritter believed that he had discovered a *central phenomenon* by which means it would be possible to explain all natural phenomena and the relationships of individual entities to nature. Convinced that he was on the track of a Grand Theory of Everything, Ritter plunged body and soul into experimental praxis, the method by which he hoped to obtain his proof. In innumerable variations of experiments, he connected his body to the circuits of friction machines, Leyden jars, and voltaic piles to test and observe the effects of strong and weak electrical charges. He wanted, for example, to find out "how strong the current is that one can endure."[28] The weak current generated by friction produced only slight changes in subjective perception, whereas connecting the body to the current produced by electrochemical processes had intense effects on the human senses, depending on its strength: "The strongest charge of a Leyden jar, which our eye can endure, is still not capable of creating a trace of those flashes in the eye that weakly charged metal piles from the well-known galvanic experiments produce in such quantities."[29] With stronger current, not only did the battery shake, but the entire body shook as well, and all the senses were affected. Painstakingly, Ritter noted down the different sensations he experienced while under the influence of electric current.[30] His main goal was to experience and explain his own body as a bipolar electrical system. Time and again he connected his body parts to the positive and negative electrodes of electrical apparatus: head, neck, nose, tongue, eyes, and other members "which are otherwise not used in experiments,"[31] to verify that each part of the body was a reflex of a pulsating Great Unity.

From the letters that Ritter wrote to his friends among the romantics, we get an impression of the dreadful effects of these experiments, which Ritter

Figure 6.9 Magnetism as universal force, which connects everything in nature, microcosm and macrocosm alike, including all branches of knowledge. This interpretation by Athanasius Kircher appears in his first published book, *Magnes sive de arte magnetica* (1641).

accepted willingly for the sake of the experience and knowledge that was of such existential importance to him. His state of health was catastrophic: his teeth fell out; after a series of continuous experiments lasting fourteen days and nights, his mouth was full of sores and he had chronic symptoms of dysentery. He began to suffer permanently from severe diarrhea. In the last years of his life, believing he could control, or at least ameliorate, his self-induced physical and mental distress with drugs, he regularly took opium. His condition did not improve after his marriage in 1804 nor after finding permanent employment—finally—in a different "foreign" state, at the Academy of Sciences in the newly founded Kingdom of Bavaria. In fact, the reverse was the case: he isolated himself even further because of his growing fascination with magical and occult practices in which he saw confirmed his view of galvanism as a cosmic formula. His espousal of siderism, belief in a subterranean electromagnetic force connected with iron, did not find favor, either, with his academic colleagues or his friends from the circle of Romantics. The once-revered cult star became a shunned crank whom no one cared to follow to the places where he had ventured. Only a few friends remained who helped him and his family; notable among them was Gotthilf Heinrich von Schubert, who took Ritter's three-year-old daughter into his own family. A colleague from the academy's geology department visited Ritter at home a few months before he died. Clearly unsettled by the experience, he wrote: "I found Ritter in a dark room in which everything imaginable was strewn around in confusion—books, instruments, wine bottles. He was in [an] indescribably over-wrought mood of bitter hostility. One after the other, he poured down his throat wine, coffee, beer, and all manner of other beverages as though he was trying to quench a fire that was raging in his insides."[32]

"Thus, at almost the self-same moment when one had thought to grasp life in its *completeness,* one *lost it completely,*" said Ritter in a lecture on "Physics and Art" at the academy in Munich on March 28, 1804.[33] In the subtitle of the lecture, Ritter states his aim: "An attempt to interpret the future direction of physics in the light of its history." He sees the ultimate goal of all thought and action in physics as "re-unification with nature to return to the former state of harmony," and in a further passage, he interprets the history of art to date as a specific positive anthropology,[34] in which the necessary interweaving of activity and time plays a seminal role. Architecture, as the earliest art form, attempts to preserve human deeds in monuments for posterity; in sculpture, the creator embodies himself; painting then partially gives back to humans the necessity of an active role, for the observer is compelled to complete the image space, which Ritter calls "half-space," with the aid of the imagination. All these three art forms, which address the eye, live off the past and are arts of memory: "Purpose of art: to render present what is absent . . . monument. The beloved, however, is more than her image."[35] In music, however, history takes a different turn. Sounds draw human actors into the very act of artistic creation; in music, action is present. Whatever comes after this, physics will accomplish: the (re)establishment of consonance between nature, which is external to humans, and their inner nature; the identity of nature and action, of life and enjoyment of living. For Ritter, this consonance is the highest form of art, and he believes that it will be realized by the physics of the future, indeed, he sees this goal as imperative, if physics is not to forfeit its meaning.[36] Like Ariadne's thread, which Ritter himself broke repeatedly or became entangled in, this notion of physics as "art within time" encompasses all Ritter's heterodoxies. Ritter's entire work is pervaded by contradictory qualities, such as extreme disregard for the body (including one's own) simultaneous with excessive celebration of it; as well as comparing, calculating, and measuring on the one hand and a highly overwrought state of mind on the other. The unity of life and enjoyment, which he propagated, naturally included the tense identity of science and art at the highest level of their praxis: the experiment. This has a name, which the physicist has no compunction in mentioning in the *Fragments:* "The longing for knowledge about things is simply the struggle for the art of loving."[37]

The idea of electricity as a *central phenomenon,* which permeates everything and keeps it in motion, initiated a change in perception: away from the gravity of mechanical physics toward a dynamic relationship between time and space. Ritter worked and researched at the imaginary boundary between the two. In one of his many texts on galvanism, he concludes by writing about oscillation as the principle of life, announces "a theory of glowing," and then makes the following generalization: "Over a long period of time, physics was concerned merely with the organization of space; however, it soon became apparent that, without history, the whole enterprise would only lead to cold petrification. A new field emerged: *time.* Time is also organized and, from the fusion of *both* organisms, time and space, all that is great and true in life and existence originates. Change is everywhere; nowhere is there stasis. All things have their own time and this does not consist in peaceful succession, which never exists anywhere anyhow."[38]

The world and all its parts, great and small, are in a state of perpetual oscillation for Ritter. That state is the present. It is realized for him in sound, an expression of the velocity of oscillations, which articulates time directly just as light does: "Above all, here there [is] inseparability of the organism of space from that of time. For in sounds, language, and music time is obviously organised and its form [Gestalt] in space is nothing less than the figure of this sound."[39]

Imaginative statements from the future, such as these, date from Ritter's final years, in which he devoted the major part of his time to two questions. The first concerned the physical interpretation of the figures published in 1787 by Ernst Florens Friedrich Chladni.[40] These were the sensational result of Chladni's early research on acoustics, which was inspired by Lichtenberg's visualizations of electrical polarity. Chladni's visualizations of sound are of extraordinary minimalistic beauty. Usually the figures are presented from a bird's eye view; the observer looks down upon visible patterns of sounds that we can hear. Chladni took thin plates of glass and metal cut into squares, rectangles, or circles, sprinkled fine sand on them, and used a violin bow to make them vibrate like a violin string. Depending on the sound frequency produced, the sand formed patterns of fine nodal lines or discrete shapes. Chladni's explanation, which still has currency today, is that the sand collects in the areas of the plate that vibrate only faintly or not at all. Ritter, however, was mainly interested in the vibrating bodies, which Chladni had used to capture the patterns on and more specifically in the plates' relation to the material used in forming the patterns. As an experimenting observer, Ritter changed the customary perspective. He crouched down so that he could view the edges of the sand-covered plates from the side. From this perspective, the rigid material of the plates appeared highly flexible: "The body is only hard . . . because of its rigidity. When there are different values of rigidity, there is also a value of electrical difference between bodies, an electrical charge."[41] In this manner, Chladni's figures became for Ritter media for concentrating or expanding time—precisely, fast and slow motion. While vibrating, the glass surface in contact with the sand rapidly changed shape, alternating between convex and concave. When the glass bent upwards, the material had a greater static charge, whereas in the concave position the sand loosened up again slightly. Therefore, there were not just two different states, at rest and in motion, but permanently changing states of varying movements. The oscillations produced an electrical charge, which could also be observed by

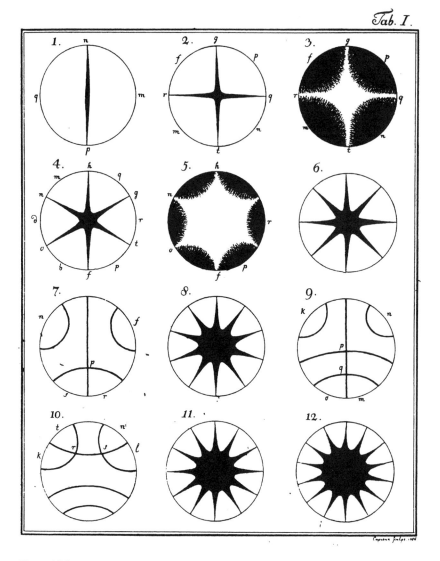

Figure 6.10 Some of Chladni's sound figures from his book *Theorie des Klanges* (1787), which he dedicated to the St. Petersburg Academy of Sciences.

the fact that the sand adhered to the insulating glass for a period of time, and most strongly in the places where the charge was the strongest.[42] For Ritter, the discoverer of electrochemistry, there could be no rest in oscillation, not even relative rest. Building upon Democritus's views of vibration as the sole higher class of movement,[43] Ritter's concept of art led him to endow the illustrations in Chladni's books with the status of frozen memories.

Turning acoustic frequencies into light by means of their extreme acceleration or the very slow oscillations in nature occupied Ritter until he died. "When bodies vibrate *extremely fast,* they *glow*" is one of his great aphorisms. Ritter saw the dynamic "light figures" or the "fire-writing," which were a source of constant delight to him, as extremely fast, high-frequency oscillations with which sound turned into the only visible phenomena of light; this represented for him "the highest degree of reality," as he wrote in a letter to his publisher in 1804.[44] When he describes the opposite pole—the transformations in the ultralong frequencies that are scarcely audible to the human ear—Ritter chooses the image of an unusual projection: "The rotation of the Earth on its axis, for example, may make an important sound; this is the oscillation of its internal conditions, which is caused in this way; The orbit around the Sun may make a second [sound], the orbit of the Moon around the Earth a third, and so on. One gets here the idea of a colossal music, of which our own poor [music] is but a significant allegory. . . . This music, as harmony, can only be heard on the Sun. The entire system of planets is *one* musical instrument to the Sun. To the *inhabitants* of the Sun its notes may appear as simply the zest for life, however, to the Sun's *mind itself,* it is the ultimate and true sound."[45] What an amazing imagination. Associations with Georges Bataille's poetic economy of the universe from the 1930s spring to mind: to the sun, which perpetually and selflessly squanders its energy, the inhabitants of the Earth can give nothing back in the form of light. The sun is the source of everything that is visible on Earth. A return gift can only be in the form of something that is perceived by a sense other than the eye—for example something that can be heard. The greatest gift to the inhabitants of the sun is for earthlings to project their *joie de vivre* onto the star.

In the short span of time left to him, Ritter could touch upon only the second subject. He probed possibilities "for the rapid conduction of sound through solid bodies. . . . This would then be a tele*language.*"[46] In all probability, this field of activity stemmed from an official request by the Bavarian minister of state, Baron von Montgelas, to the academy's members: they must do more research on speeding up communications. This request was made for reasons of state. Aiming to reconquer the former territories of the Austro-Hungarian empire in southern Germany, Austria had declared war on Napoleon in 1809. With the construction of the optical telegraph line beween Paris and Strasbourg, Napoleon had secured a strategic advantage.[47] Not only was Paris informed quickly about the movements of troops, but the optical telegraph also enabled fast retaliation using surprise defensive tactics. Thus faster communication over long distances was of prime importance for the Kingdom of Bavaria in this war, and the members of the Academy heeded the call. The anatomist and physiologist Samuel Thomas von Soemmering, who gave up his Frankfurt medical practice in 1805 to take up a professorship in Munich, very quickly constructed a telegraph that utilized the principles of electrolysis.[48] Soemmering's telegraph consisted of a voltaic pile plus a transmitting and a receiving instrument, which were both inscribed with the twenty-five-letter alphabet (minus *J*). Each letter had its own wire. A further wire, connected to the clapper of a bell, existed to signal the beginning and end of a message. At the receiving end, the wires terminated in a glass trough filled with a weak acid solution. When one of the wires at the transmitting end was hooked up to an electrical supply, bubbles appeared at the other end of that particular wire with its letter of the alphabet.[49] Electrolysis had also been a special area in Ritter's electrochemistry. Parallel to certain English and French researchers, around 1800 he had already proved by experiment that electrical current passing through liquids produces chemical changes. Thus, we are indebted to Ritter for practical proof of the electrochemical separation of water into oxygen and hydrogen. Soemmering chaired the particular committee at the Bavarian Academy to which Ritter regularly submitted his proposals on siderism, which all the committee members regularly rejected outright.

Initially, interest in Soemmering's telegraph was negligible. After a demonstration in Paris, Napoleon is reported to have dismissed it summarily as an "idée germanique." However, the idea began to take root that initiating microevents in distant places with the aid of electrocircuits might have a variety of uses. A young diplomat, who was attached to the Russian legation in Munich and whose father had been an officer in the Russian army, became very enthusiastic about Soemmering's device and took one back to St. Petersburg. A short while later, this diplomat, Baron Schilling von Cannstadt, proposed a scheme to his government for detonating depth charges from a distance using electrical wiring insulated with rubber. Twenty years later, he attracted considerable international attention with his own concept for an "electromagnetic needle telegraph."[50]

An Audiovisual Telegraph from Hungary

On January 3, 1796, the première of a one-act opera took place at a theater in Pest, the western part of the city known today as Budapest. The opera's title was *The Telegraph or, the Tele-typewriter,* and its composer was Joseph Chudy. Chudy came from Pressburg (Bratislava/Pozsony) and worked in Budapest as a master pianist and conductor for the Hungarian theatrical association. Regrettably, neither libretto nor score of the opera has survived.[51] Yet its title alone is enough of a surprise, for at that time, neither typewriters nor machines for transmitting texts over distances existed in the form of built artifacts. In the decade preceding the French Revolution, people in various European countries were working busily on concepts for fast long-distance transmission of messages. Efforts were redoubled after the French clergyman Claude Chappe presented his "tachygraph"—a device for the optical transmission of messages—to the French National Assembly of the newly constituted monarchy on March 22, 1792. The assembly, of which Chappe's brother was a member, commissioned the priest to build the world's first telegraph line between Paris and Lille, which began operating successfully in 1794. Chappe was hailed as the inventor of telegraphy. This aroused the indignation of many who had made much earlier proposals for all kinds of "writings into distance through signals and texts," the cumbersome name for telegraphy in this period,[52] but who had been unsuccessful in attracting either the attention or financial support of politicians and institutions. One of them was Josef Chudy. Using the medium of the theater, his one-act opera aimed to draw attention to an invention he had made nine years earlier. To pave the way for the opera performance, he published a slim brochure in Ofen (then Buda) in German entitled *Beschreibung eines Telegraphs, welcher im Jahr 1787 in Preßburg in Ungarn ist entdeckt worden* [Description of a Telegraph Discovered in 1787 in Pressburg, Hungary].[53] With charming modesty, the master pianist describes proposals for an optical and an acoustic apparatus for transmitting messages, which can also be used in combination as a sound and image system.

Chudy's optical telegraph was a device with five separate light elements arranged side by side, or with one light source behind a construction with five movable shutters. He chose the number five because his (Hungarian) language has five vowels and we have five senses. With these five lights, Chudy represents the letters of the entire alphabet as different combinations of their possible states—light on or off. Thus, it is a binary code, with permutations of five places. The beginning of a message transmission is signalled when all five lights are on. Chudy noted this with the wide open mouths of the capital vowel *O*— OOOOO. When the last lamp is covered, this stands for the letter *A,* which is

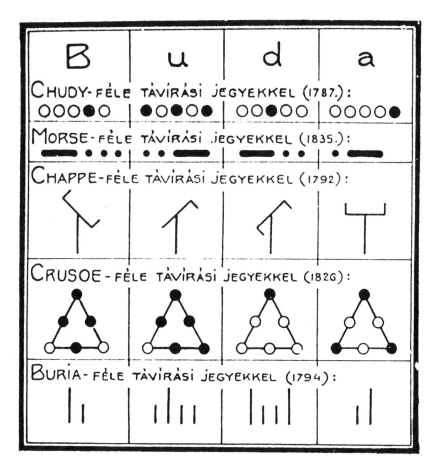

Figure 6.11 The word *Buda* written in the telegraphic code systems by Chudy (*top:* the optical binary code), Morse, Chappe, Crusoe, and Buria. (From Lósy-Schmidt 1932, p. 13).

written OOOOI; for the letter *B,* the fourth lamp is covered, OOOIO; *C* is OOOII, and so on. With the variation of the two states, on and off, and the five places, Chudy's teletypewriter was able to transmit thirty-two different signals, which is "completely sufficient for communicating sequences of letters in most languages,"[54] and it could even accommodate the special characters found in the Hungarian alphabet, such as vowels with accents. For communicating numbers, Chudy said that the code could be used analogously, in which case, the lamps that are on stand for zeros and those that are off for ones. The parties communicating have only to agree upon a signal to indicate that numbers are being transmitted instead of letters. Although Chudy mentions this possibility of representing numbers with permutations of a five-bit code, he does not give it much further thought; his interest clearly centered on communicating language over distances. By adding two more lights to the device, he was able to signal capitals and small letters as well as punctuation marks. A burning lamp positioned above the middle lamp of the row of five signified a capital letter; punctuation marks were represented by a lamp positioned beneath. For the most common punctuation marks—comma, semicolon, colon, question mark, and full stop—he used the same variations of five positions as for *Q, H, D, B,* and *A.*

For the acoustic device, Chudy proposed two types of code. The first uses the idea of his optical device, substituting lamps in on or off states with high and low notes, for example, two different drums. In his notation, a high note corresponds to a lamp that is on (O), and the low note corresponds to an off lamp, or covered aperture (Ø). For example, the word *Victoria* is noted ØOØØØ–OØOOØ–OOOØØ–ØOOØO–OØØØO–ØOOOØ–OØOOØ–OOOOØ. Of course, it is also possible to write down the system of high and low notes in musical notation form, which Chudy proceeded to do in exemplary form.[55] The second variety of code uses only one note, produced, for example, by a bell. Here, the two values are a single sound and two sounds. Obviously, gunshots or any other loud, discrete sound can be used instead of bells. Chudy was not interested in developing impractical, elaborate, theoretical designs; he wanted to develop simple methods for transmitting text messages over distances, which could be easily mastered by nonspecialists. This goal is quite apparent in his suggestion for the transmitting device: "The machine is designed to resemble a piano; it is a cabinet with five round openings, or windows and positioned beneath these is a keyboard. It can operate with a minimum of five keys . . . however, to speed up the operation . . . there can be as many keys as there are letters of the alphabet. Obviously, each illuminated opening must have a shutter, or cover; each shutter has a string fastened to it that is attached to one of the keys. . . . If there are only five keys one has to depress them using several fingers, as many as are required for the letter."[56]

The principle of a binary code can be traced back to Sir Francis Bacon, who stated that everything that can be formulated in language can also be expressed in variations of just two letters of the alphabet. For his *alphabetum biliterarium,* created in the early seventeenth century, he used the letters *a* and *b:* aaaaa represented the letter *a,* aaaab the letter *b,* aaaba the letter *c,* and so on. Although this code was Bacon's contribution to the debate on creating a *lingua universalis,*

Figure 6.12 Chudy's alphabet for the acoustic telegraph, which uses two notes, transcribed by Lósy-Schmidt (1932, p. 15).

the subject that interested him the most, as an expert in the secret affairs of state of the English court, was cryptography. He suggested that the elements of code comprising a cryptogram should be inserted in an innocuous letter or text, but in italics or another font. Bacon was also well aware that the procedure of reducing language to binary codes would be excellent for transmitting messages over distances. His *alphabetum biliterarium* offered a tool that allows friends who are far apart to send each other messages by means of signals, either acoustic or visual, providing that two distinct forms of signals are available, for example, bells, horns, fire, or cannon fire.[57]

At the end of the eighteenth century, using light to signal over distances was nothing new. In 1616, before Sir Francis Bacon published the details of his system, Franz Kessler, a portrait painter from the small town of Wetzlar in Hesse, proposed a method for transmitting texts over distances, which he called *Orts-forscher* (place explorer). Kessler operated with a reduced alphabet of fifteen letters; each letter was represented by a number of light signals previously agreed upon by the parties communicating with each other. Kessler's signalling equipment consisted of a wooden barrel lying on its side with an artificial light source within. The bottom of the barrel facing the recipient of the message was raised and lowered the appropriate number of times to signal each letter. Kessler's interest and intentions regarding his "place explorer" were of a private nature, as illustrated by the fact that he gave a successful demonstration that enabled the friends "Hans in Nahport and Peter in Eckhausen to converse."[58] At the receiving end of his apparatus, Kessler relied on the telescope, which had only just been invented. From the seventeenth to the early nineteenth century, the telescope became a central and indispensable component of all proposals for the optical transmission of text messages. It was employed routinely to bridge the distances between transmitting stations, including in examples such as Chudy's proposal, which do not mention it explicitly. This usage is illustrated by a statement made in 1782 by Christoph Ludwig Hoffmann, personal physician to the elector of Cologne. In a treatise on venereal disease, smallpox, and dysentery, he jumps suddenly to the subject of transmitting messages. The terminology is highly reminiscent of Porta: Perspectives, or telescopes, are now of such perfect manufacture that in clear weather one can see clearly the hands and numerals of a clock on a tower three miles' distant. Therefore, if someone stands on a tower holding large numbers, he can show anyone three miles away which numbers have been drawn in the lottery, as fast as light travels between the two places. When appropriate arrangements are in place and conditions are favourable, such messages can be disseminated over one hundred miles' distance and more in an incredibly short time.[59]

The many proposals published in the seventeenth and eighteenth centuries for transmitting encrypted messages and plaintext via visual contact presupposed use of the telescope as part of the transmission process. This instrument for decreasing spatial distances optically actually fulfilled the same function as fast-motion in film, for it connected places that were far apart; later electricity took over this role. Whatever was being transmitted, whether letters of the alphabet, numerals, or special characters—as in the system devised by Chappe or the comparable one by Johann Lorenz Boeckmann—relay stations equipped with telescopes significantly shortened the time taken to transmit messages. The length of time taken depended critically on how powerful the telescopes were, because this determined the number of relay stations needed. Decoding the messages and relaying them took up further time. The first optical telegraph line between Paris and Lille had twenty-three relay stations between four and fifteen kilometers apart, depending upon the lay of the land. It took one hour for a message of thirty words to reach its destination. Compared to the twenty-four hours that a despatch rider on horseback took to cover the same 212 kilometers, this represented an enormous time-saving.[60] Similarly, England's first optical telegraph line, which was installed along the south coast and began operation in 1795, had relay stations that were all equipped with telescopes. Like Chudy's, Lord George Murray's devices employed shutters; letters were represented by different combinations of light and dark. Murray's system had six positions, not arranged in a line but in two rows of three, one above the other.

Obviously, optical contact over distances relied heavily on good visibility; time of day and weather conditions constrained the time window in which messages could be sent. Small wonder that efforts were underway to find possible ways of transmitting messages acoustically. Emiland-Marie Gauthey is credited with a proposal dating from 1783 for a system of underground "speaking-tubes," which has a strong affinity to Kircher's studies on acoustics. Gauthey's idea also proceeded on the assumption that metal pipes would amplify the sound of the human voice through bouncing it off the walls of the pipes. For the human relay stations, Gauthey envisaged using men invalided out of the army: they could be trusted to listen to the clandestine communications and relay them accurately to the next station. Likewise, Johann A. B. Bergsträsser's complicated "Synematographie" of 1784—the only forerunner that Chudy mentions explicitly in his treatise—provides for an acoustic version of transmission.

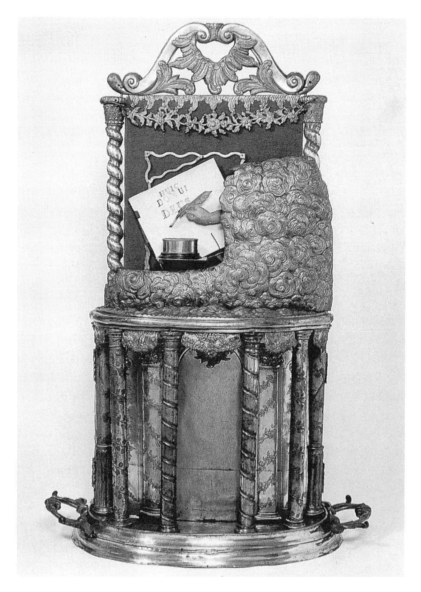

Figure 6.13 The popularization of electricity and the invention of mechanical automata proceeded in parallel. The constructors included mathematicians, clock-makers, and composers. "Self-writing wonder machines" was how E. Knauss, director of the physical-mechanical cabinet of wonders in Vienna described the automata that he built ca. 1764. The working mechanism is concealed. In this example, it consists of clockwork and several curved disks, which control the movements of the arm that writes the letters, one at a time. The automaton is one meter high. (Istituto e Museo di Storia della Scienza, Florence; photo: Franca Principe)

The distinctive feature of Bergsträsser's conception is that he elaborates various methods whereby numerals substitute for letters and compiles these in a thick *Parolenbuch* (password book). This lexicon lists all words he deemed necessary for telecommunication, together with their corresponding code numbers. Bergsträsser's concept of transposing language into numbers takes over Leibniz's binary system (1703–1705), in which the twenty-four-letter alphabet is expressed as combinations of zeros and ones. For acoustic transmission of the cryptograms, Bergsträsser suggested using the corresponding number of gunshots or rockets. There was even a proposal that suggested using bells before the idea came to Chudy: Simon Nicolas Henri Linguet, a French writer and lawyer who was guillotined in 1794 for persistently criticizing the Paris National Assembly, developed a system during his incarceration in the Bastille that is truly Pythagorean. His idea for an acoustic telegraph used five bells, each with a different pitch. The bells are struck singly, in pairs, or in various combinations, and the notes produced stand for fourteen letters of the alphabet, which Linguet estimated would be sufficient for composing short messages.[61]

Thus, many of the elements that Chudy used in his idea for a teletypewriter in the mid 1790s had already been employed in other devices or suggestions. Yet what makes his design so fascinating is not only its elegance and simplicity, but the fact that he uses a keyboard for both optical and acoustic signalling, which could also be combined. Additionally, the keyboard makes the instrument very easy to use. A master pianist from an eastern European provincial backwater, Chudy thought up a device that later became the model for the telegraphs of the nineteenth century, although the inventors of these machines in the West did not explicitly acknowledge the idea's originator. At around the same time as Chudy was developing his audiovisual telegraph, another mechanic, Wolfgang Ritter von Kempelen from Pressburg (Bratislava), was constructing his first machine, which was able to make the sounds of vowels and consonants. A bellows positioned close to the mechanical apparatus produced the sounds. The vowels and consonants were selected by operating a keyboard modeled on that of a harmonium.[62]

Belladonna and Digitalis

Living organisms must be studied as a whole, that is, the body in its entirety: "if one dissects a living creature, one kills it; it is then possible to pursue anatomical studies on the corpse, but the study of life is no longer possible."[63] Nonliving matter is the province of physics; Johann Wilhelm Ritter and his

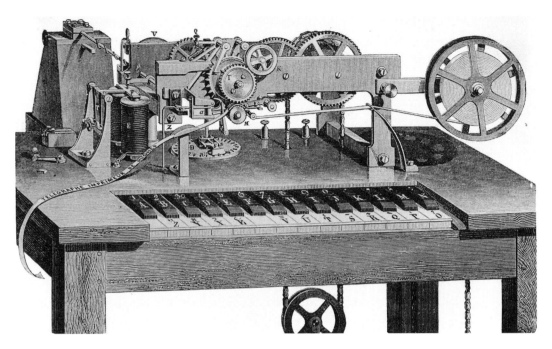

Figure 6.14 Electrical printer telegraph invented by the American David Edward Hughes. In the 1860s, it was also in service in England, Germany, and France. In principle, it uses Chudy's idea. At the transmitting end, the twenty-six letters of the alphabet are selected on a modified piano keyboard. (Carl 1871, p. 163)

"central phenomenon" was an exception. Because Ritter viewed electricity as a bridge connecting the domains of living and nonliving matter, as a physicist he was able to hypothesize connections between the two and explore these in experiments. The medium he used was his own body. He conducted experiments on his body not only to prove how physical and chemical processes interact, but also because he had begun to be interested in the body physiologically, as an assemblage that reacted to stimuli. Further, Ritter wanted to get closer to understanding the linkages between the physical and the mental. He extrapolated his idea that organisms function in a bipolar state to include the relations between the sexes, emotions, and positive or negative attitudes about the world. It was this element of his thinking that inspired his poet friends to flights of fantasy crediting the physicist with the discovery of a new *Weltseele,* a universal soul. Ritter's ultimate failure—if it can be construed as such at all— did not concern the impossibility of attaining the absolute romantic ideal, the validity of which he put aside in favor of the persuasive power of experiment. For Ritter, only what could be proved in the laboratory was worthy of being generalized. "Whereas Hegel wished to be regarded as the efficient secretary of the *Weltgeist,* analogously, one can say of Ritter that he always considered himself a marvelling guest in the laboratory of nature."[64] This fundamental conviction of Ritter's was undoubtedly one reason for his gradual alienation from the philosophers of the romantic school, particularly Schelling. If Ritter can be considered to have failed at all, it was only in his relationship with academia. The academic institutions reacted to his premature discoveries and inventions, his excessive lifestyle, and, at the end of his life, his interest in occult phenomena, by ostracizing him—a reaction born of fear. Ritter suffered immensely from this exclusion and, in turn, reacted to it by being difficult and isolating himself more and more.

In the year that Ritter died, after many years' preparation, Goethe published his *Theory of Colors,* with which he hoped to enhance the significance of the faculty of vision. The poet and natural scientist from Weimar describes a great number of his own careful observations. For example, in a darkened room, sunlight entering through a small opening falls upon a sheet of white paper. The observer looks intently at the bright spot, closes the aperture, and in the dark can still see the bright spot with the same contours and several additional colors before the light spot finally vanishes. Another example from paragraph 44: "One evening, I visited a forge, just as a mass of molten ore was being hammered. I looked directly at it, before chancing to turn and face an open coal-hole. An amazing purple image swam before my eyes and when I turned my gaze from the darkness toward the bright room, the phenomenon appeared to me to be half green, or half purple, according to whether the background was light or dark."[65]

Of course, such observations were not original. Over the centuries, many people had observed and recorded such phenomena; perhaps the most painstaking of all was the Persian researcher on optics Ibn al-Haytham writings 750 years before Goethe.[66] Yet the differences between the two men's observations are significant, particularly as they indicate the interests that drove their research. Ibn al-Haytham, an expert on optics and a physician, was interested in color and form only in the context of how far perceived images corresponded to actual phenomena in the world outside the human organs of vision. For him and his contemporaries, the distinction between internal and external reality had no meaning. For the romantics at the end of the eighteenth century, however, the distinction was of paramount importance. They sought to give back independent meaning to the human sensory organs, which, ever since the seventeenth century's theories of division and separation, were felt to be painfully lacking in the construction of individual identity. With his observations, Goethe wanted to demonstrate that the phenomena of color and color vision were relatively independent of objects external to the observer. Not least, Goethe wanted to prove Isaac Newton wrong in his statement that the color spectrum arises through refraction of white light. Goethe considered this explanation "cold." For Goethe, structure, form, and color could not arise from something that is pure, but from the confrontation of the pure and the impure, which he interpreted positively as synonymous with artistic genius.

Ritter had also turned his attention briefly to the faculty of vision. He wanted to find out how he would react to strong external stimuli. With characteristic rigor, he exposed his eyes to direct sunlight continuously for up to twenty minutes at a time by fitting a contraption to his eyelids to prevent them from closing.[67] After this extreme excitation of the eyes, colors appeared to change into their opposite: for example, when he looked at blue paper, it appeared fiery red. Ritter reported that, for days afterwards, the fire in the hearth seemed to burn "with the most wondrous blue of burning sulphur." He made similar observations of reversed color perception after passing electric current through his eyeballs. When the current was weak, he saw reddish colors; if he increased the current, the red became more intense, and at the highest voltage, red turned to magenta, which Ritter conjectured must lie at the other end of the color spectrum. These experiments represented a further attempt to prove that a living organism exists in a bipolar electrical state.[68]

Goethe's observations on the phenomenon of color made a lasting impression on a young Czech named Jan Evangelista Purkyně from the small town of Libochovice, who studied medicine and philosophy in Prague. They provided inspiration for the subject of his doctoral thesis *Beiträge zur Kenntniss des Sehens in subjectiver Hinsicht* [Contributions to the Understanding of Vision in Its Subjective Aspects], published in 1819.[69] The word "subjective" is the decisive keyword. Purkyně wanted to find the laws that govern inner visual impressions, which we all have periodically, whether we notice them or not: he called this "subjective vision." His dissertation became one of the seminal texts on the physiology of sensory perception, which in the following years was established as an independent discipline at institutions of higher education, a development to which Purkyně made a major contribution.

"The lively mind of a child revels in the colorful diversity of the outside world, which pours into it; everywhere the child gives form to what is indeterminate and delights in the repetition of these forms—each moment brings a

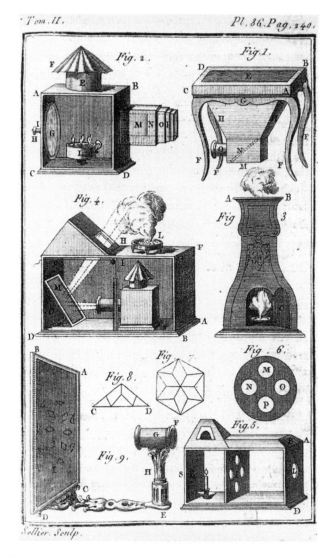

Figure 6.15 State of the optical art: More than two hundred years after Porta and one hundred after Kircher, Guyot presents a résumé of catoptric devices in his treatise "New Inventions in Physics and Mathematics." In the meantime, the camera obscura is a handy, table-top drawing aid (figure 1); the magic lantern is portable (figure 2) and able to produce many more special effects, such as projections on smoke (phantasmagoria, figures 3 and 4), kaleidoscopic and other figures (figures 6–8). (Guyot 1786, vol. 2, facing p. 240)

new discovery."[70] As a child, Purkyně was fascinated by so-called entropionic images, which are produced when the border of the eyelid is turned inward against the eyeball; then, when the eyes are closed, the images can be seen, with no direct visual reference to the outside world. The young Purkyně manipulated these images, just for fun. As a young man and scientist, he wanted to understand these sensory impressions as systematic phenomena, that is, he sought to elucidate and quantify them. For his doctoral dissertation, Purkyně determined to write a "physiography" of vision, which would complement contemporary knowledge about so-called objective vision. Purkyně did not see his physiography as a theory of deviations or exceptions because, from the viewpoint of the natural scientist, nothing is considered pathological in a pejorative sense, just as weeds do not exist for a botanist or dirt for a chemist.[71] Proceeding on the assumption that each human sense is an "individual," he ascribed to each sense organ a life of its own, both with regard to perception of external reality and its production of phenomena independent of external reality. This idea did not originate from Purkyně, nor was it first proposed by Johannes Müller[72] or the physiologists who came after him; rather, it is a fundamental concept of natural philosophy, from Empedocles to Lucretius to Porta in his *Magia naturalis.* All of these thinkers still viewed physiology as identical with the "investigation of the phenomena, forces, and laws of nature in all its domains." Only in the second half of the sixteenth century did a more narrow conception of physiology as "the science of the nature of healthy humans, their abilities and functions" gain acceptance and bring increasing specialization,[73] including with regard to the two senses viewed as most important for perception, namely, vision and hearing.

As a doctor of medicine, Purkyně conducted his research in the period that represented the transition to this specialization. His work on other sensory phenomena, such as orientation and balance, used the same systematic approach as his work on vision. His ground-breaking work in medicine, however, which assured him a place in history, was on the visual faculty. The importance of Purkyně's findings became known and was appreciated very quickly, which was certainly due in part to the fact that he not only published texts but used other media skillfully to communicate his results. His doctoral dissertation, for example, included an appendix of impressive plates illustrating observations made during his experiments. In the following years, he developed his visual presentations and also integrated new optical media for presenting rough visual movements, such as the zoetrope, into his work to popularize his research.

His spectacular success was grounded in the fact that he favored and adhered to a method, which we are familiar with through Ritter: "In this research, the only way is rigorous sense abstraction and experimentation on one's own body."[74]

The laying of the foundations for "exact subjectivism in physiology"[75] reads in many parts like a continuation of Ignatius of Loyola's *Exercises* but using the tools of science (in fact, as a young man, Purkyně had entered a monastery as a novice; he gave up his ecclesiastical career to pursue scientific research). Other parts could be protocols of trials in a pharmacological laboratory (Purkyně published his self-observations after taking opium in 1829, seven years after Thomas De Quincey published his book *Confessions of an English Opium Eater;* however, the effects of opium had played a role in Purkyně's first studies on subjective vision). Purkyně defined his method as "at the outermost limits of empiricism" and recommended that similar rules be followed as in therapeutic treatment, "namely, begin with small doses, discontinue treatment and observe the effects, then gradually increase the dose until the point is reached . . . which is the limit of all sensation and unconsciousness is imminent."[76]

In twenty-eight chapters, Purkyně elaborates his "experimental art" like a glittering kaleidoscope of inner visual experiences.[77] The majority of these experiences were induced by controlled external stimuli, because this was the only way to guarantee repeatability of the experiments, which allows qualitative and quantitative comparisons and temporal measurements. This feature distinguishes Purkyně's studies from those of Johannes Müller, published seven years later under the title *Über die phantastischen Gesichtserscheinungen* [On Fantastic Visual Phenomena], which focuses mainly on vision in dreams during sleep. In the beginning, Purkyně described simple self-observations of many variations of phosphenes, which are excited by extremely bright light or pressure applied to the eyeball when the eyelid is closed. Rectangular figures in different colors appeared; also honeycomb-like structures, and wavy lines that converged at the center, and, "when pressure on the eyeball is increased, a great many bright and fine dots appear, to begin with in the middle, then in the rest of the space, which diverge in shining lines."[78] This series of experiments led to one of the first highlights of his analyses. The forms of the phosphenes that he observed bore a strong resemblance to Chladni's figures, which had been produced by sound, and he began to experiment with these. To produce even more impressive optical results, Purkyně put lightweight and viscous fluids, instead of the sand that Chladni had used, on the vibrating plates to visualize the sound waves. His conclusions are reminiscent of Ritter's space-time inference: "Wherever continually operating forces restrict each other, in the course of their alternating ascendencies, one over the other, periodicity in time arises, oscillation in space." For Purkyně, too, there is no rest, no full-stop in the realm of the living, and for him, too, sound is a media event, which captures most aptly the sensing of the world. Subjective vision becomes the effect of a vibrating state: "Just as this actually occurs in the movement of sound, it seems likely to me that the eye, when external pressure is applied to it or it self-contracts of its own volition, goes into an intimate oscillatory motion mode."[79] In a different section of his dissertation, his profound understanding of visual experiences as processes led Purkyně to an astonishing anticipation: "When observing regular geometric lines, spirals, circles, wavy lines, symmetrical figures, ornaments, or curlicues, where laws and necessity prevail, the eye feels drawn involuntarily away from the outlines of the objects, the movements are made easier, even semi-automatic, so that they are transferred to the observed objects, which seem to acquire their own life and movement; this makes a strange impression and is accompanied by a weak tension in the eyeball. It would be worthwhile, to work on this kind of *Augenmusik* [eye music], which beckons to us everywhere in the worlds of nature and art, as an independent genre of art."[80]

Purkyně viewed experiments with galvanic current as the logical consequence of the temporal aspect of his "physiography." He constructed a battery consisting of twenty pairs of copper and zinc plates with cardboard soaked in ammonia solution between them. For the conductor outside the electrolyte, initially he used guitar strings covered with metal. He took the conductor from the zinc pole and put it in his mouth, and then with the copper pole touched his forehead, bridge of the nose, or temple (from Latin *tempus* because that was the point where the pulse, the individual time of the human organism, was measured). Purkyně varied the poles and the positions of the guitar strings, even making the two poles touch briefly while keeping the metal touching close to his eye, "whereby extremely fast discharges ensued through the winding metal wires." When he combined the application of pressure and electrical current to the eyeball, he saw the "artery figure . . . which flashed with each discharge across the sphere of vision from the entry point of the optic nerve."[81] This figure is one of the most beautiful in Purkyně's collection. Because it is similar in form to the aorta, it became a synonym for his discoveries in the field of subjective vision.

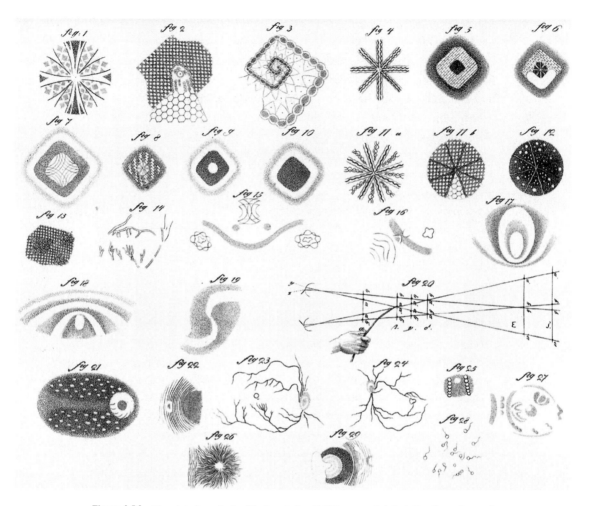

Figure 6.16 The plate from Purkyně's dissertation (1819, appendix) depicting the patterns of "subjective vision."

The "galvanic light figures" play an even greater role in his *Neue Beyträge zur Physiologie der Sinne* [New Reports Investigating the Physiology of Senses], published in 1825. In the meantime, Ritter's experiments had come to Purkyně's notice and he continued his investigation of the color spectrum. Purkyně also began to take greater risks with his own body. He reported in detail experiments where he placed both poles of the conductors, which he had replaced with the more effective silver wire, on the eyelid covering the eyeball. To lessen the burning sensation caused by contact with the electric current, he recommended moistening the skin in those places. He also intensified his work on reporting the effects of various drugs, especially belladonna and digitalis purpurea. Belladonna is a legendary extract from the deadly nightshade plant, which causes dilation of the pupil when administered as eye drops. Since classical antiquity, women had used it as an "invasive" cosmetic. In these experiments, Purkyně actually deviated from his main subject, because he tested the substances for their different effects on seeing close up and far away; he attested that the lattere was associated with a higher degree of fatigue. Although Purkyně was interested in the medical problems of near- and far-sightedness, he returned to the phenomena of subjective vision when taking digitalis, which comes from the common foxglove and whose effects include nausea. He started by taking a small dose for four days. The predominant effect was flickering vision, which he had also observed in himself after strenuous physical exertion. Several weeks later, after he had recovered, he began to take a much stronger dose of the drug. He boiled a large number of foxglove leaves for half an hour and in the early morning took more than seven grams of the concentrated extract. At first, the effects were gradual, but then very intense. After nineteen hours, his left eye began to flicker, an effect he had observed in the first experiment. Then, "Nausea, angina pectoris, weakness, and muscular tremors lasted the entire day. At midday, the characteristic flickering began also in the right eye." He was able to draw exactly the contours of his visual impression. He saw shimmering, light and dark concentric circles and roses with many petals and shining contours, patterns that he named "shimmer-roses." The effect of the drug lasted undiminished for three days; the images finally disappeared after fifteen days and with them, the very unpleasant accompanying physical symptoms. Purkyně attached importance to the fact that during the entire time "the brain was not affected in the slightest, unlike, for example, after taking opium, camphor, datura [thorn apple], etc."[82] He had written up his observations in a perfectly lucid state of mind.

The final chapter of Purkyně's dissertation, "The After-image: Imagination, the Visual Faculty's Memory," begins: "I often wondered why the blinking of the eyelid does not impair vision, as I imagined that during this, there must be a brief instant of complete darkness. On investigating this more closely, I found that the field of vision of the open eye with its entire complement of images and lights remains before the sense for a short time after the eyelids are closed."[83] Again, others had observed this phenomenon before Purkyně. Ptolemy, for example, is reputed to have made such observations and, in Lucretius's powerful poem on the world of atoms, *De rerum natura,* there are similar references; these were followed up by Egyptian investigators interested in the eye and had been taken up intermittently by others since late medieval times.[84] However, accounts explaining the first media artifacts of the 1820s and 1830s cited the effect of so-called persistence of vision, which was assumed to be a purely physical process. The image of whatever was being observed impressed itself briefly on the retina, and this image then fused with following images to give an overall impression. This effect was also referred to as "sluggishness" or a "defect" of the eye. Simple devices, such as John Ayrton Paris's Thaumatrope of 1824 or Joseph Plateau's Phenakistoscope,[85] were manufactured and successfully marketed, yet a satisfactory explanation of the effect was lacking. In his interpretation of this phenomenon, Purkyně took a completely different tack from his predecessors, which also turned out to have surprising consequences for his explanation of subjective vision as a whole. He discarded the deceptive notion of "persistence," which Goethe and adherents of the defect theory had espoused. Persistence of vision would inevitably cause a short-term, flat visual impression that corresponded to the external stimulus. However, to explain the sensation of the afterimage, which is certainly perceived as stereographic, Purkyně said that after perceiving an object the "the sense of touch of the eye" continued "to situate the afterimage outside of the organ. . . . The vividness of the afterimage differs according to mood. It is particularly vivid when the emotions are aroused, after imbibing alcohol, or taking narcotic substances or when one is particularly interested in something: in feverish states, particularly those affecting the brain, it is often heightened to the point of ineradicable objectivity."[86] Purkyně thus refutes the idea that the effect is a simple physiological process or defect and declares it to be the result of psychological activity and neurophysiological processes. The reception of an impression is no longer the decisive factor but, instead, imagination and memory "become active themselves in the sense organs." The "animal within the animal," as Purkyně also called the senses, thus become "mediators," media for feelings as well as for the unifying action of consciousness.[87] Essentially, the production of the afterimage is a constructive act.

Goethe was very impressed with Purkyně's dissertation. He is reported to have studied it for a whole year and he also wrote a long and detailed review. Goethe interpreted it as confirming some of his own ideas formulated in the *Theory of Colors.*[88] Thus the famous man of letters was all the more piqued that this obscure young Czech scientist did not mention him at all. Purkyně made amends with an effusiveness that today seems most bizarre. [New Contributions to the Understanding of Vision in Its Subjective Aspects], published in 1825 under the main title *Neue Beyträge zur Physiologie der Sinne* [New Reports Investigating the Physiology of Senses] begins with a dedication to "His Excellency, Johann Wolfgang von Goethe, Grand Ducal Privy Counsellor and State Minister, Grand Cross of Weimar Order of the White Falcon, Imperial Russian Order of St. Anne, and Officer of the Royal French Order of the Legion of Honour," plus the remark: "If I . . . have been so fortunate as to have made a few discoveries in the subjective realm of vision, and hope to make more, this should merely be regarded as a day's work that, commanded and guided by yourself, enters reality." For all that, in the tribute he restricted the inspiration derived from Goethe to paragraph 41 of the didactic section of *Theory of Colors,* in which Goethe had made time measurements of how long the afterimages of different color impressions lasted. Although this observation seemed reliable enough to the experimental scientist to merit a mention, in the rest of the work, Purkyně does not refer to Goethe at all. However, Goethe's vanity had been gratified, and his letter of thanks to Purkyně was likewise effusive in the extreme.[89]

This short digression concerning the hierarchies and sensibilities of the learned world in Purkyně's era is necessary. Purkyně had an independent mind, and the rigor with which he pursued his self-experiments was mirrored in his uncompromising stance toward the officialdom of the Hapsburg monarchy in his country. For example, he refused to accept German as the decreed language for teaching at university and insisted on lecturing and teaching in Czech. Despite the great deal of attention attracted by his doctoral thesis, none of the centers of learning in the Austro-Hungarian empire—Prague, Vienna, Budapest, or Graz—offered him a professorship. To begin with, he took up the post of an assistant in the anatomy department at Prague, where he began his sensational experiments on dizziness. To observe the interplay of physiological and neurological processes in the perception of actually or apparently moving objects, he strapped mentally ill patients in rotating machines and performed extreme experiments on himself, including testing the effect of strong drugs on the perception of movement. Of prime interest to media archaeology in this connection are Purkyně's observations concerning the perception of stationary objects as moving: "If the gaze is determined too often in a certain direction by moving objects, such as a line of soldiers or the rim of a wheel, this movement becomes fixed and, for a time, involuntary, so that even stationary objects appear to be in motion."[90]

Eventually, Purkyně accepted a professorial position at the University of Breslau (now Wrocław) where, in spite of the persistent resistance of his Prussian colleagues, he attempted to set up a physiological institute within the faculty of medicine, the first of its kind in the world. However, the resentment directed toward this unwelcome foreigner was so great that progress was very slow. The department was only established in 1839 and officially opened as an institute three years later. Ultimately, Purkyně was able to realize his work ambitions to a greater extent after accepting in 1850 a professorship in Prague, which he had wanted for so long. By then, he was sixty-three.

Purkyně is recognized as a truly outstanding scientist, and not only in his era. He discovered large branching nerve cells in the cerebellum, which are named

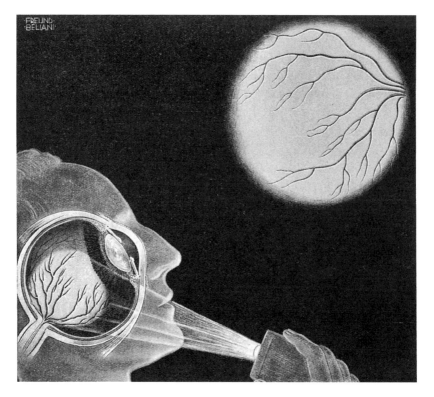

Figure 6.17 Purkyně's "artery figure": In a darkened room, the eye is illuminated from the side by a strong light beam. The image of the illuminated retina shimmers purple in the subjective perception, and the shadows thrown by the blood vessels in the retina are seen as a branching structure. (Kahn 1931, vol. 5, p. 61)

after him, and Purkinje fibers conduct pacemaker impulses to all parts of the heart. His name was also given to the effect that short-wave rays are better visible longer in semidarkness, which later became significant for astronomical observations.[91] Visual phenomena, their physiological basis, and systematic investigation appear to have held the greatest fascination for Purkyně. One could call him an "astronomer of the inside," as John Frederick Herschel has done.[92] However, as a doctor of medicine working in the field of experimental physiology, Purkyně focused on the unique human body with its complement of individual insignia and sense faculties of expression. In 1823, shortly after his first contribution to subjective vision, he developed a system that is still used today to ascertain the identity of a person: after innumerable observations, he defined nine recurring variations of patterns of lines in the skin of the fingertips. It does not do justice to Purkyně, however, to see him as one of the protagonists of the criminological identification of offenders. The motives that drove Alphonse Bertillon or Francis Galton at the end of the nineteenth century in their theories of the *signalements* of criminal elements were completely foreign to Purkyně. His approach to the practice of medicine was as an art of individualization. Paradoxically, it was his concern for the individual patient that led him to search for external indicators of uniqueness and not any intention of classifying characteristics in a catalogue of deviance.[93]

The Discovery of a Pit, a Camera Obscura of Iniquity: Cesare Lombroso

This rocket's going nowhere. It is travelling so fast.
—PETER BLEGVAD, *IN HELL'S DESPITE*

"The age of progress and steam is an age of crime, but also of noble, lofty, philanthropic aspirations," wrote the Heidelberg jurist A. von Kirchenheim on August 5, 1887, in his introduction to Cesare Lombroso's most famous book, *L'uomo delinquente,* which was published in Germany (and England) with the sensational but incorrect title *Der Verbrecher—The Criminal Man.*[1] At this point in history, a feverish excitement accompanied the emergence of a new species, *homo industrialis;* moreover, this was also the founding era of the new media. Photography was already looking back over a sizeable and impressive history. Electric telegraph lines connected metropolises, nations, and—via submarine cables—even continents. Telescopes were electrified and had also been modified for the first mechanical models of tele-vision. The human voice is now immortal, captured as a reproduction on Thomas Alva Edison's cylinders and Emil Berliner's records. The chronographic images by Ottomar Anschütz, Etienne-Jules Marey, Eadweard Muybridge, and Bertalan Székely successfully fixed bodies in motion on a two-dimensional plane and projected them. A few short years later, Wilhelm Conrad Röntgen would make fluorescing rays pass through his hand. Insights that had formerly been gained only through the bloody process of dissecting dead bodies were now presented in transparent black-and-white images of living bodies. Inner organs were also visible to the eye; on these images, only golden wedding-rings resisted the newfound force for rendering bodies transparent.

In the preceding decades, human and animal organisms had been thoroughly investigated under the microscope and hooked up to machines and instruments. Ritter and Purkyně had made laboratories of their own bodies to investigate what happens in a living organism when it interacts with the material external world, including what happens on the inside to the psychophysiological system of an individual contained within a permeable mantle of skin. Apart from verbal reports, texts, and sketches in which they recorded their self-observations, Ritter and Purkyně had no means at their disposal whereby they could objectivize their passion for these inner vibrations, which disturbed and fascinated them to such a degree that, in their experiments, they exceeded by far the limits of prudence or normal endurance.

In the mid-nineteenth century, physiological research became established at universities and academies. Physiology no longer had any pretensions to being an all-inclusive investigation of nature; it was now a specialized field, alongside medicine and anatomy, which concentrated on how the healthy human body functions. Ever more sophisticated "experimental systems"[2] proliferated to observe, measure, and classify all phenomena and processes of living organisms. The objects observed were externalized, so that physiology now meant experiments on other bodies. A priority field of research among young physiologists and physicists was the microvibrations of organisms. Hermann von Helmholtz aptly referred the measurement of the "smallest units of time" in which movement and processes unfolded to as "microscopy of time."[3] It was now possible to make graphic representations of how the blood circulates, how the field of vision is laid out, how the eardrum vibrates, how bodies react in free fall or rotation. At the same time technical systems, such as the telegraph, became models for body functions as well as instruments to investigate them. When interpreting the data gathered, physiologists at first exercised caution with regard to physical explanations. What counted was facts, insofar as these were measurable, that is, capable of being expressed in numbers that could be related to spatial and temporal parameters. In spite of all endeavors to maintain a distance for idealism and its universalistic ideas, nevertheless, on the basis of a new materialist-positivist view of the world, a notion of an ideal body took shape, which could be divided up into characteristic features and processes. For this, the body did not need to be dead; it was only necessary to protocol physical movements. Facts and measurements were idealized. Just as the discoverers of central perspective in architecture and painting did not care at all what the solid bodies in front of their eyepieces were smelling, tasting, or feeling while they were drawing them, the experimental physiologists were indifferent to the sensations of the individual body. Its motions and irregularities were primarily of interest as statistical values, as orders of magnitude in the context of functions, which needed to be understood, repaired if necessary, and possibly rendered more efficient.

Then they dared the attempt to measure things that previously had not been measurable. There were more than enough challenging subjects: the effects of introducing industrial production devices into all corners of everyday life, the massive changes in the organization and perception of time, the nascent tensions in the relationships between social classes and between the sexes, and the declassing of considerable sections of the populace to human work-machines and lumpenproletariat, plus their concentration in the big cities, became visible. Phenomena appeared at the boundary between mind and body, which had not been experienced to this degree or at this intensity before, or had carefully been kept at a distance from the attention of scientists. Alcoholism, social decline, criminality, prostitution, and hitherto unknown disturbing disorders combined to constitute a fabric of deviance with which societal authorities and their collaborators in science had not learned to cope.[4] Empirical experimentation concerning body functions had begun by examining the extremities. Now the investigators' attention turned to states where the mind exhibited extremes: in overt deviation from social norms, madness, hysteria, epilepsy, crime, and the equally misunderstood "genius."

A fair number of scientists from several countries, who were also early media experts, were engaged in this study of pathological phenomena. Building on earlier work by Franz Joseph Gall, Johann Gasper Spurzheim, and Alexander Morrison, the style of investigation included strategies from the natural sciences, statistical methods, and elements of theatrical performance, with a conspicuous leaning toward the artistic. Hugh Welch Diamond's most important photographic work in a psychiatric context was done in the 1850s, during his time as superintendent of the Female Department of the Surrey County Lunatic Asylum.[5] At the Salpêtrière in Paris, Martin Charcot and Paul Richert began their series of publications on a new iconography of the abnormal with a study on the demonic in art—in the same year that the German translation of *L'uomo delinquente* appeared. Lombroso was one investigator among many. However, with his bold theories he succeeded in polarizing both the positivists and specialists who adhered to a factual and objective approach. His ambitious goal to rewrite the human sciences as anthropology was a provocation to intellectuals outside the field. He was also polarizing because, although he dogmatically propagated the contemporary empirical methods of taking measurements, he took these methods to absurd lengths through gross exaggeration and subjective overstatement. His exaggerated explanations rattled the confidence of the positivist school because his work effectively caricatured their noble aspiration of organizing social progress through observing, measuring, and classifying.

Lombroso admitted that he would rather have been an artist—specifically, a poet. Originally, he chose to study medicine simply to earn a living and, at first, found it boring and unpleasant.[6] He came from a family of Venetian merchants with strong Jewish traditions, who, in the early nineteenth century, became impoverished under Hapsburg rule in northern Italy.[7] Lombroso's parents sent him to a pro-Austrian Jesuit school in Verona, which he experienced as a nightmare: the education was one of mindless obedience, a "violation of all independence."[8] At the time, Lombroso, who would later be a passionate collector of measurements, was much more interested in deviations from the norm than in conforming to educated mediocrity. During his studies at the University of Padua, he wrote an essay on the physician and mathematician Girolamo Cardano, who, together with Porta, had influenced sixteenth-century natural philosophy, was imprisoned for a year in Bologna for heresy, and declared a lunatic. The essay, "Su la pazzia di Cardano," was published in 1855 in the specialist journal *Gazetta Medica Italiana* when Lombroso was twenty years old. In the following year, he focused on cretinism at the University of Pavia and made long field trips through northern Italy, where he found the condition particularly frequent in Lombardy and Liguria. He carefully investigated the affected persons' conditions of life—a course of study not at all usual in medicine of the period—and concluded that the physical and mental abnormalities were due to regular intake of contaminated drinking water and dysfunction of the thyroid gland. At that time, little was known about hormonal secretion, or its links with sexuality, for example. Lombroso's academic colleagues and officialdom reacted to his indictment of these unhygienic conditions with a complete lack of understanding. This reaction grew into outright rejection when Lombroso investigated a further phenomenon that deviated from the norm, a disease known as pellagra that was widespread in northern Italy. The symptoms of the disease are severe dermatitis, gastrointestinal disorders, and neurological impairment. Little work had been done on this disorder, not least due to the fact that it was restricted to the poor and thus was not at the top of the research agenda for the scientific establishment. Lombroso's diagnosis caused a scandal. He correctly observed that the disease was confined to country areas where there was hardly any variety in the people's diet, which consisted mainly of spoiled maize that they cooked and ate each day in the form of polenta. Although Lombroso concluded incorrectly that the disease was caused by corn toxins, his sharp criticism of the conditions was right, pinpointing the deprived social component of the disease, which is caused by a dietary deficiency in niacin. The big landowners sold the best of their maize crop and gave their workers the unsaleable, low-quality, or spoiled remainder. Among the conclusions that Lombroso drew from his study of the pathology and epidemiology of pellagra were the urgent need for land reform and redistribution of wealth.[9] When his conclusions led to an outcry among the rich and their political lobby, Lombroso lost his medical practice and his position as lecturer at the University of Pavia, where he had been teaching since 1863.

This image of a socially committed doctor does not fit with the cliché of Lombroso as an archconservative specialist in forensic medicine, a fanatic who, behind his metal-rimmed spectacles, furrows his brow and thinks up ways to stigmatize others. Thus, such aspects are rarely mentioned in the literature on Lombroso. In 1859 he volunteered as a medical officer in the Italian war of liberation against Austria and he fought alongside his countrymen for an independent Piedmont. He served in the army for more than five years before resuming his work in forensic medicine and psychiatry. His first position was as a doctor in the "madhouse" at Pavia,[10] before he was appointed extraordinary professor of psychiatry at the university when he was just thirty-one years old. For a time, he also worked as director of the lunatic asylum in Pesaro. His career really took off in 1876, when he was appointed professor of forensic medicine at Turin University, where in 1896, he also became professor of psychiatry. In 1900, he accepted the chair of criminal anthropology, a field of which he was cofounder.

"It is an unhappy duty that we must fulfill. For our analysis, we have to take the scalpel and dissect and destroy, one after the other, the fine and various tissues and skins that Man in his vain triviality and stubborn self-deception is so proud of. Moreover, as compensation for our work of destruction we are not even able to offer new and more noble ideals, or more lovely and peaceful dreams: we can only respond to the cries of those who have been thus robbed and exposed with the icy smile of a cynic!"[11] These are the first lines of Lombroso's first monograph, *Genio et follia* [genius and madness], published in 1864. Here, Lombroso sets out the issues that would drive his research for the next forty years: deviance as exhibited in extreme expressions of human individuality—on the one hand, the ability to cross boundaries in art and science, which transcends the constraints of bourgeois society as do criminal acts, particularly violent ones, and on the other hand, unfathomable evil.

One of Lombroso's favorite authors was not a scientist but an Italian poet, the great Dante Alighieri, author of the *Divinia commedia.* For Lombroso madness and exceptional artistic talent were inextricably linked. Quoting Democritus, he agreed that "a free spirit and a healthy mind are [not] suited to creative writing."[12] Lombroso read Dante's masterpiece as if it were a clinical report. In his first book, there is only a brief intimation of his fascination with the *Divine Comedy,* but in his later *Studien über Genie und Entartung* [Studies on Genius and Degeneration] (1894), Lombroso characterizes it as replete with "signs of nervous overstimulation and degenerative anomalies of character." For him, the book was a treasure trove of indicators of epileptic attacks and mental disturbances, which he interpreted as phenomena connected with the mind of a genius. Throughout, he found signs of "premature passions and enormous sensitivity, perpetual highly strung eroticism, a tendency to mystic symbolism, constitutional melancholy, extraordinary wrathfulness, thirst for revenge on political and literary opponents, arrogance, megalomania."[13]

A delinquent is someone whose behavior is not in accord with accepted behavior or the law, a person who offends. Lombroso's characterizations of *homo delinquens* with reference to the criminal differ from his characterization of a genius more in nuances than in substance: moral insanity, epilepsy, extreme vanity, easy excitability, early sexual maturity, and base drives—atavisms, which he derived from his studies on "crime and prostitution in savages and primitive peoples,"[14] all categories that are highly compatible with his characterization of extremely creative people and revolutionaries. The two-volume *Der politische Verbrecher und die Revolutionen* [The Political Criminal and Revolutions], which he coauthored with Rodolfo Laschi in 1891, served above all to rescue the revolutionary in a moral sense, for, compared to the rebel, the revolutionary was a genius to Lombroso. "Between development and overthrow" there existed for Lombroso, as a convinced evolutionist, the same difference as "between normal growth and a morbid tumour."[15] A primary negative figure among political deviationists is the anarchist, to whom Lombroso also devoted a long study.[16] As

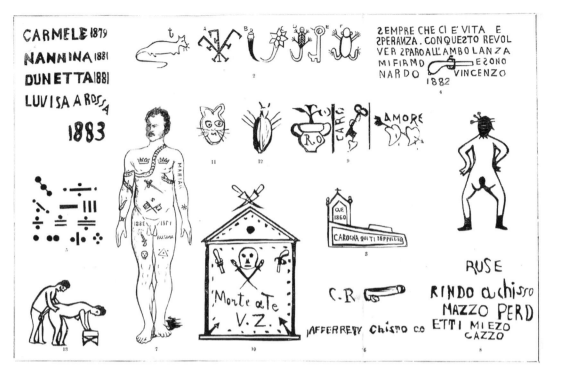

Figure 7.1 Lombroso was particularly interested in "prison epigraphics," a term he used for grafitti on the prison walls, whatever was scratched on prison objects, and the tattoos that prisoners etched on their skin. In his book on "prison palimpsests" (1899), he interprets numerous examples in connection with the individual biographies of the inmates or their social milieu. This illustration shows epigraphic examples created by members of Naples' *Camorra,* who had links with the tradition of the Spanish brigands, *Gamurri* (the name derives from *gamurra,* a particular kind of jacket that they wore). The Camorra dates from the Middle Ages when, together with the regular forces of the monarchy, they formed "to persecute all those who were not in favour because of political or religious reasons." (Lombroso 1909, p. v f; illustration in Lombroso 1896a, vol. 3, plate 53)

a committed socialist, he finds that anarchists are erratic, unpredictable, and outside the law—unlike the criminal, who is in conflict with it.

Criminal Anthropology

The prefix *hetero* produces derivative words where the meaning is different, other than usual. The Greek adjective *heteros* can also mean "deviant." *Heterogeneous* means "consisting of dissimilar or diverse constituents," and *heterologous,* "derived from a different species." Georges Bataille's *Hétérologie* was a project that attempted to amalgamate manifestations of the different, other, and alien in such a way that these elements would not relinquish their uniqueness and robustness. He needed to create a language and form of representation that would not kill deviation through fixing it, but would, on the contrary, help it to develop within the medium of poetry and philosophical reflection as "inner experience." This project was and is paradoxical, an impossibility. When they founded the journal *Documents* in 1929 to present heterologous themes,[17] Bataille and Pierre d'Espezel were not the only ones to fail in this endeavor. Other authors who followed the trail they had marked out, like Gilles Deleuze, Michel Foucault, or Pierre Klossowski, also failed—but honorably— in their attempts to write a *hétérologie.* Artists working with images are more likely to succeed in communicating an intimation of the other and otherness, such as in Balthus's paintings, Klossowski's drawings, or Pierre Moliniere's photographic compositions.

Lombroso, too, was attracted to delinquency and ensnared by it. However, he took the diametrically opposite path from Bataille, who sought to liberate the deviant phenomenon as confirmation of its autonomy. Using positivistic methods, Lombroso tried to apprehend it, to put it behind bars in a cell. His goal was to render all its manifestations utterly readable and then to convert them into data. In this way, heterogeneous phenomena became quantities of information, amenable to calculation. To this end, Lombroso used the entire range of tools provided by positivist science. Together, separately, and in tandem, he utilized physiological measuring methods (including electrical systems of registration), craniometry (for measuring criminals' skulls), the entire array of anthropometry (measurement of ears, eyes, hair growth, and other physical traits), photography (systematically as a recording and storage medium and for analysis), identification techniques (such as fingerprinting), the anatomy of brains and reproductive organs, embryology, and graphology.[18] The data thus gathered he wove into a dense fabric by constructing analogies between the various fields to produce a seemingly indisputable body of facticity. Transformed into texts and images, these unsettling phenomena could at least be laid to rest in this form. Particularly influential was Lombroso's variant of criminal pathology, which tended to refer all mental facts to biological causes.[19] In this, the doctor and psychiatrist with a strong awareness of social issues became severely entangled. Lombroso described going into criminal anthropology as an "accident."[20]

"The development of a scientific discipline as an enterprise to seek truth rather than to proclaim it, is dependent above all upon its inner freedom from any ties to power and authority."[21] In the last decades of the nineteenth century, criminal anthropology emerged as a discipline and there was no doubt about its affiliations to power. It claimed "to know the nature of the criminal completely," as Lombroso put it. In essence, its basis was "a science of what is evil," which aimed to identify and isolate evil, thus rendering it harmless. Earlier approaches directed more toward social reform, such as that of the Italian Cesare Bonesama or the Englishman Jeremy Bentham, were either repressed or forgotten.[22]

Before criminology (in its earlier guise of criminal pathology) embarked upon the path leading to its status as an allegedly objective science, it alternated between two images as explanations for the syndrome of criminality. First, there was the myth of the "totality of the beast": all aspects of the person of the criminal represented evil, and evil was manifested in all of them. However, this image was implicitly dangerous to society because it meant that the criminal, and thus the domain of crime, was territory separate from normality. The second model held evil to be ultimately inexplicable: in spite of the spectrum of explanations deriving from social circumstances, individual biography, or character traits of the criminal, the evil deed remained an enigma; it indicated something outside and beyond the individual concerned, that is, it harbored a transcendental element. The birth of modern criminology was based on the attempt to escape from both models, the myth and the empty space of transcendence. Neither could be proved empirically nor documented with facts. The solution to the problem was to transform the myth into experience and to do away with transcendence by simply reducing the complexity of both criminal and criminal act. Henceforth, criminologists would be interested only in those dimensions of crimes defined as legitimate reality by policy makers. Everything else was not part of the criminologist's brief. The central aim was to dispense with everything that could render the criminal a person "who is imbued in a disastrous way with autonomy." Yet the endless suffering, to which the modern system of justice subjected criminals, was actually a spectacular affirmation of this autonomy.[23]

The construct of *homo delinquens* as a counterpart, not a subspecies, of *homo sapiens* by Lombroso and others, particularly some of the manic encyclopedists working in Italy at the end of the nineteenth and beginning of the twentieth centuries, actually resulted in a naturalization of the phenomenon of crime. The autonomous being became a heteronomous individual situated firmly in the

Figure 7.2 "Italian criminal types": In addition to his innumerous indexes of photographs, Lombroso had many drawings done for a large album that emphasized specific physiological characteristics of deviance. (Lombroso 1896a, vol. 3, plate 25)

realm of natural objects. It was a radical solution to the problem of the other who is difficult or impossible to explain. The delinquent no longer had any importance as an independent personality (which was dangerous for the investigator, because this approach cast doubts upon him, too), but was now an object within nature. Only by making deviant persons "things" was it possible to count, measure, and file them away. Lombroso and his colleagues made a point of stressing that they only used scientific methods, conducted empirical-inductive studies of the criminal personality, and that their analyses could always be checked. An apt label for this type of modern criminology is "social engineering" (Strasser). On the one side were the delinquents and, on the other, delinquency-measuring devices and apparatus plus instruments to catalogue and prevent delinquency. For this was the consequence: once the criminal had been declared "natural," he or she did not even need to commit an infringement of law; such persons could be arrested beforehand because inescapable biology had determined that they were evil.

Photography was the master medium for this variety of criminology. A picture of a criminal taken with a camera was assigned the same degree of truth as the measurements of skulls, ears, or other body parts. At their institute in Turin and later in the Museum for Criminal Anthropology, Lombroso and his collaborators collected and catalogued portraits of criminals from all over the world as well as pictures of faces that they regarded as exhibiting deviant characteristics. These technical pictures were used as an index of violent crime and served as evidence for the fundamental assumption of its biological origins, as expressed in the photographed faces. Composite photography, which Francis Galton had introduced into the criminological discourse, had a special status. By laying a number of different portraits on top of one another, criminologists obtained portraits of characteristic "types" of criminal. Hailed as a method for defining conceptual typologies of criminality, composite photography was statistics translated into images.

Lombroso and his collaborators' views, methods, and conclusions were by no means shared by all contemporary actors in criminological research. At three large congresses, 1885 in Rome, 1889 in Paris, and 1892 in Brussels (to judge from the records that have been preserved, these were the Olympic Games of criminal pathology), there were vehement arguments between the representatives of the various schools. Decided opponents of the Italian positivists were a group of French physicians, psychiatrists, and jurists, who neither accepted biological determinism nor their Italian colleagues' methods of examining and

Figure 7.3 "For the first time, photography enables us to capture the features of a person clearly and for the duration. The history of detection began at this point because the decisive conquest of a person's *in cognito* was assured. Since then, efforts have been unceasing to pin him down in words and actions." (W. Benjamin, 1972, vol. I, p. 2; illustration in Lombroso 1896a, vol. 3, plate 32, from the *Berliner Verbrecheralbum* [Berlin Album of Criminals])

presenting findings. For the French group, crime was a phenomenon that was primarily rooted in social conditions. Although they also made use of measurements and media strategies, such as photography, these were employed rather to better understand crime than to explain it. A major theme of debate at these congresses, which were accompanied by exhibitions (the most opulent one in Rome), was the degree of truth that photographic images represented.[24]

The *Fossa occipitalis media*

Lombroso took immeasurable numbers of measurements. During his time as director of the asylum in Pesaro and as professor of forensic medicine and director of the psychiatric clinic in Turin, he examined and dissected thousands of bodies, particularly brains and skulls, examined handwriting and tattoos (which he found especially fascinating), read palimpsests and engravings by criminals, and even scrutinized the last confessions of the condemned for signs of deviance.[25] Many of the objects he archived meticulously, conserving and exhibiting them. Shortly before his death (he died on October 19, 1909), he willed that his head and neck should be preserved for posterity in a special tincture in a glass receptacle.[26]

His most spectacular discovery, however, was a kind of medium. It was an anatomical pit, a dark chamber, on which he tried to shed light through his interpretation. Deep inside the skull of an infamous Italian brigand, he found an anatomical detail of dimensions that had not been observed in humans before, only in "lower" mammals: the then so-called *fossa occipitalis media.* In his measurement statistics from the examination of 383 skulls of criminals, Lombroso notes: "The *fossa occipitalis media,* found in 16%, in 11 cases exhibited normal proportions, found in 5% of the specimens. . . . In 6 of the aforementioned 11 cases, they belonged to thieves, and 5 belonged to murderers. . . . One of them, an inhabitant of Bologna, had a fossa that was twice as large as normal, and Villela from Calabria, an enterprising thief, still exhibited open sutures in spite of his 70 years of age and an extraordinarily large fossa: 34 mm long, 23 mm wide, and 11 mm deep."[27]

The *fossa occipitalis media* not only became a leitmotiv for Lombroso, but also assumed the qualities of a medium. This feature communicated to him the biologically determined character of the born criminal or deviant, which social institutions or private circumstances might tame for a while, but was innate and liable to break out at any time. Upon this discovery, Lombroso constructed his hypothesis that born criminals are throwbacks to a previous evolutionary stage of development.[28] Later in life, in a remark recorded by his daughter Gina Lombroso-Ferrero in the introduction to the American edition of *L'uomo delinquente* (*The Criminal Man,* 1911), Lombroso recalls in trivial metaphors of enlightenment and with unconcealed fascination the encounter of the "slave to facts" (as he frequently referred to himself) with this obscure phenomenon:

This was not merely an idea, but a revelation. At the sight of that skull, I seemed to see all of a sudden, lighted up as a vast plain under a flaming sky, the problem of the nature of the criminal—an atavistic being who reproduces in his person the ferocious instinct of primitive humanity and the inferior animals. Thus it became understandable in terms of anatomy: the enormous jaws, the high cheekbones, protuberant brows, singular palmar lines, extremely large size of the eye sockets, ears that are hand-sized or close to the head as found in criminals, savages, and primates, indifference to pain, extreme acuity of sight, tattoos, exceptional sloth, predilection for orgies and irresistible love of evil for its own sake, the desire to not only extinguish the life of a victim but to mutilate the body, tear the flesh, and drink the blood.[29]

The violent criminal as a raging cannibalistic creature, who not only threatens *homo sapiens,* but is capable of tearing him to pieces and devouring him: here, making crime a natural phenomenon signifies animal instincts erupting into civilized bourgeois life, the menacing coexistence of genetically backward human types with those representing progress. How to explain this result on the basis of natural laws? How could Lombroso, a Darwinist, evolutionist, and determinist render this comprehensible? Unrestrained desire, destructive tendencies, and tearing others to pieces do not appear to be reconcilable with humans as the positive outcome of the selection process in the struggle for existence. By adhering to biological determinism, Lombroso was obliged to ascribe to nature an innate tendency to violence, destruction, and excessive expenditure of energy. However, in contrast with his contemporary Nietzsche and, obviously, in contrast with Bataille and his circle later, who celebrated these transgressions poetically as high values of an antinorm, Lombroso's assessment was that of a forensic scientist serving the perspectives of law enforcement. The only way for him to resolve this dilemma was to place these phenomena in the realm of the Dionysian: lives that were wasteful and expended themselves recklessly and ruthlessly were not consistent with the notion of socially desirable cultural development within the framework of law and order. Demonizing the deviant elements of nature seemed the only solution. However, the problem with this solution is that the critique of criminological "production" of evil through positivist means[30] then no longer holds, for demonization implies an elevation to the rank of myth. At the same time, Lombroso needed this elevation within his hierarchical structure of ideas in order to define and salvage its central configuration of criminality, madness, genius, and revolution as a privileged territory of the male, that is, his own territory.

One of Lombroso's earliest published texts, written when he was eighteen, investigates the relationship of sexual and cerebral development. The subject is not apparent in the title—"Di un fenomeno fisiologico comune ad alcuni nevrotteri" [On a Physiological Phenomenon Common to Some Neurotics] (1853). (The strategy of concealing sexual themes in ponderous academic language was not unusual. The Austrian Baron Richard von Krafft-Ebing, who wrote his dissertation on "Die Sinnesdelirien" [delirium of the senses] a few years after Lombroso's early text and of whose works Krafft-Ebing thought very highly, formulated many passages of his famous *Psychopathia sexualis* in Latin. Latin continued to be a hermetic code for Western scholars well into the nineteenth century, an effective means of excluding the majority of people from knowledge, particularly where taboo areas were concerned.) The construct of the criminal as a "psychosomatic setback for progress"[31] is the unspoken but constantly resurfacing focus in Lombroso's texts. The unexplainable other, the enigma, which aroused profound fear in him and drove him to manic activity, is articulated for Lombroso most directly in the other sex, in the difference between the sexes and their deep conflicts.

"When we embarked on our collection of facts, we often felt as though we were groping in the dark, so that when a bright and clear goal appeared our joy was that of a hunter whose enjoyment of success is doubled when he catches his prey after much toil and trouble." This quotation is from the preface to the nearly 600-page-long book *La Donna delinquente, la prostituta e la donna normale* [The Female Offender, the Prostitute, and the Normal Woman] (1893), which Lombroso coauthored with his collaborator and later son-in-law Guglielmo Ferrero.[32] In the first part of the book, the two authors set down in a kind of statistical delirium their notions of a "normal" woman. Weight and height; covering of hair; and proportion of fat, blood, skull, brain, urine secretion, menstruation, sensorium, sexual sensibility, sense of shame, sensitivity to pain, cruelty, and many other parameters are expressed in statistics with numerous cross-correlations to distill the main results. Thus armed, they then move on to the "criminology of woman" and the pathology of female criminals and prostitutes:

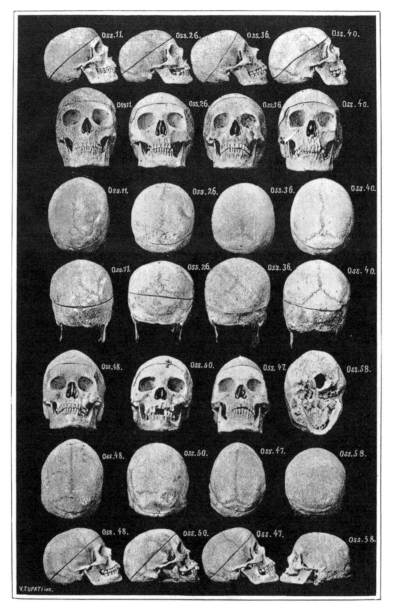

Figure 7.4 "Skulls of Italian female criminals." (In *La donna delinquente, la prostituta e la donna normale,* Lombroso and Ferrero 1894, (Italian edition) plate 3).

normal woman is "an underdeveloped man," that is, a backward being who is only superior to man during the short prepuberty phase. An expression of this brief period of superiority is the "prematurity" of woman, which of necessity leads to later "inferiority," deficient sensitivity, and sexual frigidity. Combined with the female's lower intelligence, these are the principal traits that lead the psychiatrist and forensic scientist to their central conclusion, which succeeds in saving the status of the male genius and universal superiority of men: there are hardly any born female criminals except for a few rare examples that correspond to the male atavistic type.[33] Real criminal creativity is the exception rather than the rule in women. The same applies to the positive side of genius: "in the world of the intellect, women are completely absent. Female geniuses are an alien phenomenon in the world," Lombroso writes in another publication, which he co-authored with Rodolfo Laschi.[34] Delinquent behavior most appropriate to women is prostitution, which Lombroso does not see primarily as a crime but rather under the aspect of its social use-value, as a "safety valve for morality and public order"—as a useful lightning conductor for the "depravity of the male" so that "one can say that woman, when she sins or becomes brutalised, is still of use to society."[35] Perhaps as a concession to the feminists in the Italian socialist movement, to which Lombroso was committed, he is at pains to condemn certain specific forms of discrimination against women: "Not a single line contained in this work . . . can serve to justify the many forms of tyranny to which women were and still are subjected, from the taboo forbidding them to eat meat or touch coconuts to prohibiting women from learning a profession and, what is worse, from forbidding them to exercise what they have learned in a profession. Through such ridiculous and cruel restrictions we have undoubtedly contributed to maintaining the inferiority of women, indeed, even to increasing it in order to exploit it for our own advantage, even when we hypocritically shower the credulous bondmaid with flattery, which we do not even believe ourselves."[36] For all that, the vanishing point of Lombroso's position is that he defines the relationship between the sexes as production for the benefit of the male, which, unsurprisingly, he also applies to his own daughter: "And most of all you prove it to me, dear Gina—the last remaining and only tie that binds me to life, the strongest and most fruitful collaborator and inspiration for my work."[37]

Hyper

This is the era when the new media were founded, the end of the nineteenth century. Cesare Lombroso's books are highly elaborate and bizarre products, which

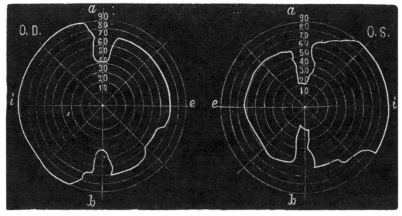

Fig. 16.
Gesichtsfeld der jugendlichen Diebin J. M. im ruhigen Zustande.

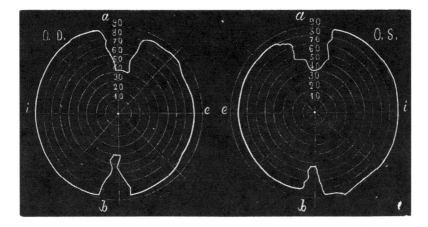

Figure 7.5 Irregularities in delinquents' field of vision were strong indicators for Lombroso and Ferrero of unstable mental states. The top diagrams are from "a young female thief . . . in a quiet state." The bottom diagrams are "the field of vision of the same person during a presumed mental epileptic fit." (Illustrations in Lombroso and Ferrero 1894, p. 379)

appear to have left the medium of the traditional book far behind them. They exhibit, on the one hand, the voluminous phenomenological variety and mania for analogy found in Porta's *Magia naturalis* and, on the other, the complex, connected arbitrariness of the coming media age. In presenting his evidence, the scientist Lombroso, a poet manqué, utilizes the entire repertoire of techniques for registering, archiving, and visualizing that were available in the second half of the nineteenth century plus an inflationary use of statistics as a method of description. Similar to his contemporary Galton, who in all seriousness proposed and carried out statistical surveys of the efficacy of prayers, Lombroso found divine significance in quantified comparisons.[38] The vast quantity of pieces, which he produced in the form of graphs, images, tables, diagrams, and text particles, were woven into networks of argumentation that appeared highly complicated and, like in a chamber of horrors, were juxtaposed generally without any hierarchical order. A characteristic of the books is a fondness for divisions and structuring. *Das Weib als Verbrecherin* [The Female Offender], for example, is divided into two volumes and nineteen chapters, which are subdivided further into over 250 sections. The impregnable single fact is carefully orchestrated to the point of irrefutability. This approach dictates the superficial structure and rhythm of the texts, which are invariably backed up by the impressive numbers of cases that he investigated: the "observation and study of 23,602 mentally deranged persons"[39] led him to the absolute certainty that going mad coincided with rising temperatures in certain months of the year. One of the verbs that Lombroso uses most frequently is "to link," and he is in fact a master of conjunction: he connects and correlates everything, even the most disparate conjectures and data, most of which he raised himself. In *Der politische Verbrecher und die Revolutionen* [The Political Criminal and Revolutions], there are maps of France showing the distribution of "races," political parties, and the "number of geniuses per 10,000 inhabitants."[40] Also with reference to France, in a detailed diagram he classifies the various *départements* according to the "genius index" in combination with the "orographic composition" and the "geological soil composition."[41] In the second chapter of the first volume, Lombroso links the frequency and quality of revolts and revolutions with a variety of factors, including climatic conditions, such as air pressure and barometric swings, or the "inhibiting influence of very high mountains."[42] The validity of any single claim is not an issue. This strategy overwhelms by creating the impression of factual complexity: the individual fact becomes irrefutable because linking it to numerous other facts endows it with an unassailable status. In an obsessive manner, Lombroso cultivates a positivism of associations. The opponents of this method, who viewed social grounds as more decisive than individual ones, objected particularly to his explanations of deviancy. Kurella gives a very good description of Lombroso's method:

If they . . . had summoned Lombroso to the prisons of Liebau, Riga, Dorpat, and Reval to establish the causes of the Latvian-Estonian jacquerie, he would have ascertained the specific meteorological conditions prevailing at the time of the revolt, investigated the racial origins of the prisoners, established the signs of degeneracy in their physiognomy and body form, particularly of the skull, sought for the number of epileptics, hysterics, madmen and dipsomaniacs, subtracted the vagabonds and those with a criminal record, established how many of the women involved were menstruating at the time of the uprising, counted how many young men were completely influenced by fanatical doctrines . . . and it is more than doubtful [after this], whether many of the accused would be left over for a representative of the materialist view of history, who had been invited along with Lombroso, to establish a purely economic determination of the incriminating acts.[43]

One of the meanings vibrating in the prefix *hyper* is "excessive," or "excessively." The excessive degree to which Lombroso calculated and measured deviancy to pin it down exceeded even the narrow frame of the deterministic and Darwinian method he favored. In a self-induced delirium, the manic encyclopedist became bogged down in the mire of facts, data, and signalments he had collected and pitted against each other. The text that he worked on for his entire life is superficially one that the police, judiciary, and society attempted to make use of, but the subtext, which is discernible in all of his books on the various types of deviancy, conveys a different story. It tells of a suffering and passionate—and in this sense, pathological—hunter and gatherer, who seeks (and probably finds for the greater part of his work) in the positive school of science a way to compensate for and pacify his own fears and desires. "[T]he passions of a genius's mind are violent," he wrote in *Genio e follia*. "They give color and life to the mind's ideas. And when we think to have discovered the one or other genius whose passions do not flare up with unrestrained force, this is only because they have subsided to make way for one overruling passion: the insatiable longing for fame, or the thirst for science and knowledge."[44]

The born criminal and the genius: Lombroso must have felt the existence within himself of these diametrically opposing poles of existence that strive for autonomy. One can also read his books as a confused trail, which he laid down for others that they might achieve some idea of how this conflict feels.

Dante, who like Petrarch was closer to the dark and enigmatic nature of medieval *Minnesong* [love songs] than to modern poetry, made love for the first time when he was nine years old, Lombroso tells us. On the same page he quotes some lines by Raphael, "How sweet is my yoke, how sweet the chains of her dazzling white arms when she puts them around my neck. I suffer pangs of death when

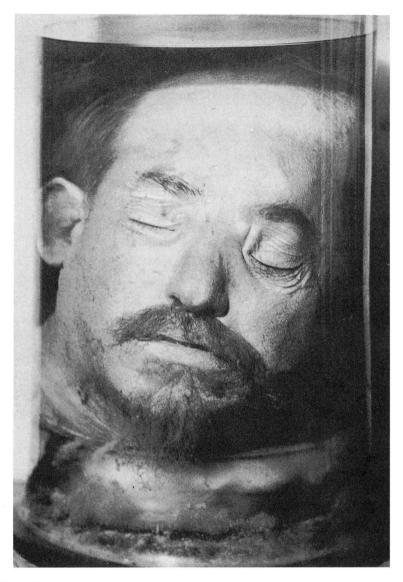

Figure 7.6 "A life dedicated to science": Lombroso's preserved head in a glass vessel, which was kept in the Museo di Antropologia Criminale di Cesare Lombroso in Turin according to the terms of his will. (Colombo 1975, p. 57)

I tear myself away from them. I shall remain silent about a thousand other things, for a surfeit of delight leads to death."[45]

Lombroso's last work, which was published posthumously in 1910 in Turin, was entitled *Fenomeni ipnotici e spiritici* [Phenomena of Hypnosis and Spirituality). He was particularly fascinated by the extraordinary states of trance and mediums' profound absorption, and thus ended his research in an area that is hidden from the eye. As an academic outsider, Lombroso was frequently the butt of ridicule by colleagues and institutions throughout his entire career. Many thought that he was crazy. This ridicule gave him a heightened sensibility, which made him avoid discriminating against forms of knowledge other than those that he practiced professionally. Lombroso said: those of us who laughed about different, deviating views of the world were often "incapable of noticing that we were wrong, and like many madmen, we laugh while darkness conceals the truth from us about others, who are already in the light."[46]

In the most famous expressionist film of all, *Das Cabinet des Dr. Caligari,* the psychiatrist's medium is called Cesare. There is only one other silent film that can compare with the expressiveness of Wiene, Mayer, and Janowitz's masterpiece: an Italian film version of *Dante's Inferno,* made in 1909, a magnificent preexpressionist film poem, full of eccentric archaic effects and a delirious portrayal of deviance and decadence. Lombroso could have written the script. Sadly, few people have seen the film or even know of its existence.[47]

The Economy of Time:
Aleksej Kapitanovich Gastev

The realm of things remained poor and paltry as long as mortal
humans were the measure of all things.
—FÜLÖP-MILLER, *FÜHRER, SCHWÄRMER UND REBELLEN*

From 1907 to 1909, Aleksej Kapitanovich Gastev worked as a tram driver in
St. Petersburg. He took great pleasure steering the heavy vehicle through the
streets of what was then the Russian capital with its leisurely pace of life (in his
subjective impression). The gently flowing Neva with its extensive network of
canals, the magnificent parks of this city that Tsar Peter I had laid out on a
grand scale, and his slow and ponderous tram passengers formed a stark contrast to the dynamic iron and steel machine he was driving, which had been
responsible for speeding up travel immensely. This experience inspired Gastev
to write in a short story, composed in 1910 while he was in exile in Paris: "with
the motors whirring mightily, you cut through the crisp air saturated with the
fragrance of fresh greenery. Slowly, quietly, as if on velvet, you drift onto the
Stroganov Bridge and rein in the car on the slope. A Stop. And after it, ignor-
ing the protests of the overdressed passengers and disregarding safety pre-
cautions, I switch on both motors at once. With a terrible lurch and bursts
of sparks I take off as if I had been stung by a bee down the twisting lane of
Kamenoostrovski Prospect."[1]

Gastev's father was a teacher and his mother a dressmaker. The family
came originally from Suzdal, a small provincial town in Central Russia. At first,
Gastev also wanted to become a teacher. At the age of sixteen, shortly before the

turn of the century, he began teacher training in Moscow. At eighteen, he joined the social democrat party (RSDRP) and, at the age of nineteen, was sent down from the Moscow college for coorganizing a demonstration. He was arrested for the first time in 1902 for distributing illegal socialist propaganda to textile workers and was exiled to the province of Vologda for three years. These are just a few details from the biography of a young man whose life was fraught with tensions. Gastev grew up in a sheltered family situation that in general was positively disposed toward intellectuals. While still very young, he was active in the nascent worker's movement, earned a living doing various jobs in different trades and industries, helped to set up the metal workers' trade union in St. Petersburg, and, under changing pseudonyms, wrote articles and short stories for the newspapers and journals of the left-wing movement. By the time of the October Revolution in 1917, he had been arrested several times by the police and sentenced to a total of ten years' banishment. However, he managed to escape and spent several years among the large group of Russian exiles in Paris. There, he came into contact with both the French left-wing movement and western European avant-garde artists. Around 1907, he left the RSDRP and did not join a political party again until 1931, when he was forty-nine years old. At the time, Gastev bore the heavy responsibility of trying to transform the organization of industrial labor in the Soviet Union. However, his ideas were increasingly being discredited, and he did not resist absorption into the Communist Party. Seven years later, he was arrested again; this time, on Stalin's orders. Sentenced to ten years' hard labor, he was murdered soon after his trial. The official date recorded for his death is October 1, 1941. In 1955, the Soviet administration began tentatively to rehabilitate him owing to pressure from friends and relatives. In the mid 1960s, some of his theoretical essays on labor were republished and, in 1971, a selection of his poetic texts appeared in print again. To date, this committed Russian Taylorist and radical poet of "machinism," as he named his overall concept for an ideal new world, is little known, either in western Europe or in his own country.[2]

One of the words that is most laden with meaning for Gastev is the Russian word *opyt*. Depending on the context, it means either "experience" or "experiment." In his understanding, life is a laboratory, and he viewed his own biography as a sequence of different states in the laboratory. "In my life, I was for long periods a revolutionary, a metalworker/design engineer, and an artist," wrote Gastev, looking back, "and I have come to the conclusion that the highest expression of labor, which was implicit in all . . . that I did, is the work of the engineer."[3]

Until the early twentieth century, the vast lands of tsarist Russia had an agrarian economy. The attempt to industrialize society and daily life collided violently with the traditional pace of life, which had hitherto been determined by the gradually changing seasons and the diurnal round of agricultural production. Gastev invested all his creativity and energy in the project to transform this slow-moving giant of an economy into an entity that operated according to the rhythm of a machine. In the early years, when he was an active trade unionist, Gastev still believed firmly in the workers as drivers of the needed transformations and in their capacity to learn. However, confronted by the lethargy (which he found intolerable) of the majority that he encountered in the workshops and factories, he began increasingly to place his hopes in the machine, in particular, the perfected form of the automaton. He became convinced that his ideas of social progress could be realized only by integrating each and every individual into the technical process. Thus, the effort was no longer to organize the machines; instead, the machines themselves would effect the transformation. In an essay entitled "On Trends in Proletarian Culture," Gastev wrote: "the machine guides and controls living people. Machines are not objects to be controlled any more, they are the subjects."[4] For him, this could not be an isolated or local process: the "technicalization of the world" (*texnizacija mira*) is only conceivable or feasible if it is universal. In the course of this process, the individual disappears completely, with the workers of the future becoming "cogs in a giant machine." A psychology will develop that operates in an identical way to the laws of the machine, with its two states of "switched on" or "switched off" and its closed functional circuits. The individual will develop into a semiological apparatus, a "mechanism, which executes operations at a signal or process of signals."[5] Gastev knew that such a fundamental transformation of attitudes and skills was possible only for an elite minority of the workforce. Only very few were capable of following the intrinsic laws of machines and their rhythms, which are so different from those of human beings. One had to accept these special machine-literate workers in the same way that one had to accept the machine itself.[6]

The change in the way that Gastev regarded the revolutionary masses and his direction toward the concept of collectively mechanizing social performance developed in symbiosis with the forms and themes of his literary oeuvre. His first

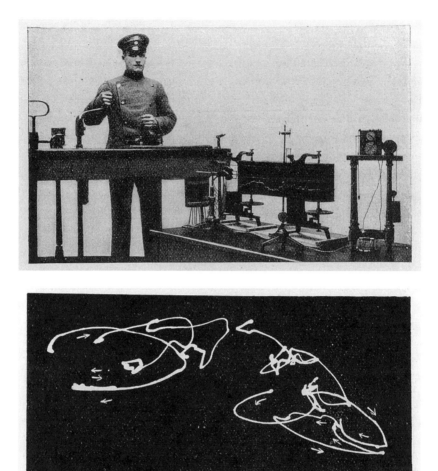

Figure 8.1 Dissecting the labor process in a Berlin institute for time and motion studies: Experiment to measure the time taken for a tram's emergency stop. Times are stopped by two registering devices (top); the motions of the tram driver are recorded photographically as light trace (bottom). (Tramm 1921, pp. 108 and 113)

published work in 1903 consisted of reports on his daily work situation and precise observations of the processes of production and exploitation under which he saw the workers suffering. His first poems, written around 1913, he grouped together in the 1926 collected edition of his poetry under the heading *Romantika* (Romanticism). These were resounding invocations of the struggle and successes of the rebellious workers' movement expressed in conventional lyrical prose or traditional verse forms. During the course of World War I, after his involuntary sojourns in Paris and under the spell of the October Revolution, he radicalized his writing style and began to practice a severe economy of language. The lines of his poems became increasingly condensed; indeed, they are rather like the output of a machine. "Write fast," he exhorts the reader, in a text addressed to factory fitters, "always carry a notebook and pencil with you. Obviously, it would be good if everyone could do shorthand."[7] Gastev's poetic masterpiece is a collection of poems entitled *A Packet of Orders*. Written in 1919–1920 and published in 1921 in Riga, this thin volume was the last thing Gastev published that had explicitly artistic aspirations. His idea of universal "machinism" has obvious religious overtones.[8] He saw himself as a missionary of the mechanization of all areas of life—on a global scale. Thus, it is plausible to interpret his *Packa orderov* as ten commandments for a new world; in fact, they are numbered from 01 to 10—and the numbers are the titles. In the "technical instructions," Gastev directs that the poems be recited in "uniform batches . . . as if one were operating a machine." The listeners should receive the impression of "a libretto of crucial processes." Again, he develops motifs that are in evidence in his earlier texts. Music, vibrating sounds, and their dynamism serve as metaphors for the process of transformation, and orchestration is a symbolic representation of the universal. For example, Order "06":

Asia completely on the note of D.
America one chord higher.
Africa in B minor.
Radio concert-master.
Overture: cyclone-cello.
The bow across forty towers.
Orchestra at the equator.
Symphonies around the 7th parallel.
E-strings down to the centre of the Earth.
. . . Eruption of volcanic fortissimo crescendo:
Hold at volcanic tempo for half a year.
Slow down to zero.
Close orchestra pit.[9]

Order "07" is a hymn devoted exclusively to the new media of telegraph, telephone, and radio. When Gastev wrote the poems, radio existed as a public medium only in the United States. Here, too, Gastev's mood is omnipotent and globalistic—"Turn off Sun for half an hour." With regard to form, however, Order "02" is the high point of the collection. It begins with the instructions, "Chronometer: ready. / To the machines," and ends with lines that consist of only one-word orders in Russian: "Get ready. / Switch on. / Automatic operation. / Stop." Economy of language becomes economy of time and vice versa. Both converge in the idea of the poem as a manual, as operating instructions for Gastev's concept of machinism. Poetry of this kind met with a very mixed and contradictory reception in the young Soviet Union. Some vehemently rejected this idolization of machines in futurist language because it was too apolitical; others hailed Gastev as "the Ovid of the miners and metal workers."

Velimir Khlebnikov, one of the leading futurist writers in Russia, criticized its religious connotations but also praised Gastev's poetry, not least because it accorded with his own prewar ideas: "This is a fragment of the conflagration of the workers in its purest essence; it is neither *You* nor *He* but the firm *I* of the great blaze of the workers' freedom. It is a factory siren, where one hand reaches up out of the flames to pluck the wreath from Pushkin's weary brow—its leaves of pig-iron melt in the fiery hand."[10] Khlebnikov also addressed a thorny problem, which had triggered much controversial discussion in the Proletarian Culture Group: he asked whether such poetry of daily industrial work experience could be written only by those who were themselves part of the technical production process. Besides having artistic ramifications, this question was also an existential one for, according to how it was answered, artists would be either included or excluded. For Khlebnikov, whose socialization was exclusively that of an intellectual, Gastev was the proof that it was possible to generate relevant artistic designs for a future world from a position outside the factory floor. People with experience of the production system, who at the same time maintained a certain distance from it, could write programs in an artistic form.

Gastev's poetry was the outcome of his identification with technology, and the radical aesthetic of his ten commandments represented the culmination of a long developmental process rather than a revolutionary new departure. After the turn of the twentieth century, particularly in the capital St. Petersburg, a great number of painters, writers, musicians, and theater people who had turned their backs on traditional art organized in fluctuating, loosely knit groups. When Marinetti, a leading Italian futurist, visited the city in 1914, the eccentric activists of the Russian futurism movement, who met regularly in the Stray Dogs Café, hailed him as a satiated, boring bourgeois. In the first decade of the

Figure 8.2 The sound of "machinism" for a mass urban audience and the city as a gigantic "music box": On November 7, 1922, a symphony composed by Arsenij Awraamov for hooters, factory sirens, fog horns, machine guns, and steam whistles was performed in the oil-producing city of Baku; it was repeated exactly one year later in Moscow. The conductor stands on the roof of a tall building next to a chimney (bottom) and conducts the concert performance by signaling with flags. (Rumantsev 1984; images: Fülöp-Miller 1926)

new century, Khlebnikov began a rigorous analysis of language and constructed poetry based on mathematical and astronomical principles. Around 1910, Alexei Eliseevich Kruchonykh wrote poems with one-word lines. The composers Alexander Scriabin, Nicolai Roslavets, and Michail Matiushin explored music that invoked ecstatic states, ritual, and ceremony. During World War I, the Polish writer, painter, photographer, and art theorist Stanislav Ignacy Witkiewicz, also known as Witkacy, experimented with multiple exposures of figures on a single photographic plate. His *Multiple Self-portrait* as an officer in the Russian army, for which he invented a sophisticated arrangement of mirrors, was created between 1914 and 1917 in St. Petersburg.[11]

An important highlight of the combined artistic activities of the St. Petersburg *Budetlyane,* (those who "know what the future will bring")[12] was the premiere of the opera *Victory over the Sun,* in December 1913. Khlebnikov wrote the prologue, addressed to "You, the people who have been born and are not yet dead," which ends with the exhortation: "The theater is a mouth! / Spectators, be an organ of hearing (be all ears) / And be observers." Matiushin composed the score and Kruchonykh the libretto. First, two actors described as anonymous sing,

The fat beauties
we have locked up in the house.
Let all the drunkards
Run around there naked.
We have no songs,
rewards of the sighers
Who amuse the mould
Of rotting naiads!

The opera's first stage direction describes the opening set: "Tableau 1. White with black—walls white, floor black." The stage design was by Kasimir S. Malevich. Later, Malevich said that while he was working on this opera, he painted the first version of his famous *Black Square,* which was exhibited publicly for the first time in 1915 together with many of his other suprematist paintings. The exhibition, entitled *0/10,* was in fact the Russian cubo-futurists' last.[13]

Gastev's machinism was but one element in a complex constellation where artists in prerevolutionary Russia sought to communicate with each other and, at the same time, to position themselves in an international context. Machinism was not a unique phenomenon. The cubo-futurists fell voraciously upon the findings of the scientific and technological avant-garde and attempted to translate these into artistic praxis or to find autonomous forms for them in art. "Science has its ions, electrons, and neurons. So, art must have its attractions," wrote Sergei Eisenstein in his self-reflective text *How I Became a Film Director*.[14] Khlebnikov had studied mathematics at the University of Kazan, where Nicolai Lobachevsky taught in the early nineteenth century. Together with the Hungarian Janos Bolyai, Lobachevsky is hailed as the cofounder of non-Euclidean geometry. His famous essay on a new, "imaginary geometry" was first published in 1829 in the *Kazan Courier* newspaper.[15] Through a point, which lies beyond a given line, more than one additional line can pass that does not cross the given line. This theory was postulated in 1824 by Karl Friedrich Gauss, also one of the early protagonists in the development of telegraphy, but Lobachevsky and Bolyai were the ones who delivered the mathematical proof in their early writings. It marked the beginning of exact calculation of the dynamic relations of time and space, which radically changed the worldviews of mathematics and physics. There was, however, a time lag of decades: initially, the theories of Lobachevsky and Bolyai were neither understood nor accepted. In the 1880s and 1890s, they underwent a revival and sparked much interest in the fruitful intellectual climate in which Einstein formulated his general and special theories of relativity. For the mathematician and poet Khlebnikov and his fellows in the cubo-futurist scene, Lobachevsky was a symbolic figure in the revolt against the old, static conditions of society. It was imperative that the ends of the lines through the given point should bend away from the line with which they must not intersect.

St. Petersburg, known as "Venice of the North" because of its magnificent palaces and waterways, was the powerhouse that drove new technological developments and scientific dreams well into the 1920s. Under Peter the Great, a wealthy and lively academy of sciences was founded in the 1720s. From 1728, its first periodical, *Commentarii Academiae Scientiarum Imperialis Petropolitanae*, appeared regularly and was read avidly in other European centers of learning.[16] Many distinguished non-Russian scientists of the eighteenth century were corresponding members, among them, the acoustics expert Chladni, who in 1794 gave a demonstration in St. Petersburg of his sound figures and Euphonium, a sound instrument he had invented.[17] Proposals for calculating machines, for example, are also in evidence from the early nineteenth century. In September 1832, Semen Nikolaievich Korsakov published a paper describing a new statistical method to investigate "the comparison of ideas." His proposal included a machine that would classify logical operations.[18] The data for this formal intelligence device was represented by holes on punched cards, a process that had been invented a century before by the French mechanic Falcon for the weaving loom and subsequently refined by Joseph Marie Jacquard in the early nineteenth century for weaving patterns on his fully automatic loom. It was not until fifty years after Korsakov's paper that Hermann Hollerith, founder of the firm that gave birth to the IBM empire, introduced punch cards for mechanical writing.[19]

An integral part of this early scene of enthusiasts for all things technical was a small physics institute at the Konstantin School of Artillery, which became a nucleus for the worldwide development of electronic television. Boris L. Rosing, professor of electrochemistry and electrophysics there, had studied mathematics and physics at St. Petersburg's Technical University where, in 1893, he submitted his Ph.D. thesis on "the effects on matter during magnetization." Magnetism and electrolysis were the twin foci of his early work. In addition to developing a new system for accumulators, at the artillery school Rosing concentrated on technical solutions for transmitting images over distances. In his first experiments, he projected symbols in baths of silvery electrolyte, whose bases were connected by five wires, with a different image signal assigned to each terminal—a visual electrochemical telegraph. At the turn of the century, Rosing began to experiment with the Braun tube, which was available as of 1897 as an electronic display device.[20] In 1902, Rosing was already using the tube with its stream of electrons as a receiver, while still utilizing electrochemical components for the transmitter. In 1906, he designed a complete system using electron tubes to transmit simple stationary images—the same year that the Strasbourger Max Dieckmann also proposed using the Braun tube as a transmitter—but Rosing continued to experiment with the electromechanical Nipkow disc at the receiving end. Until 1907, Rosing refined his apparatus, adding drums with mirrors as scanners and modifying the light intensities of the tubes, to the point that he was able theoretically to transmit half-tones and rudimentary movements at a resolution of twelve lines. He patented this system in Germany—patent no. 209320.[21] In 1911, he used his system to transmit successfully the first moving television pictures.

In the 1920s and 1930s, two Western countries drove the development of electronic television to the level of a viable commercial product—Great Britain and the United States. Isaac Schoenberg and particularly Vladimir Kosma Zworykin are considered the main proponents of TV's technological development. Both men had studied with Boris Rosing in St. Petersburg before the Russian Revolution. Schoenberg emigrated to England, where he became responsible for developing television for Thorn-EMI. Zworykin worked with Rosing from 1910 to 1912, during the period when Rosing was experimenting with practical improvements to his system of cold cathode ray tubes. In 1917, the Russian Marconi Company in St. Petersburg commissioned Zworykin to set up a laboratory for electronic television. After an interim stay in Paris, he emigrated to the United States in 1918, where the Radio Corporation of America (RCA) provided him with excellent facilities to develop his "iconoscope," the first functioning electronic camera.[22]

Perhaps the most remarkable thing about Gastev's world of ideas is the rigor with which he attempted to translate his ideal of life as a functioning machine into a universal mechanism, from poetry into social reality. He was a passionate advocate of scientific-technical management, which had already been tested in Russia before World War I under the acronym NOT (*Naucnaja organizacija truda*) and had exerted an enormous influence on culture, music, and language. Gastev saw scientific-technical management as an ideology-free method for increasing productivity and, what is more, as a concept for radically reforming individuals—the components of the great automaton known as society. The hero of Gastev's ideal came from the class-enemy camp: Frederick Winslow Taylor. In 1881, Taylor had both won the U.S. National Tennis Championship in doubles and begun his spectacular studies and experiments on more effective use of labor-time in industry. Taylor's *Principles of Scientific Management* was published in 1911. With its central theme of "the necessity of transferring intelligence not only to machines but also to workers,"[23] it became programmatic for a new conception of industrial productivity—worldwide: "In the past the man has been first; in the future the system must be first."[24] Taylor's vision was also all-embracing. Originally a lecture addressed to the American Society of Mechanical Engineers, the text, when published, stressed the validity of Taylor's "principles" for all areas of human activity, as an "energetic imperative," as Wilhelm Ostwald phrased it in 1912.[25]

Gastev took Taylor's ideas much further. From the rigorous application of mechanical principles to human activities he envisaged that perfect worker-machines would emerge as well as systems of experts at all levels, "directorate systems, administrator mechanisms, or works-manager regulators."[26] From this, Gastev evolved the idea of a new "art of combining" (*kombinirovannoe iskusstvo*) for artistic praxis, which required an entirely new set of qualifications: "Culture does not mean literacy nor philology; we have enough literate, educated people but they are helpless, they live a life of contemplation, they are skeptics. The

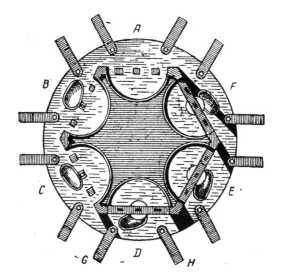

Figure 8.3 We should take work to the people, not vice versa. Henry Ford saw this maxim fulfilled with the introduction of the production line. Gastev was also of this opinion and shared with Ford (and Marx) a fascination for the clock as metamachine. His ideal of a new spatiotemporal organization of labor and structure corresponded to a clock-face. As an example, Gastev used this eighteenth-century diagram from Denis Diderot and Jean Le Rond D'Alembert's *Encyclopédie ou Dictionnaire Raisonné des Sciences des Arts et des Métiers* (1751–1780) for the proposed manufacture of needles. (Gastev 1966, p. 313 ff.)

culture for today, which we need to transform our country, means above all skill; the ability to model, arrange, select what goes together, assign, and install; the ability to assemble, to order masterfully what is disordered and scattered into mechanisms, into active objects."[27]

To dream of integrating such a system into the backward, predominantly rural economy of contemporary Russia was both anachronistic and recklessly bold. Moreover, sections of the organized left-wing movement were violently opposed to the idea. As late as 1914, Lenin was still branding Taylorism as "enslavement of the people by the machine": "Recently, the supporters of this system in America implemented the following method: a small lamp is fastened to the hand of a worker. The worker's movements are photographed and the movements of the lamp analyzed. They found that certain movements are superfluous so they forced the worker to avoid making these movements, that is, to work harder and not to waste a second on rest. . . . The cinematograph is employed systematically."[28]

Critics of machinism from the ranks of traditional industrial psychology and *Psycho-technik,* including followers of Hugo Münsterberg's ideas described in his book of the same name, insisted upon identifying the individual as the "subject of labor."[29] They rejected Gastev's functionalization of humans as mere executors of mechanical principles because they considered this a drastic reduction of complexity and ultimately unproductive. They also cited political reasons: the mechanization of labor would lead to even greater exploitation of the workers. However, contemporary economic pressures were overwhelming. There was a willingness to try anything that might facilitate the process of reformation and so, at first, Gastev was given a free hand. On August 27, 1920, the executive of the trades unions' Central Council founded the Central Institute of Labor (CIT—*Centralnyi Institut Truda*), which was directly responsible to the council. Although its headquarters were in the former luxury hotel Elite, initially, it began its work under the most impoverished and provisional conditions. Within a short time, however, the institution grew into a broad movement with branches all over Russia.

The CIT was, in Gastev's own words, his last "work of art." Indeed, its brochures and books look rather like the products of a publishing house devoted to promoting constructivist-futurist aesthetics. Avant-garde artists, like Krinski,

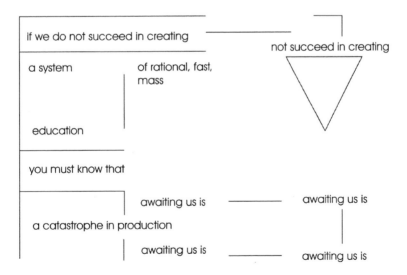

Figure 8.4 Diagram from a brochure of the CIT in Moscow. (Translated by Gloria Custance from Baumgarten 1924 [p. 18], graphically reconstructed by Nadine Minkwitz)

created a kind of visual corporate identity for the institution, and the language of the texts points to the poetic tradition from whence Gastev came. ". . . not *maybe* an exact calculation / not *somehow* a well thought-out plan"—instructions such as these in the style of his ten commandments are examples from the masses of printed words addressed to the "new man," whom Gastev sought to create with the aid of CIT training programs: "he must observe with the eyes of a devil and listen with the ears of a dog." Gastev's entire system was to resemble a "biological mechanism."[30] Paradoxically, the concept of work that is articulated here is far from alienated industrial labor. In actual fact, Gastev celebrates complete identification with the job as exercising an artistic skill: "We shall not only feel love for the machine, the machine that is often a mere theoretical fantasy for us, but also for tools. We shall give priority to the hammer and the knife. He who wields a hammer skillfully has grasped the principle of the rotary press. He who uses a knife skillfully understands the secret of all cutting machines. . . . Enough of all those grand designs, become artists of labor!"[31]

Gastev's enthusiastic relationship with technology did not develop against a background of vast machine parks awaiting rational organization, but exactly the opposite: machines were in very short supply, indeed. As Fülöp-Miller writes, "with very little industrial production anyway and then war destruction, to begin with there were but few examples of those steel marvels with cogs and wheels in the country. Thus, for orthodox Bolsheviks, every telephone, every typewriter, in fact everything connected in some way with technology, was an object of rapturous admiration."[32]

Gastev and his colleagues employed two closely related strategies in their struggle to win the battle for a new human subject. Confronted with the slow and ponderous Russians, the "objective hygiene of brain activity" must consist of installing "a key to the economy of time" and getting them moving.[33] It was impossible to realize any new economy based on the old leisurely pace. Parallel to the work of the CIT, in 1923, Gastev also founded a "Time League," which developed into a popular movement. Within the space of a few months, branches sprang up in St. Petersburg, Charkov, Kiev, Kazan, Rostov, and Tbilisi. Factories were declared the cells of the Time League. Flyers and propaganda leaflets spread its message in permutations of three words: "Time—System—Energy!" The members were named "elvists," a neologism formed from the initials of the league's name in Russian, *Liga Vremeni*. "Stop your time, check it! Do everything within the time, to the exact minute! . . . The

Figure 8.5 "Let us take the storm of the revolution in Soviet Russia, unite it to the pulse of American life, and do our work like a chronometer!" Text by Gastev, drawing by Kriki. (Fülöp-Miller 1926)

The Economy of Time

Time League is a collective means of propaganda for introducing Americanism in the best sense of the word: Our work is our life!"[34] As no more hours could be added to the day, they must be used more effectively, during both work and leisure time. To train people to the new time structures, each *elvist* was given "chronocards" on which they recorded not only the day's work in detail, but the worker's entire daily routine, from getting up to going to bed. It was a tool for self-regulation. Time, which in agrarian production had appeared cyclical and continuous, was transformed into the arrow of progress and broken down into its component parts. The clock, with its stop and go (also the principle that made the cinematograph run), became a metamachine with symbolic character.

The breakdown of complex processes corresponded to the development of a code for mechanical motions. From his analysis of the range of different industrial tasks, Gastev distilled two elements, which he viewed as the basis of it all: strike and pressure. In his book *The Emergence of Culture* (1923), he explains: "Striking is a motion performed in the labor process where most of the movement takes place external to the work-piece; it is a movement that is fast and hard. Pressure is a motion to which the work-piece is subjected the entire time, a soft movement. . . . All labor-motions are either strikes or pressure. Thus, work in the foundry, such as riveting, hammering, and cutting, are striking movements, and filing, cogging, and grinding are pressure."

Similarly, Sergei Eisenstein's "bimechanics" and Vsevelod Meyerhold's "biomechanics," which he developed for stage acting, used the breakdown of time into microstructural units in order to render the dramaturgy of movement programmable, also utilizing two of time's basic elements. Acknowledging the new economy of time, Meyerhold sums up the ideas of bi- and biomechanics in an essay on the future of acting: "The Taylor system will enable us to play in one hour what we currently take four hours to present."[35]

This was an ideal field for media machines. After the motion studies of René du Bois-Reymond and Ernst Mach, and the physiological experiments of Wilhelm Wundt and Etienne-Jules Marey, the institutes and departments of physiology and medicine involved in this research had developed into advanced media laboratories by the end of the nineteenth century. They pursued the idea of noninvasive, "nonbloody anatomy,"[36] which brought experimental physiology an astounding number of devices to record and analyze movements of all kinds. The series photographs of Anschütz and Muybridge paved the way for the sensational spaces where bodies were seen in motion: cinema—for the time

Figure 8.6 *Top:* The Centralnyi Institut Truda, in Moscow, founded by Aleksej Gastev in the early 1920s. *Bottom:* Cinematographic study of a man working with hammer and chisel.

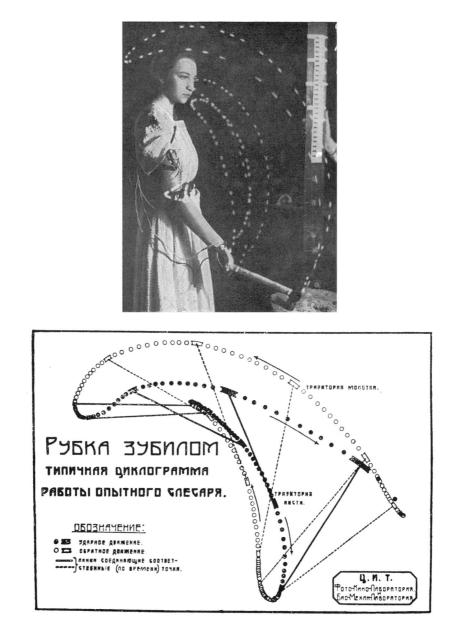

Figure 8.7 *Top:* Strike and pressure: A demonstration of the chronocyclographic method in Gastev's Moscow institute by a female worker with an artificial arm. Two phosphorescent dots are fixed on the hammer so that the movements can be recorded photographically as curves. On the right side of the picture an assistant holds a measure to provide a scale. *Bottom:* Diagram of the movements of a worker wielding a hammer.

Chapter 8

244

being, the media-historical vanishing point for generating illusion commodities and off-the-peg fantasies. In the field of experimental physiology, its traditions led to the development of a different strategy for registering, recording, and utilizing bodies. The focus was to analyze the microelements of motion and translate this into data, diagrams, statistics, and graphs. The (chrono)graphic method and its notation became a universal language of physiologists, which was understood and compatible all over the world.

Some of the most meticulous studies in this period were performed by a pair of German investigators who were associated with the mathematics and physics section of the Royal Science Institutes of Saxony. Wilhelm Braune was a member, and Otto Fischer, his closest collaborator, published their joint work after Braune's death. They experimented in a specific way with the members of the human body. Obsessive about detail, they analyzed the movements of arms, elbows, legs, thighs, and calves in terms of both motion sequences and their interdependent relationships. *Bestimmung der Trägheitsmomente des menschlichen Körpers und seiner Glieder* [Determination of the Moments of Inertia of the Human Body and Its Limbs] was published in 1892, shortly after Braune's death. The text's passion for detail and mathematical calculations make it rather bizarre reading, but at the same time highlights the mathematical possibilities of creating an exact reconstruction of complex movements by reducing them to a few parameters. The data can then be used to generate a graphic image. Braune and Fischer identified two modes whereby one could pin down any movement of a body, translation and rotation: "One refers to a movement as translation, when all points of the body describe parallel straight lines. Rotation, on the other hand, is when during a movement, the points of one of the body's straight lines, called a rotation axis, do not change their position while all other points describe circles with centers on the rotation axis and planes perpendicular to it."[37]

Using just these two parameters, the physiologists developed a complete system of the movements of the human body and its subsidiary systems. In *Der Gang des Menschen: Versuche am unbelasteten und belasteten Menschen* [The Human Gait], they describe their method and its application with an impressive illustration.[38] The male test person wears a tight-fitting, black-knit suit and, for safety reasons, thick leather shoes. His entire body appears to be wired, including his head. The electricity supply is connected to the wires at his head, which enables the test person to move relatively freely. Technically, the thin white lines running down the sides of the body, called Geissler tubes, are the decisive

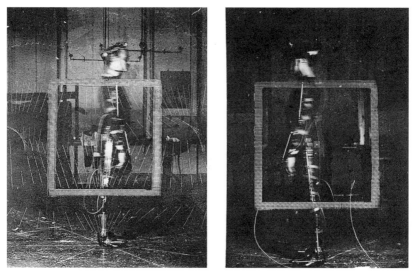

Figure 8.8 In the (photo)graphic representations of the cabled man the position of the limbs, head, and feet are clearly recognizable as discrete lines. Even the joints and principal positions for the movements, which particularly interested Braune and Fischer in the context of their binary code, are visible, as dots. At these points, the Geissler tubes were painted over with thick black paint, which appear in the photos as breaks in the white lines. Like Marey, the two researchers include the spatial and temporal parameters of the experiment in the photos: the lighter frame of dots, which is also visible in the frontal view of the male test person. (Braune and Fischer 1895)

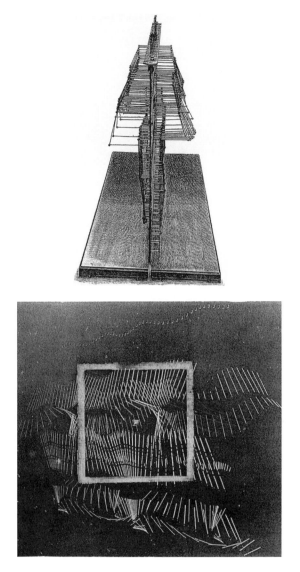

Figure 8.9 Braune and Fischer were not only interested in two-dimensional studies of human locomotion; they were well aware of its spatial dimension. From the graphic data of two views of the same movement (*bottom*), they constructed three-dimensional models of figures (*top*). To study bodies in locomotion, one walked around these sculptures, built from physiological data. (Braune and Fischer 1895)

feature of the suit. Around the turn of the century, these tubes were used in a different experimental context by Anschütz—to illuminate the glass slides in his rotating electrical *Schnellseher,* or tachyscope, to produce moving images— and in other early experiments with mechanical television. Braune and Fischer's experimental design sought to solve a problem associated with Marey's method. The test persons in Marey's experiments had white or shiny metal strips attached to their limbs when he photographed them in motion, but these strips left light trails in the photographs and tended to blur the images. Exact registration and reconstruction of movements was not possible. The Geissler tubes in Braune and Fischer's experiment ran parallel to the rigid parts of the limbs and were held in place by leather straps. In all, there were eleven tubes. The advantage of this experimental design was that, since the Geissler tubes used induction current, they could produce short flashes of light in quick succession. Further, when the thin tubes filled with nitrogen flashed, they emitted much photochemically active light and, in a darkened room, it was possible to take photographs in which the individual limbs appeared as separate lines.[39] This equipment also enabled precise recording of how the movements of arm and leg related to each other or how the head related to the feet. Even twenty-five years later, the Berlin engineer K. A. Tramm still enthused over Braune and Fischer's method in his book *Psychotechnik und Taylor-System* [Psychotechnology and the Taylor System] because it enables the study of walking movements to a precision of 1/1000 mm. Braune and Fischer's research is still exemplary and, so far, nothing better has been devised.[40]

An important part of the CIT was the "photo cinemas." In Moscow and other towns in Russia, these cinemas were equipped with cinematographs, chronophotographic instruments, and other devices for registering the movements of muscles or pulse frequencies and translating them into graphic representations. Gastev used a method of recording that he named "cyclography"; it is reminiscent both of Braune and Fischer's method and that of Marey. The various tools and the test person's clothes carried light-colored spots at strategic points, which were important for the particular movement being studied. Double exposures of the test person's actual movements while using the tool or machine made it possible to analyze the labor process precisely and to determine whether reorganization was necessary. A strength of Russian Taylorism was that the photo cinemas—or media labs—were part of a broad network of experimentation and research wherein many vastly different approaches to the physiological and psychological dimensions of labor processes flourished.

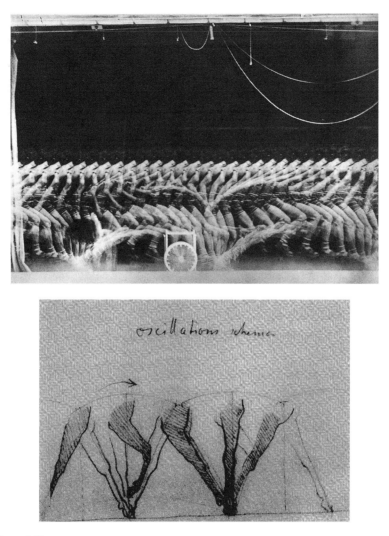

Figure 8.10 *Top: Etude de la course,* a single-plate exposure by Etienne Jules Marey (1886), captures the body of a runner moving through space and time. At the bottom, rather faint, is a measure in meters and in the center is the chronometer. The images of the runner look rather like pictures of a rotation that has been stretched horizontally. "[I]t occurs to me that originally, photographic material belonged to the techniques of cabinet-making and precision engineering: essentially, the devices were clocks for looking at [things]" (R. Barthes 1985, p. 24).

Bottom: Parallel to Marey, the Hungarian artist Bertalan Székely avidly studied movements, particularly of the horse. The physiologist Marey greatly admired Székely's precise and dynamic drawings and corresponded at length with the Hungarian artist, who is virtually unknown in the West. (Székely 1992)

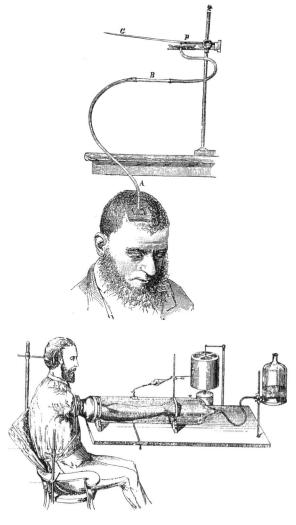

Figure 8.11 *Top:* Measuring pressure against the skull while the test person listens to music. The illustration originated at the Technical University of Kazan in the 1920s. (Mitrofanova 2000, p. 172)

Bottom: Apparatus for measuring the volume by which the arm expands when different kinds of music are heard. When blood pressure rises, the liquid in the chamber presses against the device connected to the needle of the recording instrument. The illustration is from a study by Tharkanov entitled "Effects of Music on Humans," which was published in 1898 in St. Petersburg. (Mitrofanova 2000, p. 172)

Among them were sophisticated studies on sound, noise, and music. An anthropological study by Karl Bücher at the turn of the century held these to be significant categories for labor productivity. He understood rhythm as "the ordered structure of movement in its temporal sequence" and saw its function in a close relationship with pleasure/motivation, which made work easier.[41]

The central institution in Russia for theoretical and practical experiment in this field was the State Institute for Reflexology and Brain Research at the Technical University in St. Petersburg, one of the first neurobiological institutes. The director of this facility was the charismatic psychotherapist, physician, and psychologist Vladimir Bechterev, who worked in close proximity, but also in competition, with the psychophysiological laboratory headed by Ivan Pavlov. Bechterev named his research field "psychoneurology." His institute focused on reflexology, studies on the ability to concentrate, and labor hygiene. Particular interest centered on experiments using musical structures, theories about intervals and rhythms, and the physiopsychological and therapeutic significance of music. In addition to psychotropic drugs, treatment of neurotic and hysterical patients included musical elements that influenced brain processes. Bechterev and his coworkers attached great importance to the therapeutic power of harmonious musical structures. Research areas included the "effect of combinations of major and minor keys on the excitation and inhibition of the human cerebral cortex"[42] and how specific music compositions affected fatigue. In a collection of essays on "labor reflexology," published in 1926, Bechterev contributed a paper entitled "Geistige Arbeit, vom reflexologischen Standpunkt aus betrachtet und Messung der Fähigkeit zur Konzentration" [Mental Work As Seen from a Reflexology Perspective, and Measurement of the Ability to Concentrate], in which he compares the different effects of Beethoven's *Moonlight Sonata* and the overture to Charles Gounod's opera *Faust* on intellectual work.[43] Bechterev's conjectural hypothesis was that the rigorous physiological and reflexological method can be compared to the way a piano functions: the brain performs the role of a fixed arrangement of notes, the reflexes play the keys, and the totality of the whole organism's reactions is the musical piece. Obviously, atonal and twelve-tone music were entirely incompatible with the concept of a healing therapy that utilized simple consonant structures. In later years, the experiments and theories in neurophysiology became an important weapon levelled against radical artistic experiments with novel musical structures.[44]

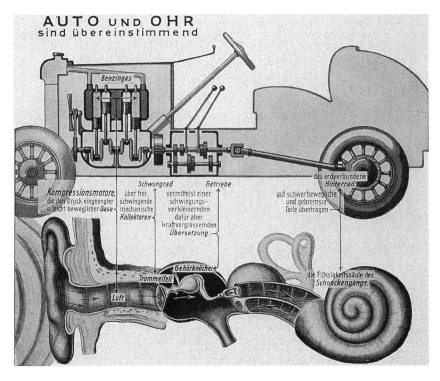

Figure 8.12 "Concurrence of car and ear": "[O]ur primary task is to investigate the wonderful machine that is closest to us: the human organism. This machine possesses many mechanical parts, an automatic and fast gear shift. . . . The human organism has a motor, gears, shock absorbers, perfect brakes, sensitive regulators, and even a pressure gauge. . . . There should be a special branch of science, biomechanics, which could be pursued under select laboratory conditions." (Gastev 1923/1978, p. 245; illustration: Fritz Kahn, vol. 4, 1929)

As the ideas of a harmony-obsessed socialist realism became increasingly established politically, the more the small band of people who sought to translate avant-garde artistic aesthetics and scientific ideas into social practice were discriminated against and isolated. Their socialist idealism, including enthusiasm for technological utopias, collided dramatically with the dictatorial reorganization of the Soviet Union under Stalin—who as the son of a peasant farmer was hostile toward technology—and his party bureaucracy. Paradoxically, Bechterev was one of the early victims of the Stalinist purges, even though his methods and theories certainly complied with the state rationale. In 1927, he was called upon in his professional capacity as a doctor and physi-

Figure 8.13   Portrait of Aleksej Gastev

ologist to examine Stalin's brain. Courageously, he diagnosed the party leader's mental state as paranoia. Twenty-four hours later he had died under mysterious circumstances.[45] (The doctors in the Kremlin claimed he had died of food poisoning.) Nor did it help Gastev that he had recently joined the Communist Party; in 1938, he was arrested as a counterrevolutionary and sentenced at one of the show trials, like Meyerhold. He was probably killed shortly afterwards. The avantgarde's project to create the "new man," a flexible, excellently functioning biomechanism, had failed, not least due to the dogmatic rigidity of a strictly hierarchical, ponderous administrative system and its conservative protagonists.

Conclusions: Including a Proposal for the Cartography of Media Anarchaeology

The things are here—why invent them?
— JEAN-LUC GODARD, *ÉLOGE DE L'AMOUR*

Developed media worlds need artistic, scientific, technical, and
magical challenges.

For the generation that began to work imaginatively in and with media worlds at the turn of the twenty-first century, it is of vital importance to know that a magical approach toward technology continues to be possible and to be reassured that investment in it is meaningful. Photographic and cinematographic apparatus, highly differentiated and automatized forms of imaging, electronic tools, local and networked computing machines—these devices are not simply awaiting discovery by today's media activists, as was the case in the avant-garde movements of the 1920s, the postwar pioneers of fluxus, action and concept art, video, or the early NetWorkers. On the contrary, they are hemmed in on all sides by standardized technical devices and systems, yet access to the functional bases has become enormously complicated and expensive and is only available to a privileged minority. To find one's own way through all this and arrive at original creative expression is not easy, assuming that merely reprocessing what already exists for the new channels of communication is not an option. Many art and design activists choose to create something original by establishing unusual connections between existing means of expression and/or material; such work stands out significantly from the media products we encounter every day. For example, they may cooperate on a casual and temporary basis with the club or dance scene. Combinations of DJing and VJing create amalgamations of sound and image processing in real time—a contemporary equivalent of *expanded cinema*.[1] Hooked up directly to music machines, the body's sensory functions, or even brain waves, generate self-constructed or annexed worlds of images that are projected to the rhythm of technomusic in dilapidated buildings abandoned due to industrial affluence. The event locations in Germany, for example, are old factories, called *E-Werk* in Berlin and Cologne and *Stahlwerk* in Düsseldorf. Few activists, however, take the more daring path of exploring certain points of the media system in such a way that throws established syntax into a state of agitation. This is poetic praxis in the strict sense that the magical realist Bruno Schulz of Poland understood it: "If art is only supposed to confirm what has been determined for as long as anyone can remember, then one doesn't need it. Its role is to be a probe that is let down into the unknown. The artist is a device that registers processes taking place in the depths where values are created."[2]

In the 1930s, the Polish writer and artist Bruno Schulz corresponded briefly with his more famous colleague Witold Gombrovicz. Gombrovicz wrote to Schulz that he had met a lady on a tram, the wife of a doctor, who said that in her opinion Schulz was either mad or a poseur. This provocation, which Gombrovicz published in the avant-garde journal *Studio,* was intended to challenge his younger colleague to an intellectual duel. But Schulz refused: "Actually, I don't believe in the sacred codex of arenas and forums at all; I despise it." At the end of his written reply to Gombrovicz, however, Schulz is driven into giving an opinion, one that strikes straight to the European heart: "You have the makings of a great humanist; for what else is your pathological sensitivity to antinomies if not a yearning for the universal, a yearning for the humanization of *non-humanized* spheres, a yearning for the dispossession of minority ideologies and their conquest for the sake of the grand unity."[3]

Schulz came from Drohobycz, a small town in Galicia that is now in the Ukraine. In his collection of short stories, *The Cinnamon Shops,* he charges the now forgotten things and figures of his hometown with the new energy of magical fantasy. Schulz, who also created one of the most fascinating and bewildering books of the twentieth century, the *Book of Idols,* taught art at the local grammar school. On November 19, 1942, he was shot dead in the street. Schulz had attempted to survive the ghetto by painting and drawing for an officer in Hitler's Gestapo. This officer had killed the protégé of another German officer, and the murder of Schulz was retaliation for this killing. In 1936, three years before the Nazis marched into Poland, Schulz had written a text while in Warsaw that, like so many of his works, remained unfinished: *The Republic of Dreams.* In dreams, Schulz writes, "Hunger for reality" is summed up, "a demand that commits reality to mature imperceptibly into credibility and into a decree, a bill of exchange that falls due and must be paid." Schulz proclaimed the republic of dreams to be "the sovereign territory of poetry," where one could live "a life of adventure, of perpetual dazzlement and amazement." His personal paradise is not so different from the realm of flowing honey that Queen Kyris protected for Empedocles. Schulz conceives it as a haven and, especially, as a place of boundless hospitality. Anyone "pursued by wolves or robbers" who manages to get there is in safety: "His reception is triumphal, his dusty garments are removed. Festive, blissful, and happy, he steps out into the wafting Elysian winds, the sweet rose-scented air" that pervades the garden and its "cells, refectories, dormitories, libraries, pavilions, balconies, and belvedere."[4]

Bills of exchange that fall due and must be paid, defense of antinomies versus universalization of the remaining heterologous remnants, and politics permeated by the poetry of hospitality: these are ways of describing the subterranean currents of energy that course through the deep time of the media. Schulz, a poet from a tiny village at the back of beyond, whose texts and drawings have been catalysts for the work of many artists and scholars in the latter half of the twentieth century in their labors to remodel reality in their favor,[5] was also an inspiration for this study. Throughout this expedition through the deep time of media worlds that my protagonists thought up or actually constructed, I have made no attempt to conceal my partiality for a magical relationship to things and the relationships between them.

In a superb essay titled "Form and Technology" (1930), the philosopher Ernst Cassirer, who came originally from Wrocław, examines the historical relationship between the methods employed by magical natural philosophy and by experimental physics—from the viewpoint of committed and enlightened inquiry. He concludes that, in principle, the partitions dividing the two are as permeable as Empedocles envisioned the interfaces of his active organs to be. At the same time, Cassirer challenges the notion that the magical arts should be regarded as the direct precursors of scientific experiment,[6] with reference to a decisive question for the modern inquirer: "[This idea] ascribes a significance to the magical approach and lays claim to achievements, which belong by rights to the later technical approach. Yet magic does differ from religion in that with magic, man emerges from a wholly passive relationship to nature: the world is no longer received as the mere gift of a superior divine power but man seeks to take possession of it and impose a particular form upon it."[7] The magical arts also contrast with the systematic investigation of things and their interrelationships by experimental science, implemented by technology, in that their dreams reveal enhanced wishful thinking: the "Allmacht des Ich" [all-powerfulness of the self].

From the standpoint of an archaeology that assigns special importance to the poetic permeation of media worlds, one can take Cassirer's idea a step further: the operations of the magic arts cannot be tied absolutely to a particular purpose, and their prerequisite is a specific mental attitude. This way of looking at things should not be understood as an underdeveloped precursor of the experimental approach to the world of things and their relationships that flourished and died out in premodern times. Historians of science like to classify the magical way of thinking as "primitive," when judged on the criteria of the extent and degree of certainty of its knowledge: "The field of observation is too limited, the methods of observing fluctuate and are too unreliable for the outcome to be any statement of tenable empirical laws."[8] This is precisely why the magical approach to technical media worlds holds such potential for new impulses and inspiration. Science, which seeks to establish general laws, cannot afford to concentrate obsessively and passionately on one area of observation any more than it can allow fluctuations and uncertainty in experimental proof. Such obsessive focus is essential, however, for the kind of experimental thought and practice that can afford to fail and is not afraid of including the possibility of failure in its calculations. The determined pursuit of a single idea and its investigation until all possibilities are exhausted will likely stir up unrest among firmly established structures and procedures. In most cases, established firms and institutions react to this method of study with discrimination. Yet such discrimination may be of only short duration—one task for anarchaeology of media. Magical, scientific, and technical praxis do not follow in chronological sequence for anarchaeology; on the contrary, they combine at particular moments in time, collide with each other, provoke one another, and, in this way, maintain tension and movement within developing processes. When heterogeneous approaches meet, openings appear that, in the long term, may even result in relatively stable technical innovations. Porta's experiments to sound out the media possibilities of the camera obscura in staging his theatrical performances of moving images with sound or his rotating cryptographic devices are examples of this as well as Kircher's combinatorial boxes for mathematical calculations and musical compositions or Ritter's discoveries about electricity and chemical processes.

Cultivating dramaturgies of difference is an effective remedy against the increasing ergonomization of the technical media worlds that is taking place under the banner of ostensible linear progress.

The collision of diametrically opposing concepts of creative work with and in the computer-centered media has come to focus on the operation and design of the interface. This boundary between media users and media devices simultaneously divides and connects two different spheres: that of the active users of the machines and that of the active machines and programs. In the 1990s, both technological developments and dominant media concepts were oriented toward making the boundary between the two imperceptible. The vision was to use a computer and be unaware that it is a machine based on algorithms for calculating and simulating. The user would be immersed in a so-called virtual reality of images and sounds without noticing the transition and, what is more, without knowing that one was dealing with a precisely prestructured, calculated construction of visual surfaces and temporal sequences. Computers were, and still are, designed for their users like a camera obscura; one works with them, enjoys the effects they produce, and has no access to their mode of functioning. A number of artists, together with programmers, conducted experiments to challenge this smoothly functioning technological and semiological ergonomy with the aim of facilitating and sustaining possible dramaturgies of difference, including with the most advanced technology. As heirs of classic film and of the video avant-garde, they insisted that access to computer worlds must remain as access to artificial contructs. Interfaces to these worlds must be designed to maintain an inherent tension with the worlds outside the machine; this would enhance, rather than diminish, enjoyment of both of them.

Brecht's concept of "thinking as intervention" was envisaged as an alternative to thinking as an option, which the real world—as the world of commodities—supports and engenders. His *Short Organum for the Theatre* (1948) is a theoretical and practical plea for operational dramaturgy—that is, for a dramatic art, that does not invite its audience to illusion and catharsis but that encourages thinking to continue during pleasure. The senses and reason are not construed as being in opposition; rather, they are forces engaged with each other in an exciting social game that we may call art. A comparable *Organum* for the interface does not yet exist.[9] However, powerful artistic work is being done on the dramaturgy of difference, both in and outside the Net. Interestingly, particular groups are engaged in this project, like the Critical Art Ensemble (CAE) from North America or the German-Austrian trio Knowbotic Research, who have been working in this area for the last ten years. Logically, their projects are located between the disciplines of art theory and practice; the critique of technology policy aimed at standardization is an important element.

Perry Hoberman from Brooklyn, New York, is one of the few individual artists currently working in the production of art with and through media outside the Net who brings off the balancing act between fascination with technology and thinking as intervention. At one level, his installation *Cathartic User Interface* (1995) makes use of the need to vent the frustration and aggression that arise from dealing with the external interfaces of personal computers and their manufacturers. As in earlier works,[10] Hoberman uses a simple experience from everyday culture, in this case, throwing soft balls at stacks of tin cans, familiar from the fairground sideshow where direct hits win prizes, such as small toys or gadgets. In the *Cathartic User Interface,* computer keyboards replace the cans. If a ball hits one of the active keys, the reward is not an artifact from the world outside the game; the prizes here are technical images projected onto the screen showing the keyboards, and they all derive from the world of machines and programs: ironic user instructions or error messages, satirical shifts in the user interface graphics, or faces of computer-industry agents. The physical act of throwing things at these objects of anxiety and aversion has a short-term liberating effect. However, hoped-for catharsis does not take place. The prizes on offer are merely from the world of machines and programs, and physical action attacks only their visual images. A short in the cybernetic system—one cannot get the better of this programmed and standardized world by machine wrecking; that course of action was already doomed to failure in the century before last. The only effective form of intervention in this world is to learn its laws of operation and try to undermine or overrun them. One has to give up being a player at a fairground sideshow and become an operator within the technical world where one can work on developing alternatives. For artistic praxis with computers in particular, this means learning the codes they function with. In the 1920s, Gastev demonstrated to the techno-avant-garde that taking up this position does not necessarily have to be identical with that of the programmers. An important part of *Cathartic User Interface*'s concept is that several users take part at the same time, a common feature in many Hoberman works. The presence of several people performing actions together within a twilit space invariably leads to interaction between the visitors, something that is not possible in this way in the dark cube of the cinema. This aspect makes it expanded cinema of a special kind.

Establishing effective connections with the peripheries, without attempting to integrate these into the centers, can help to maintain the worlds of the media in a state that is open and transformable.

Modern audiovisual mass media first became established in industrial centers: cinema and television were introduced as innovations in Berlin, London, New York, and Paris. Although we are accustomed to think of, write, and see media history from the perspective of these metropolises, this way of looking at the subject leads to a dead end, not least because decentralized and networked media systems no longer need the industrial and financial capitals in the way that mass media did. The entrance of Japan into Western markets has already brought about considerable shifts. With their focus on mobile and electronic media artifacts, Japanese manufacturers after World War II began to change the geographical conditions of the media economy. And this trend continues. The People's Republic of China with its hundreds of millions of potential media users is entering the world market, at great speed and with enormous power—and will change established hierarchies in the foreseeable future. Seoul and Singapore have only just begun to impact hegemonial conditions from the Far Eastern periphery. Although many current software solutions are still marketed lucratively by American corporations, they are no longer primarily developed in the United States. Automobile production, energy, and heavy industry, the pillars upon which Western Europe's and the United States's economic strength developed and grew, are not suitable models for producing the highly ephemeral products of the service industries. The media worlds of telematics have become as ubiquitous as their designers are mobile and nomadic.

One conclusion of this quest through the deep time of seeing and hearing using technical devices, with its additional focus on combining (*ars combinatoria*), is to advocate a two-fold shift of geographic attention: from the North to the South and from the West to the East. This shift has nothing to do with commercial markets. Stated oversimplistically: both the philosophical and the practical foundations for the construction of modern media worlds stem originally from the Far East, particularly ancient Chinese culture, and from regions adjacent to the Mediterranean, such as Asia Minor, Greece, and Arab countries,

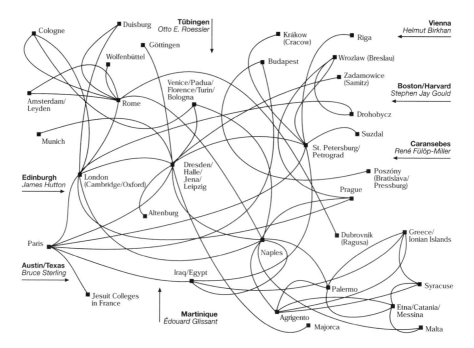

Figure 9.1 A suggested cartography for an anarchaeology of the media.

including their outposts in southern and southwestern Europe. We have followed and traced this movement broadly, for example, in the development of optical concepts and artifacts. From roughly parallel beginnings in China, the Greek islands, and Sicily via the reactivation and expansion of this branch of knowledge by Arab researchers around the turn of the first millennium, these activities intensified and gradually moved northwards. In early modern times, the southern Italian city of Naples was an incubator for various attempts to investigate phenomena and acquire knowledge about them magically. Further north, and to the northeast, Tuscan cities appeared on the scene and, before the turn of the seventeenth century, the Prague of Rudolf II, as hubs of astronomical, mathematical, and technical knowledge, with links to London, Oxford, Cambridge in England, Paris, as well as Krakow in Poland. With the Jesuit order's network as intellectual avant-garde and the Vatican as supreme authority, in the seventeenth century, Rome became the center where knowledge acceptable to the Catholic Church about the new media worlds was collected, analyzed, evaluated, and redisseminated worldwide. Rome and the

Vatican reduced the South more and more to the level of the periphery. Championing a different worldview and in competition with Rome, there emerged Paris, with its Catholic minimalists and early rationalists of the Enlightenment; the classical university towns Oxford and Cambridge; London; and the strongholds of liberal thinking in the Netherlands. Heretics fled from the persecution of the Inquisition and left their mark on the places that gave them temporary shelter. From this perspective, *Electricorum* by the Roman professor of rhetoric Mazzolari, represents a brilliant highlight, but also the turning point that heralds the end of this geographic order. In his Latin hymn of 1767 to all things electrical, Mazzolari recapitulates all that is known about this new realm, which will be so fundamentally important to media, culminating in its reification in his proposal for an electrical teletypewriter. However, the new protagonists in the poem now hail from other places: Dubrovnik, Philadelphia, Leyden.

At the turn of the eighteenth century, with Ritter, Chudy, and Purkyně, a region comes increasingly to the fore, which until then had only attracted notice when its outstanding teachers and scientists (who could afford it) moved to the north Italian universities, Rome, or Paris to study and to teach: present-day Poland, Hungary, and the Czech Republic, with their extremely checkered history of conquest and foreign domination. Scientists and engineers of this region

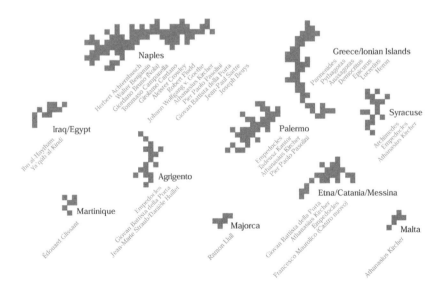

Figure 9.2 Cartography detail 1.

still went to study and teach in the academic centers of Austria, but also increasingly to eastern German universities in Thuringia and Saxony: Dresden, Halle, Jena, Leipzig. World War II and its aftermath resulted in a brutal caesura whereby the links and lines to these places and archives in the East were interrupted for decades. As though in defiance of this history, since the mid 1990s, the media faculty at the Bauhaus University in Weimar, situated at the border between eastern and western Germany, has developed into one of the country's most advanced institutes for teaching and research.

The breaks were even more profound for St. Petersburg, Russia's former center of science, technology, and art. They lasted for over eighty years and were of a twofold nature: first, there was the rigorous political and ideological turn away from everything that had to do with the West, and second, the geopolitical reorientation within Russia. Moscow became the center of political power and thus also the focus of national and international attention. With the founding of the Theremin Center more than ten years ago at the Moscow State Conservatory, there is again a laboratory for experimenting and testing new forms of art and media. It is named after the inventor of one of the first electronic musical instruments, which is played without touching anything: moving the hand or fingers between two electrodes influences electromagnetic waves. First demonstrated in 1922 to Lenin at the Kremlin, the melodic instrument was invented by the physicist and musician Lev Sergeivich Termen in 1920 in St. Petersburg, while he was director of the Physics and Technical Institute there. Through Brian Wilson's legendary composition "Good Vibrations" for the Beach Boys, the theremin's strange wailing sounds have found a place in the history of pop music. With his project *Forgotten Future,* the Theremin Center's director Andrei Smirnov has started to bring the power of these older inventions to the contemporary artistic game of creating technical illusions. Outstanding young artists, such as Anna Kuleichov, combine in their work the aesthetic ideas of the Russian kineticists, cubo-futurists, and suprematists with modern electronic concept and performance art.

Familiar geography is gradually changing again, but not only as a result of the new Moscow laboratory. In the years of perestroika, the nascent art scene in St. Petersburg began to reforge intellectual links with the legacy of their city's techno-avant-garde of the 1920s. Housed in a vast building in a courtyard off Pushkinskaya Street are the provocative "neoacademic" school, founded by the eccentric artist Timur Novikov, and the Techno-Art Center. Directed by Alla Mitrofanova and Irena Aktuganova, the Techno-Art Center realized media art projects in the 1990s in its Gallerie 21 under the most difficult infrastructural conditions and cultivated debate. At the Budapest Academy of Arts, the Intermedia section commenced work in the autumn of 1990, before the majority of such academic initiatives in the West. In cooperation with the associated Center for Communication and Culture, this institute has since achieved an international reputation for exceptional media projects. Miklos Peternàk, director of both institutions, is untiring in his efforts to reestablish connections between the new media worlds and technological and cultural gems from Hungarian and Eastern European history. *Excavating the Future* was the title of a symposium organized in 2001 by various institutions in Prague, which focused on Jan Evangelista Purkyně and his discoveries connected with technical envisioning. For decades now, Jan Švankmajer's brilliant animated films have ensured that advanced media worlds remain connected with the deep time of the Prague alchemists, magicians, and mannerists. *Alice* (1987), *Faust* (1994), and *The Conspirators of Pleasure* (1996) are but three masterpieces from the *mundus animatus* of this Prague surrealist to whom, lamentably, German cinemas remain closed.[11] In the 1970s, Poland already had its own school of video artists centered on the film school in Lódz.[12] Under the dictatorship of Jaruzelski, several of these artists—like Zbigniew Rybczynski—emigrated to the West, where they enriched the experimental film scene as well as the commercial world of music video. Others, like Josef Robakowski, elected to stay on and work in Poland under politically and technically difficult conditions. In the 1990s, the Biennale in Wrocław established itself as an important nexus of East–West relations in the art world of electronic media. WRO2000 took place in a building belonging to the old university whose tower once housed one of the first astronomical observatories in Europe. The media activists of the Russian, Polish, Czech, Slovenian, and Hungarian scenes are beginning to link up the most valuable elements of their archives and museums with advanced technical and media know-how of the West or to develop them further independently.

At the end of the 1960s, two exhibitions took place almost simultaneously with the objective, each in its own way, of investigating the interdependent relations between science, technology, art, and media. At the New York Museum of Modern Art, the Swedish curator Pontus Hultén organized *The Machine,* a retrospective devoted to various avant-garde movements in the past age of mechanization. In a brief appendix, artists and engineers were invited to present contemporary collaborations and experiments with electronic instruments and computers. In the exhibition catalogue, which has a metal cover and weighs

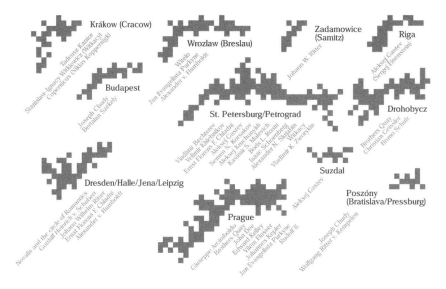

Figure 9.3 Cartography detail 2.

kilos, these electronic projects, described in blue print on white at the end, are in danger of being overlooked.[13] However, together with Jasia Reichardt's less spectacular exhibition at the London ICA (Institute of Contemporary Arts) in 1968, it was a beginning.[14] Solving the title of Reichardt's exhibition is a welcome gift for anarchaeology: *Cybernetic Serendipity.* Serendip is an old name for Sri Lanka (Ceylon), and *serendipity* was coined by the English author Horace Walpole after the title of a Venetian fairy tale, *The Three Princes of Serendip,* the heroes of which "were always making discoveries, by accident and sagacity, of things they were not in quest of."[15]

These two exhibitions and their publications are now legendary, but most of the pioneers of rudimentary digital graphics and computer-generated installations are nearly forgotten. The most ambitious project of this early period took place three years later, in the Galerie Grada in Zagreb, Yugoslavia. Under the title *Dijalog sa strojem* [Dialogue with the Machine], for the first time artists and scientists from eastern and western Europe, the United States, and Japan met to discuss their approaches to computer-programmed art. Among others, Marc Adrian from Vienna presented a program that he had developed in collaboration with Gottfried Schlemmer and Horst Wegschneider, a computer programmer. A 162-II IBM computer was used to rearrange syntactically text fragments taken at random from popular magazines. The texts, which were put together according to the rules of the program, were performed by three actors, a, b, and c:

c: shouldn't intersections fall in love? body-fresh telephone lines chat about heaven. you must get this! but who really tells their child what this is: heaven? embarassments liquidate tired stomachs or perhaps not?

a: drink embarassments! women discover independent tastes. martinis know no boredom.

b: shouldn't women fall in love? tongues cut budding intersections. what happened to the prettiest cover girl in the Soviet zone?[16]

   In Tuscany, northern Italy, the media theorist and activist Tommaso Tozzi and the network-duo 0100101110101101.ORG are fighting a rather solitary battle for greater integration of advanced technology into political and academic culture. In Venice, for decades Fabrizio Plessi has been building his lovely baroque video sculptures. It is difficult to imagine that southern Italy will have a renewed shift in that direction. The immediate reasons are severe economic problems and poor infrastructure, but also that the most recent innovatory thrusts in the area of media technology are the result of social and cultural processes that are relatively alien to southern European societies. Work and play on PC workstations hooked up to the Net are solitary pursuits and still dependent to a large extent on enclosed architecture equipped with ports connecting it to global data networks. Imaginary excursions in the World Wide Web still do not hold a great attraction for members of societies whose culture is traditionally oriented on the public sphere of streets and piazzas and oral communication. The exception is the cell phone, which enables southern Europeans to practice their favored form of exchange with others in a technically expanded form. In South America, the situation is very different. Particularly in those countries where the users of telecommunicative systems were subject to control under dictatorships, such as Argentina or Brazil, the Internet has made considerable inroads into urban everyday life. In the more affluent districts of Buenos Aires or São Paulo there are veritable supermarkets for accessing the World Wide Web. These retailers of time provide the infrastructure and terminals, which the customer uses to connect to the Internet. The customer pays for admittance just as for the fantasy-machine, cinema. Naples, one-time center of the magical researchers of nature and inventors of fantastic media worlds, plays no role in contemporary media geography. That which the Internet strives to deliver as a simulation—multifarious identities, ordered and restricted as little as possible, side by side and overlapping—is everyday reality in Naples, with all its incompatibilities, disasters, and surprise promises of pleasure. In this way, the city of Porta retains the same status that it has had for many generations of intellectuals: it is a place of desire and longing, particularly for those who hail from the sleek and well-organized North. In Herbert Achternbusch's film *Das Andechser Gefühl* [The Andechs Feeling] (1974), the grammar school teacher (played by the director) has just received the security of tenure from the Free State of Bavaria. This achievement plunges him into depression and desperation. Instead of the usual family lunch, the kitchen is the scene of a violent quarrel between him and his wife, in which his mistress is also involved. Attacked by his wife with a huge carving knife, he sinks to the tiled floor at the feet of his mistress and breathes his last with the words, "See Naples and die." A priest, who is also present (played by the film director and later director of the Berlin Film and Television Academy, Reinhard Hauff), is sent off shortly before this into the garden.

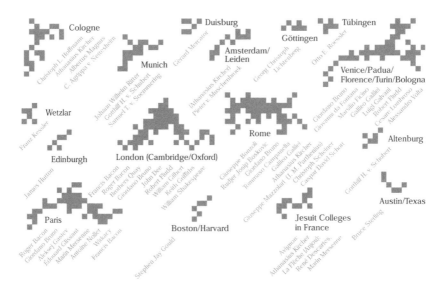

Figure 9.4 Cartography detail 3.

The most important precondition for guaranteeing the continued existence of relatively power-free spaces in media worlds is to refrain from all claims to occupying the center.

In the deep time of media, two models can be distinguished that correspond to the tensions, that Georges Bataille began describing in the 1930s in his proposals for a general economy, written under the deep impression made by fascism and Stalinism. An economy of adjusting and shaping that is committed to the paradigm of productivity, which first culminated in the project of industrial cinema and television and then, for the time being, in the postindustrial[17] phenomenon of the Internet, confronts an economy of friendship. The first model serves to effectivize systems, to protect them, and to ward off attack by rival, competing systems. The second has a subversive relationship to the first and is a luxury. It requires no legitimation, just as pleasure and art require none. It either develops and expands or it is nonexistent. It exists in and alongside the hegemonial economy. Even in the media worlds that developed critically close to the spheres of power—telecommunications and cryptography being cases in point—this other economy was also present: inventive, ingenious, and imaginative. From Trithemius's *Polygraphia* and Porta's proposals for communicating over distances to Kessler's place-finder and Bozoli's electrical teletypewriter, the ideas were, down to the last detail, motivated by concern for the friend who was obliged to remain in an inacessible place.

In the first half of the 1990s, the Internet enjoyed a short phase of euphoria. Anyone with access to a computer and a telephone line could send and receive messages that were largely uncensored. Political and artistic utopias of free exchange and freedom from the straitjackets of markets and power structures were projected onto the new networks. From the outset, this young scene of international NetWorkers did not accept the old boundaries of competing ideological and political systems. Quite the contrary: a high priority of their activities was to establish connections with the relatively few machines in eastern Europe via the few lines that could be freely accessed. This effort was both an affirmation and a test of the democratic potential that they presumed to be inherent in the data networks, which in the meantime were being expanded and made accessible to a mass public of users. The denizens of big cities in the countries where political and economic change took place relatively peacefully, under the banner of the global marketplace, connected up most quickly with the marketplaces in the West. For others—for example, in Albania or Kosovo—who were suffering wars and persecution, these lines of communication were the only ones which censorship could not reach for the time being. In 1996, the Syndicate Network came into being as "a translocal network that is based upon personal relationships and a healthy mixture of disagreement, respect, and solidarity, which characterizes all good friendships. . . . The Syndicate has its roots in the tactical media associations linking individuals and groups on either side of the Iron Curtain, through it and under it, far away from the eye of the mass media."[18] These relations also became established ones. A highlight of the Syndicate's activities to date was the *Deep Europe Workshop,* which took place in 1997 in the Internet platform of the Documenta X exhibition in Kassel. In a forum open to world opinion, fifteen syndicalists from various countries in eastern and western Europe discussed their ideas for a central Europe linked through networks of mutual respect.

At the beginning of the 1990s, the subscene of media worlds connected through networks, of which the Syndicate is but one example among many, underwent a transformation process.[19] The establishment of the World Wide Web as a global provider of audiovisual services, which knows very well how to commercialize anything that meets with massive interest among its users—including services that are barely legal or blatantly illegal—led to justifiable doubts about the existence of a highly intelligent lumpenproletariat. At the same time, the activities of committed interest groups made an important contribution to keeping open the option of a nonhierarchical public sphere of heterogeneous relations, also with regard to technologically advanced media worlds. They have not failed because theirs did not become the central model for the so-called information society. The economy of friendship is not amenable to generalization; it develops in and laterally to established relations and, as a rule, it is of short duration. It has to be set up time and again anew.

The problem with imagining media worlds that intervene, of analyzing and developing them creatively, is not so much finding an appropriate framework but rather allowing them to develop with and within time.
Photographs are flat surfaces showing a detail that contains visual information. The monitor of a television, video, or computer has a standardized frame with a ratio of 4:3 or 16:9; the cinema screen and electronic projection screen merely expand the same ratios. When watching media constructs, we have become used to viewing them as larger or smaller framed images. Particularly when the frame is filled with something that claims artistic merit, the misconception has arisen that all media productions that we can perceive audiovisually are pri-

marily concerned with the production of images. Monitors embedded in sculptures or piled up to form monuments in museums, with or without loudspeakers, only deepen this misapprehension under which art critics and art historians continue to labor. It was always difficult for them to get a handle on forms of processual art praxis, such as the happening, performance, or action; these forms of art eke out a very meager existence at the outer limits of their attention.

Literary studies, on the other hand, opened up to the aesthetic disciplines at a much earlier date. Dramatic texts, phonetic poems, and oral lyric poetry are all media forms that operate within time. At the former Institute for Language in the Age of Technology in Berlin, as beginners we learned our theory from Friedrich Knilli in his seminars on Gotthold Ephraim Lessing's *Laokoon.* Among other things, Knilli also introduced us to film and radio text analysis. Political science, sociology, and psychology were other disciplines that were quick to include individual media phenomena and structures as subjects of study, as were physiology, physics, chemistry, and medicine. Media apparatus has accompanied the experiments and studies of the latter since the founding of these disciplines.[20] Mixing, linking, rhythm, frequency, assembly, processes, impacts, and collisions are some of the basic modes that the sciences, which deal with bodies of all kinds, operate with on a large and small scale.

For this reason music and sound plays a significant role in media archaeology. Arts that operate with and through advanced technical media are arts of time. This applies equally to producing the illusion of moving pictures with a succession of still photographs or dynamic graphic structures. This story is told, day in, day out, by established channels of distribution for industrial media products. Yet images that appear to move are but one phenomenon among many that the arts create in time. Robert Fludd studied the construction of harmonious structures as well as arithmetic and meteorological processes. For Athanasius Kircher, composing and combining were artistic praxes that were just as important as creating sensational visual effects. However, it was the physicist and galvanist Ritter who really ushered in a paradigm shift, also for art theory, when he crouched down to get a horizontal view of Chladni's sound figures vibrating. Electricity breathed a new soul into the media worlds. From that point onward, they could no longer be thought of exclusively in static terms; they began to dance, to oscillate, to vibrate, to come alive. At the same time, they entered precarious territory, now operating in close proximity to the phenomena from the "museum of sleep" (Robert Walser) that people call life. With his projections of the real outside world inside the artificial space of the camera obscura,

Porta had already drawn attention to this dilemma. Henceforth, observers of and participants in media events found themselves in a permanent reality test. As various realities began to compete with each other for attention, the possibilities of deriving enjoyment were enhanced, but insecurity also increased. Which worlds should be considered true and which false? Perhaps Ritter sensed the turbulence into which this question plunged thought and perception when he defined the new praxis of art that was needed as physics. It is still not too late to listen carefully to the oscillations emanating from his lecture, given over two hundred years ago, which would help to prevent misunderstandings arising in the academic disciplines concerned with aesthetic phenomena.

Media worlds that need electricity as energy are synonymous with artificially created, processed, and rhythmic time. The nomadic existence of contemporary alchemists working in the electronic arts has both logistical and economic grounds: they go to the places that offer well-equipped laboratories for their experiments and spaces of freedom where, for a time, they can install and present their unstable or ephemeral works. A considerable number of the excellent first-generation artists to engage with these rather unreliable techniques do not come from a fine arts background. Nam June Paik began his career as a musician and fluxus artist. Steina Vasulka is a fine violinist and gave many excellent concerts before she turned to film, video, and combining violin-playing with electronic image worlds. Among the many identities he has assumed, Peter Weibel performed as a rock musician and continued to live out his identity as an action artist at the same time he was director of leading media art institutions. Among other activities, Alluquére Roseanne Stone worked on Jimi Hendrix's stage shows before she began performing media studies discourses. Perry Hoberman constructed projections for Laurie Anderson's multimedia shows before he began to build his own complex installations. Such artists have problems with the traditional venues for exhibiting artworks—and galleries and museums have difficulty with them—because their background is one of concert venues, clubs, tours, and the street: not primarily the places where art is quietly comtemplated and collected.

Kairos poetry in media worlds is potentially an efficacious tool against expropriation of the moment.

Under the pseudonym Heinrich Regius, Max Horkheimer published his *Dämmerung: Notizen zu Deutschland* [Twilight: Notes on Germany] in 1934. The notes contain a short passage entitled "Time is Money." Horkheimer remarks that this phrase raises the question of establishing a criterion for how much money a certain amount of time is worth. He continues:

A worker who rents a car to get to work on time is stupid (viz. the cost of this means of transport compared to his daily wage); an unemployed person with only five marks in his pocket, who uses a car to save time is crazy; but someone from middle managment will start to show lack of talent if he does not make his calls with a car. A minute in the life of an unemployed person has a different value than a minute in the life of a manager. Time is money—but what is the life-time of most people worth? If one is not bothered by speaking in such platitudes, then time is not money—money is time, in the same way as it is health, happiness, love, intelligence, honor, tranquillity. For it is a lie that whoever has time also has money—just having time won't get you any money; however, the reverse is true.[21]

Against the background of an experience of time that flowed at a leisurely pace, following the seasonal cycle of agricultural production, Aleksei Gastev made the audacious suggestion more than eighty years ago that temporal perception and praxis should be connected with the tempo of what he referred to as "machinism." Gastev's aim was not so much absolute acceleration but rather an alternative structuring of time, which would make labor time more effective, and thus save time. Linking this new time structure to mechanical apparatus on the basis of a binary code of motions in the labor process was intended to eliminate, or at least reduce to a minimum, frictions between biological and technical bodies. Gastev hoped that machinism would result in a new and sovereign individual, that the conscious act of union with a completely different other would lead to a new unity: the proletarian human machine or, what was the same from Gastev's perspective, the mechanical human. For the poet–industrial organizer, this construction possessed both a utopian and an elitist nature. He knew that it was impossible to achieve such a symbiosis as a permanent state and was reconciled to the fact that only a few, highly qualified and flexible combinations of manual and "mental" workers would actually get anywhere near this goal.

On the basis of the digital computer's binary code, at the beginning of the twenty-first century, information can now be billed to users per bit; it is immaterial whether the information comes in the form of numbers, images, texts, or sounds. The basic digital unit is becoming a new abstract currency. As the smallest techno-instant, it is the basis for calculating in an economy of providing services—services that take the form of producing symbols and programs, which prospectively include art that is realized exclusively within the framework of media networks. The development of such modalities of invoicing pleasure/leisure and work-related services is concurrent with the peaking of a trend in the mass media that Jean-François Lyotard described most aptly in one of his early texts: "Our culture valorizes and stages the only performance that, in its eyes, constitutes an event: the moment of exchange, the immediate and direct, the block-buster, real time—this is the only kind of time that is alive and vivid for our culture. One can name this instant, in which accumulated dead time is realized, obscene."[22] Thus, the power of disposal over time, as the ability to make decisions instantaneously, is under threat from two sides: from the obscene, culture-industrial compression of life-time into a staged and acclaimed "blockbuster hit," and from the installation of a universal measure of time and economy, which is indiscernible to human perception.

The temporal behavior of technical processes may be described as follows: even the qualities that affect such processes from the outset, such as monitoring, checking, and control, are time-dependent. They are re-formed by the technical process. On the output side of any machine-machine or human-machine system, we find time-dependent qualities of experience. These may also be called dynamic processes. The least that artists and engineers who engage with such processes can do is to ensure that the re-formation, which takes place in the course of the process, sets marked differences between the qualities operating on the input and the qualities of experience operating on the output. This would indeed be efficacious work on the interface; that is, its dramatization. Designed or formed time must give back to people something of the time that life has stolen from them. This is one of Jean-Luc Godard's finest thoughts on cinema, and it can also be expanded to include technical media worlds. If media activists fail to effect this transformation, then processed time is wasted time. We should not always be lagging behind the capabilities of machines.

"We wander around in circles and are consumed by fire": this is how the situationist Guy Debord described the activity of "wandering about," which he considered the only dignified course to take, confronted by a society of the spectacle.[23] The first known devices for measuring time from ancient China were square, oblong, or round metal reliefs with labyrinthine structures. In the depressions, a powder was scattered that burned slowly. The burning powder in the labyrinth marked the passing of time. Debord offered his body and imagination as material for measuring the passage of time during the period in which he lived. What alternatives for time-conscious praxes are there to this situationist, self-consuming identity? Theoretically, one possibility is to be fire, and not burning powder. However, this position can be assumed only by those who want to play God, for we are of a substance that time wears out. A course of action that is open to us is to intervene in the rhythm of the fire's burning and help to organize its intervals. From this perspective, pro-active media policy would mean committed action to preserve independent use of time and how it is organized. To be prepared and aware of the risk of loss and failure, in the sense of Debord's "consuming" and Bataille's "expenditure," are necessary caveats. Yet in this case loss would not be a category of a fatalistic economy if one succeeds in turning it into an enrichment for others. Otherwise, the act of consuming would be religious, and expenditure ideological. Both positions have had horrendous consequences in the recent past.

This expedition through the deep time of media-technological thinking and operating has focused on protagonists from very different historical epochs and configurations. They all contributed to the transformation process, whether by condensing existing knowledge, adding to it, tweaking it in a different direction, or courageously opening up other, riskier pathways than those offered by the established wisdom of the day. In an approximation of Hölderlin's epithet for Empedocles, they can be called Kairos-pilots. Each in his unique way has demonstrated that the fortuitous moment does not exist to accomplish something for us, but that we must grasp the moment.

The Quay Brothers, who are originally from Philadelphia but live in London, are directors of film and theater. Their special passion is giving a soul to lifeless objects through cinematography's tricks—that is, creating animation films. With a unique poetic power, their films wander through forgotten and banished places, often from eastern Europe. They collect and assemble old signs, discarded objects, things that exhibit a resistance to the commonplace, rhythms and melodies that seem to come from time dimensions to which we no longer have access or on which we have closed the door. With consummate precision and sensitivity, they animate their found material, combining it with the power of imagination to create minimalistic orgies of momentary sensations. One of their quests led them to Drohobycz. An early masterpiece is their *Street of Crocodiles* (1985), a willful filmic interpretation of Bruno Schulz's short story of the same name. The film is Kairos poetry par excellence. In the back rooms of the shops in the Street of Crocodiles, where mannequins play out secret obsessions and frenetic bustle reigns, a small boy searches for manufactured objects that will satisfy his curiosity and desire to play. Rusty screws twist elegantly out of dusty floorboards of their own volition, scatter across the floor, and wind back down again in different places. The boy halts one of the screws, twists it counterclockwise out of the floor, and carefully adds it to a little pile of collected things. A figure made of various metal parts with a dim light bulb for a head rubs an iron plate, the bulb's filament shines briefly, and the boy catches the light in a pocket mirror and deflects the beam onto a mechanical monkey, who rewards the boy with a furious, but very brief, roll on his drum. Later, the boy is seen with the metal figure by his side. He takes it in his arms and pulls his own little hat over its glass head.

> *Artistic praxis in media worlds is a matter of extravagant expenditure.*
> *Its privileged locations are not palaces but open laboratories.*

Media art is a strange *mixtum compositum*. On the one side, the compound noun denotes two things that are very close, rather obviously so. All art requires media for it to be perceived by others. On the other hand, media art has been developed over recent decades to describe a specific concept of cultural praxis. From this perspective, the *mixtum compositum* contains two elements that are far apart and strives to fuse two different worlds into one. The origin of the compound was strategic—not so much for media, but certainly for art. Similar to the terms "film art" and "video art," which preceded it, the prefix *media* was designed to facilitate its delineation of new artistic praxes as opposed to traditional "old" ones; its association with "art" staked its claim for tapping into historically developed markets, distribution channels, and discourses. The strategic concept of "media art," however, went even further. Since the mid 1980s at latest, the prefix *media* could count on a high rate of acceptance, both politically and economically. The designability of what would come in the future was securely tied to the media. At the same time, this acceptance was one reason media art was rejected more vehemently in the traditional institutions of art than previous concepts involving other media.

In this compound, *media* stand for several paradigms where the connection with art is not at all self-evident, including their remit of boundless popularity. The technical media of the late nineteenth and twentieth centuries no longer addressed the closed circles of social elites, but rather reached out to possible audiences that were nonspecific and unlimited socially, regionally, and nationally. Telephone, telegraph, cinema, radio, television, videorecorder, CDs, and DVDs developed as cultural technologies that could function worldwide. Their tendency to cross all borders is inherent, a trend that telematic media enhanced still further. Their users no longer saw themselves primarily as spectators and listeners but as participants in a global event, players in an interactive context, which we have learned to call "communication." In this, our world, we are confronted not with individual technical artifacts, but with technical systems built of multiple elements and, in the exact sense of the term, with technology. Telematic communication is no longer a question of individual objects and forms in which technology is articulated but a complex structure that includes technical capabilities, the training of engineers and computer scientists, technology policy and economy, its social and cultural meanings, and, naturally, the arts and sciences and their institutions. Technology is connected in a specific way with what is called progress and, thus, also with power. At present, computers and their worldwide networks are the focus of attention. Both the individual machine for processing, storing, and sending data and its transnational connections are systems for calculating. They are still systems in the mechanical tradition although they operate with sophisticated electronics and programs. For a hallmark of mechanical systems is that the processes they run must be capable of formalization—it is immaterial whether these are digital or analogue.[24] Art, too, has various dimensions that can be formalized. These can be taught and learned; they can be expressed in language or in other systems of ordered symbols; and they can be developed strategically and tested. In this sense, one can speak of artistic experiment. Thus, one can also refer to a studio with mainly technical equipment and focus as a laboratory. In a laboratory, research, development, and tests are undertaken; results are discarded or gained. Such work is connected with a peculiarity of artistic praxis, which it shares with science and industry. The difference is that it possesses far greater significance for art; indeed, for some it is the very essence of art: intuition, the specific way of looking. It is inextricably linked with the most important source of energy for artistic praxis, namely, the imagination.

Formalizability and computation on the one side, and intuition and imagination on the other, are the two poles of the *mixtum compositum* media art with regard to the actions of the subject. To understand these poles as two ends of a scale that can be played in both directions is an alternative to a dualistic view, which is an easy option but also fatal, if one remains trapped within this kind of thinking.

The spectrum of what is currently still referred to as media art is a training ground for mixtures of the heterogeneous. It is, therefore, a chaotic space, if one understands chaos to mean that dynamic linkage of multifarious[25] elements, of chance and necessity, which is by nature opaque and out of which arise phenomena and processes that we can understand. At least this is how the chaos heuristicians from the fifth to the third century B.C.—Anaxagoras, Empedocles, Democritus, and Epicurus—understood chaos. Why should we lag behind them?

Artistic occupation of media worlds requires locations where chaos is understood in this way. In these places the work of mixing and separation, dissecting and combining, is viewed as work that is valuable and worthy of support. In premodern Europe, such places were called alchemists' laboratories. The only people who could afford them were the rich nobility, queens and emperors, in Prague or London, or under the control of the Vatican, in Rome. Irrespective of their backgrounds, such patrons invited the most original minds so that they might experience at close hand their work with the impossible. In the long way from the division of the *prima materia* via various processes of mixing to the final projection, the last step in the alchemical process where the base is transformed into the noble was no less than the attempt to make the impossible more possible. These places were not permanent or enduring; they were not installed for posterity like the halls of the academies or universities. They were localities of passage, of surprises, of new departures, of abrupt cessation, but they were also havens. Yet, if the patron's money ran out, or the sorcerers' apprentices proved to be mere charlatans or mountebanks, an ear might be cut off, or the alchemists might be thrown into a dungeon. If they were more fortunate, they would be sent back onto the precarious highways where they would journey to the next promising destination that might take them in.

The situation has changed for contemporary laboratories of experimental research on media worlds in Cambridge, Berlin, Karlsruhe, or Cologne, in Japan's Ogaki-shi, Barcelona, Budapest, or Moscow, but the reasons they were set up and expanded have not changed. Behind the modern, well-equipped research-and-development facilities of today lies the hope of those who set them up that the contemporary sorcerers' apprentices, engineers, programmers, and artists will succeed in turning the digital into gold. Once established, the institutions developed their own dynamic. The people who worked there were not on the whole amenable to merely providing ergonomic designs for what the politicians were fond of referring to as "the future of the information society." When the room for maneuvering is curtailed for everything that is unusual or foreign, that is unwieldy and does not quite fit, then the attempt must be made to confront what is possible with its own dimensions of impossibility to render the possible (or reality) more vital and worthy of investment. With the advent of Web projects, exploration of new forms of cinematographic and video narratives, the opening of experimental acoustic spaces, the shifted space of artistic praxis and experience into the machines themselves—to the limits of physical endurance—or the development of new apparatus for which there was no demand and combinations of different performative forms for which there are no stages as yet, the activists from the institutions entered into a relation of friction and tension with their intended assignment. Because an explicitly formulated "mission statement" does not exist in most cases, they have freedom of action. The experiment cannot really fail. The alchemists were well acquainted with failure—not because they sought such experience for its own sake, but because, time and again, they embarked on projects that were of such significance that there was honor in failure.

In the 1970s, a bizarre economy was described by the French painter Pierre Klossowski, whose trilogy on the laws of hospitality also make him well known as a philosophical author. The text, with a letter by Michel Foucault as an introduction, was not published until 1994. In it, Klossowski proposes a possible solution to the conflict inherent in the *mixtum compositum* media art. He turns around the cultural pessimists' lament about commercialization and the resulting mechanization of the body, and declares the human body an object of exchange, "live currency." Liberated from the constraints of unwanted reproduction and the pressure to produce offspring, the body conceived of in this way is free to become a supremely confident actor. Klossowski assigns a special role to experiment in his economy. The manufacture of appliances is confronted regularly with its own "periodic infertility," which "becomes more apparent because the accelerated tempo of manufacture perpetually forces prevention of inefficiency (in the products) and this inevitably drives it in the direction of wastage. The experiment, the precondition of which is efficiency, presupposes the wasteful mistake. To explore in experiments what may result in profitable production is geared to the elimination of infertility in the product but at the price of wasting material and human labor (production costs)."[26]

This anarchaeological quest should also be understood as a plea for maintaining and continuing the right of access to those places which offer hospitality to experimenters and experiments, and for setting up more of them. Such places are able to function not only because they have a generous host or patron, for whom artistic waste is not synonymous with failure but a sign of independence and strength. My plea also includes the type of guest: guests for whom artistic praxis with and in media worlds is more than a cleverly packaged affir-mation of what we know already, what fills us with ennui, and what merely serves indolence and to harmonize what is not yet harmonious. It is a plea for guests who understand the invitation to experiment as a call to continue work-ing on the impossibility of the perfect interface of Empedocles of Acragas. In this sense, it is even meaningful to speak of the virtual world. The willingness to engage in wasteful activity oneself is the least that this economy should demand of its guests; such willingness is also the trick that makes it work.

Artistic praxis on the Internet is superfluous. Those who can afford this superfluity earn their living offline or through a second nonartistic identity in the form of productive work in the Net. The establishment of Linux, the free-ware operating system that was conceived as an alternative to industrial Microsoft, originally followed the logic of such an economy. The people who worked on improving and updating the software for all Net users did this in their spare time, parallel to their secure jobs as academics or well-paid pro-grammers. For the generations of artists who exclusively realize their works as processual, ephemeral media worlds, the situation has become increasingly nar-rower toward the end of the last millennium. Olia Lialina from Moscow has a background in journalism and experimental film. She could not make a living from her sophisticated and committed work on the Net, although her art has achieved recognition worldwide. The same applied to a host of other artists, in-cluding her fellow-Russian Alexei Shulgin; Vuk Cosic from Belgrade; Know-botic Research from Zürich; David Link from Cologne, inventor of the *Poetry Machine;* or Texan science-fiction author Bruce Sterling, just to name a few ex-amples. They all live at least double existences now, alternating between work to earn their daily bread and artistic work on inventions to invigorate the inter-national data networks of information and communication or for presentations at art forums. Bruce Sterling summed up the situation at the beginning of the new decade admirably during a discussion in Cologne in 2001: he would not be able to invest the requisite time and energy in his Internet projects if his internationally successful sci-fi books had not provided him with the financial basis. As a further seriously wasteful activity, artistic praxis on the Net may have a brilliant future.

Notes

Chapter 1

1. See Sterling 2000; the project can be visited at: www.deadmedia.org (last accessed 29 August 2004).

2. In discussions, this was one of Dietmar Kamper's favorite rejoinders to metaphysicists of media technology.

3. See also Thompson 2002.

4. Fülöp-Miller 1934, p. 330 and 275.

5. On Hutton's discoveries, see Trümpy 1996 (in German); citation p. 79f., and Repchek's wonderful new monograph *The Man who Found Time* (2003).

6. See Gould's chapter on Hutton, pp. 61–98 (1987, reprint 1991).

7. Ibid., p. 3.

8. Ibid., p. 63f.

9. See Gould 1991.

10. Gould 1997.

11. I cite important ideas from Gould's extensive writings that are relevant to my theme, which appear in various works listed in the bibliography. The text on his cancer diagnosis "The Median Isn't the Message," (the title is a play on a well-known book of Marshall McLuhan's), is found in Gould 1992, pp. 473–478.

12. Gould 1998, pp. 266f.

13. This is the title of the final chapter of the English translation of *Audiovisions* (Zielinski 1999).

14. Segalen 1994, p. 48 (in the section "Der Exotismus der Zeit" [The exoticism of time]).

Chapter 2

1. For a full description see, for example, the conference proceedings Johnson and Haneda 1966, particularly the chapter on firefly bioluminescence, p. 427f.

2. On view in the aquarium of the Stazione Zoologica in the Villa Communale Park, Naples. This marine biological station was founded in 1870 by the zoologist Anton Dohrn from Stettin (Szczecin).

3. For a description of the biological mechanisms, see Marchant 2000, p. 34f.

4. Bataille 1985, citations p. 289 and 291 (Bataille's italics).

5. Ritter paid Schubert back later in installments; on the relationship of the two men, see also Klemm and Herman 1966.

6. Wagner 1861, p. 12; for biographical details, see also Schneider 1863.

7. *Rosa ursina sive sol* [The Bear Rose, or the Sun] is the title of the principal work of the astronomer Christoph Scheiner, who worked at the beginning of the seventeenth century in Galilei's shadow.

8. Schubert cited here according to the 1818 edition, Lecture 1, pp. 1–25.

9. See, for example, Wagner, 1861, p. 38.

10. Title of chapter 2, cited here from Schubert 1840, p. 6f.

11. Ibid., p. 40.

12. Krebs, Die Anthropologie des Gotthilf Heinrich von Schubert, inaugural diss., University of Cologne (Cologne: Orthen, 1940), p. 16.

13. Quoted in Röller 1995, p. 46.

14. Kamper, 1990, p. 275f.

15. A reliable source is the essay by Middleton and Knowles (1961), which cites the original texts of all the authors discussed.

16. Kircher describes these experiments in *Ars magna lucis et umbrae* (1646), p. 888f. In the eighteenth century, there were pocket lighters that functioned with miniature burning-glasses (Heydenreich n.d.).

17. Morello 2001, p. 179.

18. "De horrendis terrae motibus anno 1638 . . . ," Vol. 1, book 2, Kircher 1664–1665, p. 3f.

19. [Kircher] 1901, pp. 40–48.

20. [Kircher] 1901, p. 43 and p. 47f. In the second quotation, Seng refers to Brischar's 1877 rendering of Kircher's text, in which Kircher is referred to in the third-person singular.

21. Kircher, *Mundus subterraneus,* 1664–1665, citations translated from the Latin, book 1, p. 1.

22. Novalis 1802/1987, p. 86f.

23. The title of this section, "Mittel und Meere" [The Means and the Seas], is the title of a lecture Vilém Flusser gave in Naples, and is a play on *Mittelmeer* [the Mediterranean]. In this lecture, Flusser spoke of the intellectual and cultural "Mittel," "thanks to which the Mediterranean penetrates other seas in order to give them form" (Flusser 1988, p. 12).

24. Glissant 1999, p. 84. This essay is a summary and poetical distillation of his earlier writings on creolization.

25. Ibid., p. 87.

26. Ibid., p. 88.

27. Quoted in Ibid., p. 14.

28. This is the field of research of Carlo Ginzburg; see Ginzburg 1995.

29. The idea is Otto Roessler's (personal discussions).

30. Ernst (1996) uses the term "Anarchäologie" [anarchaeology] in a different and interesting sense: as an activity counter to that of digging up and exposing to view. It appears in one of his essays concerning a Roman suggestion to "make the equestrian statue of Marcus Aurelius disappear underground by means of a lift and only bring it up on certain occasions for exhibition." My suggestion, with respect to this image as metaphor, would be to let Marcus Aurelius's horse gallop away with him on occasions.

31. Visker 1991, p. 309.

32. Musil 1968, p. 16.

33. Wittgenstein n.d., vol. 1, manuscript VII., p. 24.

34. This is the title of a teaching and research project that I began in the late 1990s, in collaboration with Hans Ulrich Reck and Silvia Wagnermaier at the Academy of Media Arts in Cologne. The project is devoted to the search for Kairos poetry that is in keeping with our times. See also Zielinski 2000 and 2001b.

35. Marx, 1974, p. 251f. English: *Capital,* vol. 4, chapter 21, p. 852.

36. Brand 1999.

37. See the study by Filseck (1990).

38. For an excellent review of the definitions of *Kairos,* see Kerkhoff 1973.

39. See Roessler 1992 (including the preface by P. Weibel, pp. 9–12) and Roessler 1996b.

40. See Zielinski 1985, particularly the final chapter on the "audiovisuelle Zeit-maschine" [audiovisual time-machine].

41. Roessler 1996a.

42. Thus, studies that take the trouble to analyze philologically how the concept of media is understood and used in particular historical constellations in which new in-sights and knowledge arise are of especial value; see, e.g., Röller 2002.

43. Leibniz 1998, p. 25.

44. See the notes by Specht in: Descartes 1996, p. xvi and Specht 1998, p. 13.

45. I believe that I share this approach with two other authors in the Rowohlts Enzyk-lopädie series, who stand as models for my own work: Marek [Ceram] and Hocke. In 1948, Ceram wrote not only the very popular book of the postwar era on the deep time of civilization, *Götter, Gräber, und Gelehrte* [Gods, Graves, and Scholars], but also the first *Archaeology of the Cinema*. Hocke's fascinating study of mannerism is still today an im-portant resource for archaeology of the observer's view.

46. Wessely 1981, p. 392.

Chapter 3

1. Adapted from *Ancilla to The Pre-Socratic Philosophers* by Kathleen Freeman, fragment 10 (Oxford: Basil Blackwell, 1948), in www.humanistictexts.org/democritus.htm, accessed 10 December 2004.

2. Cited from Primavesi's summary of the findings in Belz 2000, p. 33. For the com-plete research report, see Martin and Primavesi 1999.

3. In the Museum of Archaeology in Naples, there are several papyrus fragments by Empedocles on display. Looking at these, one can appreciate how difficult Martin's task was; moreover, working with photographs, he was not even able to use the material structure of the papyrus fragments as a guide.

4. Primavesi, quoted in Belz 2000, p. 39.

5. Diogenes Laertius, in *Lives of the Philosophers,* trans. C. D. Yonge (London: Henry G. Bohn, 1853); Rolland 1918, p. 15; Steinhart 1840 (p. 84) gives a very similar quotation.

6. I thank Detlef B. Linke, brain researcher, for this suggestion.

7. Hölderlin 1973, p. 7.

8. Quoted by Schroedinger 1996, p. 20.

9. Erwin Schroedinger, *Nature and the Greeks.* [1954] 1956. Cambridge: Cambridge University Press, p. 8.

10. Ibid., p. 13.

11. O. Roessler, "Edle Einfalt, stille Größe—Kann man Anaxagoras lieben?" in Belz 2000, p. 4.

12. Steinhart 1840, p. 86. In Rolland (1918, p. 15) Empedocles' expression is "royal" and the sandals are bronze. Empedocles has certainly not been neglected by researchers, and there are more recent works that contain details of the legends surrounding his biography, for example, the essay by Denis O'Brien in Brunschwig and Lloyd 2001 (pp. 535–546) or the various text editions of the pre-Socratic philosophers. I used Steinhart's text (in German) because its scholarship and wealth of (and delight in) detail appealed to me. The English translation of Empedocles' fragments that I cite follows Wright 1995, unless otherwise indicated, and Steinhart is used here mainly for biographical references.

13. Steinhart 1840, p. 85.

14. Empedocles, according to Wright 1995, p. 164; Mansfeld 1996, vol. 2, p. 75; see also O'Brien in Brunschwig and Lloyd 2000, p. 537, who likely takes up this concept because it is close to the philosophy of Deleuze and Guattari.

15. Empedocles, according to Wright 1995, p. 177; Mansfeld 1996, vol. 2, p. 83, whose edition of the pre-Socratic texts closely follows that of Hermann Diels and Walther Kranz's seminal work, *Die Fragmente der Vorsokratiker.*

16. Wright 1995, p. 228–229; Steinhart 1840, p. 98; Mansfeld 1996, vol. 2., p. 119.

17. Steinhart 1840, p. 90.

18. Rolland 1918, p. 21, here interpreting the words of Empedocles.

19. Ibid., p. 26

20. Hermann Diels, "Passages relating to Empedocles from [Hermann] Diels' *Dox-ographi Graeci* [1879]," Hanover College Historical Texts Project, pp. 226, 230–231, 1998/2001, http://history.hanover.edu/texts/presoc/emp.htm (19 January 2003); Mansfeld 1996, vol. 2, p. 129. The term "object" as applied to the entity being perceived is problematic but serves to emphasize how difficult it is for us today not to think in terms of a separation between perceiver and perceived. Empedocles, however, refuses abstraction, as is demonstrated in the fragment "So sweet seized on sweet, bitter rushed to bitter, sharp came to sharp, and hot coupled with hot" (Wright 1995, p. 231).

21. Wright 1995, p. 240; Mansfeld 1996, vol. 2, p. 107; see also the detailed account in Reiss 1995, to whom I am indebted for the first encounter with this text fragment many years ago.

22. Diels 1998/2001, pp. 227, 231; Wright 1995, p. 296; Mansfeld 1996, vol. 2, p. 132f.

23. Diels, 1998/2001, p. 232; Mansfeld 1996, vol. 2, p.133. This citation and the preceding ones are from Theophrastus's discussion of Empedocles' doctrine. I have not distinguished between the sources within my text because my concern is not to attempt a philological analysis of these classical texts.

24. Schroedinger (1956, p. 101) also includes several examples of Greek words in his list.

25. After Steinhart 1840, p. 91.

26. Ibid., p. 98; also see Wright 1995, pp. 230–231.

27. Democritus was a pupil of Leucippus and he fused Leucippus's ideas with his own in his writings.

28. Democritus, according to Mansfeld 1996, vol. 2, p. 319.

29. Ibid.

30. Lucretius, translated by William Ellery Leonard, http://classics.mit.edu/Carus/nature_things.4.iv.html (7 February 2003).

31. Simon 1992, p. 23.

32. See Rothschuh 1957.

33. Schroedinger (1956) discusses this problem specifically by contrasting Democritus with Epicurus, whom he reproaches for naively taking over the paradigm of the soul (p. 103f.).

34. Wright 1995, p. 282.

35. Empedocles, "Von der Speise des Jammers [sollt ihr] euch enthalten!" Fragment 186 in Mansfeld 1996, vol. 2, p. 155.

Chapter 4

1. I am not referring here to the first edition of 1558 (*Magia* I) but to the greatly expanded edition of 1589, which I cite in the following as *Magia* II. In della Porta's works, there are a number of different spellings of his name. Throughout, I use the same spelling as the Neapolitan publisher Editione Scientifiche Italiane uses in their edition of his collected works. To minimize the confusion further, I have omitted the aristocratic della in the body of my text.

2. See Plötzeneder (1994, p. 34f.), who also points out that Dali dedicated one of his lesser known paintings to Porta: "Phosphène de Laporte—En hommage au physicien italien Giambattista Porta." Hocke (1957, vol. 1, p. 82) says that in the period when Dali was painting clocks like runny Camembert cheeses, which coincided with his rise to international fame, he was again engrossed in Porta's *Magia naturalis*.

3. Unless there are differences in the original Latin edition, I quote from the English translation of *Magia* II of 1658, which is still widely available in a facsimile edition of 1958 (cf. here chapter 9, p. 323). Where accuracy requires it, I cite the Latin edition of 1607, which is identical to the 1589 edition. The now obsolete spelling "magick" in the English translation was also used by Aleister Crowley as, among other things, a secret reference to the Greek *kteis*. Crowley wrote his own *Book Four* near Naples.

4. Benjamin 1972, p. 315.

5. Experts disagree about both the place and the date of Porta's birth, and it is still a popular topic of debate. I follow here for the most part Belloni (1982, p. 11f.) and Clubb (1965, p. 3f.), who elucidate two different, plausible interpretations in meticulous detective work but both fix Porta's date of birth as 1535. Porta himself was decidedly indifferent to—or deliberately evasive about—any officially fixed biographical dates.

6. Flusser 1988, p. 12.

7. Goethe 1885, vol. 6, pp. 776, 778.

8. Sartre 1986, p. 67.

9. Cited in Vollenweider, 1985, p. 109.

10. Zigaina and Steinle 1995, p. 225.

11. Also referred to as the Kingdom of Naples. Hein (1993, p. 164) discusses in this connection the publication of works by Kircher in Naples. A compact description of the history of the occupations of Naples is given in Wanderlingh 1999.

12. Syphilis is discussed with reference to the rise of prostitution in di Giacomo's study of 1899; cf. also Hagen 1901, p. 235.

13. Hagen (1901, p. 234) recognizes the text as such in his early and exceptional reference work on the relationship between the sense of smell and sexuality.

14. See my introduction in Baudry 1993, p. 34f.

15. It is difficult for us today to appreciate how tight the web of obligations and dependence was in this period. For example, Giordano Bruno refused the post of professor in Paris because professors were obliged to attend church regularly, see Kirchhoff 1980, p. 32. There are many parallels in the biographies of Porta and Bruno, who was born thirteen years later in Nola, also a town south of Naples. What Francis Yates writes of Bruno—that he never lost or disavowed his "vulcanian and Neapolitan origin" (see von Samsonow's comment in Bruno 1999)—also applies unreservedly to Porta.

16. Clubb 1965, p. 8.

17. Schroedinger 1956, p. 44; see also Helden 2001, p. 1.

18. From Belloni 1982 (p. 14), who refers here to the biographical notes by Porta's first biographer Pompeo Sarnelli (Porta 1677).

19. Porta 1611, p. 5.

20. Cf. Thorndike (1958), vol. 5, pp. 3f. The resultant neologisms, which are not found in any dictionary, and many instances where he bends the rules of grammar make translating Porta's works from the Latin a real adventure.

21. Kirchhoff 1980, p. 38f.

22. Grau 1988, p. 24.

23. Benjamin 1972, p. 314.

24. Cited in Belloni 1982, p. 18.

25. This idea is explained in the twenty-one chapters of the first edition of *Magia naturalis* (1558), which consisted of only four books yet contains in rudimentary form the entire spectrum of subjects on which Porta worked in the following decades. The first book begins with the chapter "Quid sit magia naturalis" [What natural magic is], which one could call the methodological part, and the last book is on optical phenomena. The second deals with things biological, and the third with inorganic nature, particularly alchemy/chemistry. The practice of dividing works into *libri* (books) at this time did not indicate as a rule publication in separate volumes (*tomi*), but was the first main division of a monograph after the title. The first edition of *Magia naturalis* was only 163 printed pages long, approximately 20×30 cm in size. The greatly expanded version was produced as a magnificent folio edition with gold lettering on the leather spine and was also larger (26.5×39.5 cm).

26. Porta 1558, n.p. Here again there is a striking link to Bruno, who, while Porta was writing *Magia* II in Naples, was lecturing on natural magic at the newly founded University of Helmstedt in Germany and also writing treatises on the subject. Bruno, however, set the concept more within his systematic philosophical worldview of the infinity of the cosmos, which was rather alien to Porta. "For Bruno, 'natural magic' is not so much a principle of effect as of understanding; a form of meditation and focus directed toward mental perception of the 'one,' the 'world-soul' that imbues all things" (Kirchhoff 1980, p. 46).

27. In 2002, the Biblioteca Nazionale Marciana Venezia and the Bibliotheca Philosophica Hermetica Amsterdam began research to determine the influence of Hermes Trismegistus on European thought; see Gilly and Heertum 2002.

28. However, more recent work on the life of Isaac Newton, for example, shows that the first group also practiced the magic arts to a greater extent than is commonly known. It is an aspect that historians of science have simply passed over (see White 1998, who gives special attention to Newton's relationship to alchemy).

29. For a fuller discussion of this topic, see Zielinski 1997b.

30. In the preface to his treatise on meteorology (1610), he calls them *scienze fisiche.*

31. It was only very much later that *Pneumaticorum* was recognized as an important pioneering work in the history of engineering; cf. Beck 1900, p. 254ff. On Heron, see the translation by Woodcroft of 1851.

32. Cf. Beierwaltes 1978, p. 8f.

33. Porta 1586; here 1930 edition, p. 64.

34. *De humana physiognomia* is one of the most widely distributed of Porta's works, mostly due to the illustrations. The German edition of 1930 contains the illustrations in a separate annotated appendix (p. 317 f). In this connection, see Baltrusaitis 1984 for an excellent discussion of the work in an art historical context. Baltrusaitis says that Porta's micro-universe creates "an artificial world beyond the probable" (p. 8).

35. Clubb 1965, p. xvi.

36. Barthes 1978, p. 71f.

37. Zedler (1749, vol. 61, p. 642) calls this "Characteromantie."

38. See von Samsonow 2001, p. 354.

39. Cf. Belloni 1982, p. 19.

40. On the relationship between Dee and Mercator, see Watelet 1994; on Dee in general, Wooley 2001, Halliwell 1990, and the introductory essay by J. L. Heilbron in Dee

1978; for Dee's *Monas Hieroglyphica* see also the German translation of 1982. Reichert (2001) is one of the few scholars with an excellent knowledge of Dee and his works. For aspects of Dee in relation to media, see Zielinski and Huemer 1992.

41. Porta 1677.

42. On Campanella's *De ciuitate solis,* see also the review in the journal *Nachrichten von einer Hallischen Bibliothek,* part 44 (August 1751), p. 134f.

43. See especially the study by Clubb (1965). So far, Editione scientifiche has published a volume of plays, which contains one entitled *L'Ulisse* (Porta 2000a).

44. Belloni 1982, p. 14.

45. The close links between thaumaturgy and mathematics also played an important role for Dee, who referred to thaumaturgy as "a mathematical art" (Wooley 2001, p. 13). René Fülöp-Miller, who in 1931 published a book on cinema as *The Imagination Machine,* ran away from home at the age of fourteen because he wanted to write a book titled "Thaumaturgy."

46. Belloni 1982, p. 26.

47. Cf. Zielinski (1990, pp. 229–252) in the book by Decker and Weibel on this subject.

48. For example, see <http://www.sancese.com/Cripto5.html> (Italian); Taes 2001 (French); and the seminar by Tiemann (2000) (German).

49. Cf. Benoît 1998, especially p. 329.

50. See Lippmann's essay "Zur Geschichte des Alkohols und seines Namens" in Lippmann 1913, vol. 2, p. 210.

51. See Belloni's annotated edition of the *Taumatologia* manuscript in Belloni 1982, n. 28.

52. Porta 1563, p. 1. Parentheses contain terms from the Latin original; the square brackets are the author's insertions for greater clarity.

53. Ibid., p. 2.

54. Cf. the Latin title of Porta 1593a in the bibliography.

55. Cited according to Jutta Gsöl's translation from Latin in Plötzeneder 1994, p. 42. This letter by Porta is given by Athanasius Kircher in his treatise on magnetism (1654 edition of the book from 1641, p. 284).

56. Debord 1978, p. 25.

57. Beutelspacher 2001, p. 6, and Strasser 1988, p. 19. I found Strasser's text on cryptology in the sixteenth and seventeenth centuries to be the clearest and best researched.

58. Quotation from an anonymous translation cited in Strasser 1988, p. 29, who also discusses the further development of Trithemius's substitution tables by Blaise de Vigenère (p. 55).

59. Interestingly, for this Porta uses the letters of the Italian alphabet, that is, without *j, k, x, y,* or *w.*

60. Künzel and Cornelius (1986) developed special computer software to test Llull's system. An outstanding contribution on "the Llullian art" is the study by Schmidt-Biggemann (1983). For an excellent study in Catalan, see also Artau 1946.

61. A few years ago, David Link began working on generating texts for his *Poetry Machine* with the Bible because it is an ideal example of complexity and, at the same time, semantic retardation; cf. Link 2004.

62. Künzel and Cornelius 1986, p. 28.

63. See, for example, Miniati 1991, p. 8f.; Bini 1996, p. 86f.; and Watelet 1994, for some magnificent later examples of paper volvelles dating from the 1540s.

64. Porta 1563, p. 70f.

65. See, for example, Bagrow 1951; Grosjean and Kinauer 1970.

66. The Nixdorf Museum's forum in Paderborn has a goldsmith's replica of one these instruments of Porta (Beutelspacher 2001, p. 7).

67. "Sunglasses belong to the outfit of 'the existentialists' who see any human relationship as a betrayal on principle" (Linke 2001, p. 35).

68. The voluminous collection edited by Serres (1998) also treats optics only in the form of insertions within other branches of the history of science. Cf. other standard works by Mach (1921), Ronchi, and Wilde in the bibliography.

69. See the critique by Rybczynski 2000, pp. 357–366.

70. The expression originates from Simon (1992).

71. See the superb work by Baltrusaitis (1986), my reference is here the motto by Raphael Miramis of 1582 (n.p., 6).

72. Porta 1558, book 4, p. 141.

73. Needham 1962; Hammond 1981, citation p. 1.

74. See the admirable work by Sabra (Ibn al-Haytham 1989) on the camera obscura, especially vol. 2, pp. xlix–li and liii. In Western literature, Ibn al-Haytham is often called Alhazen (e.g., Wilde 1968); he lived and worked mainly in Egypt.

75. Needham 1962, p. 26 f; Graham and Sivin (1973) offer more detail.

76. Ibn al-Haytham 1989, vol. 2, pp. lviii–lix.

77. Bacon 1928 and 1998, Lindberg 1987, Wilde 1968, p. 85 f. On Bacon's conception of science, see also Lindberg's annotated selection of texts (Bacon 1998).

78. Cf. Ronchi 1897/1991, p. 39 f., and the essay by Waterhouse (1902), which has an extract from Maurolico's *Diaphaneon* (published 1611 in Naples) in the appendix.

79. Porta 1558, book 4, p. 141 f.

80. Ibid.

81. See Porta 1607, p. 962 f.

82. See the excellent account in Mann 2000.

83. See Fülöp-Miller 1927, p. 493 f; Braunmühl 1891; for more details, Shea 1970.

84. Cf. Weigl 1990, p. 31.

85. For a detailed description, see Zielinski 1989/1999.

86. For further details, see Ronchi 1897, p. 41f.; on Maurolico and Ibn al-Haytham, p. 39. There is a good and concise account of the dispute over who actually invented the telescope in Helden 1991, p. 64 f, with illustrations of some of the earliest instruments.

87. Gould 2001, p. 50.

88. Brecht 1967, p. 1246. In connection with Galilei's rationalist conception of sensory perception, Ginzburg (1995) draws attention to his severe problems with his own sensory organs: the "contrast between the physicist Galilei, who was deaf and had lost his sense of taste and smell as a result of his work [and was blind in both eyes as of 1637], and a contemporary doctor, who ventured to make a diagnosis by putting his ear to a wheezing chest, sniffed stools, and examined urine" could not have been greater (p. 21). Like so many of the "magic" natural scientists of his era, Porta was also a practicing physician.

89. Porta 1558, book 4, chapter 10, p. 148; translation based on 1658 English edition *Natural Magick,* book 17, chapter 2, <http://members.tscnet.com/pages/omard1/jportat3/html> (3 September 2002).

90. Björnbo and Vogl 1912; for the German translation of the *"(Pseudo-)Euclidis,"* see p. 107 f, n. 107. The editors date the compilation as twelfth century (p. 155).

91. Porta 1558, book 4, chapter 10, p. 147; translation based on 1658 English edition *Natural Magick,* book 17, chapter 2.

92. Ibid.; on Freud's arrangement of mirrors, see Weibel 1991. In connection with simulacra, in Lucretius's *De rerum natura* (1973) there is a passage that is difficult to understand in which he describes similar multiple reflections.

93. See, for example, the brilliant interview of László Beke and Miklós Peternàk with Flusser, which took place in 1989 in Budapest, two years before his fatal accident on the way back to France from Prague, which he had visited in 1991 for the first time since his forced emigration. Several books that I used for this study, for example, by Bruno, Bacon, and Jakob Boehme, are from Flusser's traveling library, in which there is a striking selection of texts from so-called hard science and others belonging to the magical tradition. All publications are housed in the Vilém_Flusser_Archiv at the Academy for Media Arts [Kunsthochschule für Medien] in Cologne, Germany.

94. For a detailed discussion, see Belloni 1982, pp. 65–69.

95. This is the main title of Flusser's book on "evolution to humanity" (1998).

96. Schroedinger 2004, p. 163.

97. Mach 1921, p. 19. With regard to the subject of his book, physical optics, Mach's research on Porta's works is wretched. Not only does he refer solely to the first edition of *Magia naturalis*—the sections on optics are greatly expanded in the second edition—he makes no reference at all to Porta's other works on this subject, not even to *De refractione,* Porta's special study on the techniques of vision. Incidentally, the German edition (Nürnberg 1719) did not have the courage to take over the original title of Book 20 of *Magia* II. It was translated as "Buch von allerley untereinander" (Book of all kinds of things together). There has always been a strong fear of chaos in Germany.

98. Foucault 1978, p. 64.

99. Goethe 1885, vol. 10, pp. 412, 410.

Chapter 5

1. On tarantism in the Renaissance, see Tomlinson 1993, p. 157ff.; citation p. 169f.

2. However, Porta was evidently very interested in mechanical music devices, which is articulated particularly in his book on engineering, *Pneumaticorum,* (1601, chap. 9 and 10., pp. 60–63). Here one finds sketches and descriptions of hydraulic organs, which are only a little rougher in their execution than Kircher's of fifty years later.

3. Silvia Parigi in (La) *Magia naturale* nel Rinascimento 1989, p. 102.

4. Beierwaltes 1978, p. 8.

5. Cf. Coudert 1978, p. 114.

6. Crombie 1959, vol. 2, p. 249.

7. The term "masterplan" is from Godwin (1979a, p. 93).

8. See Godwin 1979a, p. 12, for a list of these works.

9. Schroedinger 1956, p. 48.

10. In a brilliant essay, the painter and theorist Marcel Bacic from Zagreb describes how Gothic cathedral spires were built using this combination. In Beke and Peternàk 2000, p. 253f.

11. Cited in Pauli 1952, p. 150.

12. I can give only a very brief outline here. Some opponents of the Pythagoreans, like Didymos of Alexandria, also based their arguments primarily on mathematics but used other divisions; for a summary, see Stauder 1999, p. 184f.

13. See the excellent essay on harmonics by Annie Bélis in Brunschwig and Lloyd 2001, p. 297.

14. I quote Christopher Stembridge, an expert at playing and teaching these instruments. From the cover notes to his CD: *Consonanze Stravaganti. Musica Napoletana per organo, cembalo e cembalo cromatico,* Freiburger Musikforum: Ars Musici, 1997, p. 8.

15. According to Starke 1999, p. 88.

16. For example, when tuning or calculating C, if one takes eleven perfect fifths (2:3), when one gets to the twelfth, also known as the *tritonus* or "Wolfsquinte," one arrives at a numerical relation of 177147:262144. If, then, the twelfth is tuned as a perfect fifth, "one does not produce the note one started from, or its octave, but a note that is too high by a ratio of 524288:531441" (Chladni 1827, p. 98). The division of an octave into intervals (1:2) of twelve perfect fifths fails mathematically because of the problem that "it is quite impossible that a power of 2 to a power of 3 can behave like 1:1 or 1:2." "However, this serious shortcoming can be easily remedied by a deliberate and well-ordered slight adjustment of the relations between them, which is called temperament . . . i.e., . . . through equal distribution of the deviation among all fifths" (Chladni 1827, p. 96ff.). In this connection, Mersenne speaks of the "cordiality of dissonances" (Ludwig 1935, p. 64).

17. See Ammann (1967), who refers in this context to the translation of Kepler's most famous work by M. Caspar.

18. See Röller 2002.

19. Fludd's first reply to "Kepplero" is found in vol. 2. of his work of 1619, consisting of fifty-four large-format pages under the heading of "analytical discourse." These two citations follow the translation by Pauli 1952, p. 151 and 156.

20. Cf. Zebrowski 1999, particularly p. 104f.

21. Atmanspacher, Primas, and Wertenschlag-Birkhäuser (1995) locate in their respective disciplines the debate between Kepler and Fludd in the long-standing dialogue between the physicist Wolfgang Pauli and the psychoanalyst C. G. Jung on the significance of archetypes (citation in Atmanspacher, Primas, Wertenschlag-Birkhäuser 1995, p. 232).

22. Fludd 1617, p. 26f.

23. Cited in the Kircher biography by Schneider (1847, p. 599); see also *Universale Bildung* 1981, p. 47.

24. Fletcher (1988, p. 179ff.) lists a total of thirty-two individual works. In the secondary literature, one often finds mention of a larger number, but this results from including different editions of the same work and unpublished manuscripts, such as Kircher's treatise on mathematics of 1630. The manuscripts, which also comprise many hundreds of pages, contain another twenty-three titles. Fletcher's work is also the best for researching the whereabouts of the works in German libraries. In the United States, the best collection of Kircher's works is held by the Stanford University library, where the International Research Project on the Correspondence of Athanasius Kircher is directed by Michael John Gorman and Nick Wilding. Their homepage is an excellent resource for international research on Kircher: <www.stanford.edu/group/STS/gorman /nuovepaginekircher>. For libraries in Italy that possess works by Kircher, see Lo Sardo 2001, p. 25ff.

25. See Hein 1993, who has investigated in great detail the printing and publishing landscape around Kircher (for distribution specifically, see part B, chapter 9, p. 195f.).

26. Fülöp-Miller 1927, p. 103.

27. Ibid., 1927, p. 62 f; for Loyola's biography, see the detailed study by Boehmer (1941).

28. Fülöp-Miller 1927, p. 103.

29. Lo Sardo (2001) discusses the power of the Jesuit network in connection with the position of Rome in detail.

30. Kircher 1667; citation from the introduction to the 1979 reprint edition. For a detailed review of *China illustrata,* see the journal *Nachrichten von einer Hallischen Bibliothek* (August 1751), p. 146f.

31. Leinkauf 1993, p. 203.

32. Kircher 1650, vol. 1, p. 268.

33. On Leibniz and his conception of the art of combination, see Simonovits 1968, p. 36f.

34. In this connection, see Coudert 1978.

35. Foucault 1974, p. 70.

36. Wessely 1981, p. 386.

37. Leinkauf 1993, p. 20.

38. Kircher 1650, vol. 1, p. 47. For a discussion of Kircher's concept of music, see particularly Scharlau 1969; in this context, especially p. 82.

39. I should like to thank Friedrich Kittler for bringing this connection to my attention.

40. Dee 1570/1975.

41. Kircher 1650, Liber X, vol. 2.

42. For a description, critique, and position of Kircher's theory of music, see particularly Scharlau 1969 and 1988 and his introduction to the reprint of *Musurgia universalis* (Kircher 1650) in 1970; on the fugue, see Elson 1890, chapter 23, and Krehl 1908.

43. A CD of Bach's *Fantasie der Chromatik* released in 1999 by Deutsche Harmonia Mundi has Fludd's well-tempered monochord on the front cover of the booklet.

44. The Minims, *Orde Minimorum Eremitarum,* belonged to the Franciscan order. They stood for religious minimalism and extreme erudition. On Mersenne, including in relation to Kircher, see Knobloch 1979, Ludwig 1935, and the biography of Descartes by Specht 1998.

45. Kircher 1650, vol. 2, pp. 432–440.

46. *The Spiritual Exercises of St. Ignatius of Loyola,* translated from the autograph by Father Elder Mullan, S. J., <http://www.ccel.org/ccel/ignatius/exercises.html> (22 August 2004); see Loyola 1946 and Hocke 1957, p. 63.

47. Cf. Scharlau 1988, p. 57.

48. Kircher 1650, vol. 1, p. 564.

49. Scharlau 1988, p. 58, who cites here from a treatise by Pietro della Valle of 1640.

50. An exhibition in spring 2001, curated by Eugenio Lo Sardo, attempted to reconstruct parts of the museum at the Palazzo Venezia in Rome (cf. Lo Sardo 2001). Giorgio de Sepibus, who assisted Kircher in building the mechanical artifacts, compiled and published a detailed catalogue of the exhibits (de Sepibus 1678).

51. Godwin 1988, p. 23.

52. Johnston 1973, p. 17f.

53. In *Discipline and Punish: The Birth of the Prison* (German edition: 1994, p. 259), Foucault mentions that Bentham originally planned to install acoustic surveillance in the cells of his prison but ultimately discarded the idea because the sound channel would also have been open to the prison guards.

54. Ullmann (1978) provides a brief overview of Kircher's conceptions of acoustics and his contemporaries' critique of them.

55. Translation of the "First Proposition" from the German edition of *Phonurgia nova,* Kircher 1684, p. 4.

56. Kircher 1684, p. 102f.

57. Cf. Gould 1997, p. 37f.

58. Wooley 2001, p. 15.

59. Schneider 1847, p. 602f.

60. Baur-Heinhold 1966, particularly p. 121f.

61. Fülöp-Miller 1927, p. 514.

62. Rainer Specht in his introduction to Descartes 1996, n. 18; on Descartes's dioptrics, see the exceptional essay by Authier 1998.

63. 1646 is the date usually cited in the literature as the publication date of the first edition. However, Salzburg University library possesses a copy of *Ars magna lucis et umbrae* with a publication date of 1645.

64. Kestler 1680, book 3, p. 70ff. This is the book that Hocke, in his study on mannerism, insists was published in 1624. It also contains Kircher's experiments with burning mirrors and pyrotechnics. (Kircher was allegedly a consummate pyrotechnist.)

65. On the iconography of Kircher's works, cf. Lo Sardo 1999. Wessely 1981 provides an informative study of the frontispiece to *Musurgia universalis* with numerous cross-references to other copper plate engravings.

66. On the early history of the laterna magica, cf. Mannoni, Campagnoni, and Robinson 1996, pp. 40–63; Zielinski 1989/1999; on the individual protagonists, cf. Robinson, Herbert, and Crangle 2001.

67. Kircher 1671, book 10, p. 768f.

68. Ibid., p. 769.

69. Kircher 1646, p. 782ff.

70. Schott 1671, p. 265.

71. Hocke 1959, p. 123. In Lo Sardo's catalogue (2001, p. 253), Christina Caudito refers to the apparatus as "il Proteo catottrico" (the catoptric Proteus).

72. Aristoteles 1961, p. 30.

73. Kircher 1671, p. 785, fig. 1–3.

74. For a detailed account, see Baltrusaitis 1986, p. 21 f.

75. Ibid., and Kircher 1671, p. 776.

76. Cf. Miniati 1991, p. 18.

77. Knobloch 1979, p. 266.

78. He terms these "musurgia mechanica"; Kircher 1650, vol. II, p. 185 f.

79. On this subject, see the texts by Scharlau (1969 and 1988), Knobloch 1979 (p. 265 f.), and the catalogue of the Kircher exhibition *Vonderau Museum*, Fulda (2003, p. 50 f.)

80. Kircher 1650, Book VIII, Part V, Chap. IV, p. 188 f.

81. Cf. Scharlau 1969, p. 206 f.

82. For a detailed account, see Dotzler 1996.

83. Leibniz, "Zur Geschichte der Konserven und des Fleischextraktes," in Lippmann 1906, vol. 1, p. 343. First published in *Chemiker-Zeitung,* 1899, p. 449.

84. *Universale Bildung im Barock* 1981, p. 25.

85. See the pioneering study by Künzel (1989) as well as the later works Künzel co-authored with Peter Bexte (1993 and 1996).

86. Schott 1664, pp. 478–499. Here, Kircher's successor to the professorship of mathematics in Würzburg lists forty-three groups of words for his encryption method.

87. Coudert 1978, p. 91; also see Dotzler 1996, p. 157 ff., and Umberto Eco's short essay in Lo Sardo 2001, p. 209–213.

88. Cf. Cram and Maat 1999; on Leibniz, see p. 1040 f.

89. Francis Bacon, in *The Two Books of the Proficience and the Advancement of Learning,* which appeared in 1626 in an expanded Latin version; cited here according to Aschoff

1984, p. 95f. Strasser (1988, p. 88f.) also writes that Bacon developed this bilateral alphabet as a young man in Paris.

90. Wilding, 2001, p. 93.

91. Kircher 1663, n.p. (p. 45).

92. Kircher calls this glossary *epistoliographia pentaglossa* (Kircher 1663, pp. 88–127).

93. Strasser 1988, p. 171.

94. For examples, see Kircher 1663, p. 132f.

95. See Strasser 1988, particularly p. 175.

96. "Gang" denotes here a loosely knit group with hidden connections, in the sense used by Deleuze and Guattari (1992).

97. Kircher 1663, appendix, p. 6.

98. Ibid.

99. Ibid., p.22f.

100. Kircher 1671, p. 789.

101. Schneider 1847, p. 600. Fletcher 1988, p. 5.

102. [Kircher] 1901.

Chapter 6

1. The total number of pages is 288, some of which, however, are paginated incorrectly. The printed text ends on page 188.

2. Yasmin Haskell has recently published a fine study on this type of literature (Haskell 2003). She also discusses Mazzolari's *Electricorum* (p. 189f.), particularly the aspect of the special character of the text; in this context, see also her short essay (Haskell 1999).

3. Rudjer Josip Boskovic [Roger Joseph Boscovich] was born in Dubrovnik, Serbia, which the Italians called Ragusa. He studied in Rome and became professor of mathematics at the Collegium Romanum in 1740. He was one of the outstanding polymaths in the Jesuit order and one of the most prominent theorists on issues of time and space in physics and astronomy. For an introduction to Boskovic's natural philosophy, see Boscovich 1763.

4. Franklin's text is available as a reprint, published in 1983; here cited after Teichmann 1996, p. 29.

5. Mazzolari [Parthenius] 1767, p. 193, footnote a. On Franklin's electrical plates, see, e.g., Carl 1871, p. 78 f, and in general the excellent study on electricity in the seventeenth and eighteenth centuries by Heilbron (1979, part 4, The Age of Franklin, p. 324f.).

6. Mazzolari 1767, pp. 32–35. The ellipses denote the position of footnotes in the original.

7. Ibid., footnote a, p. 33f.

8. In his comprehensive study on the history of telecommunications, Aschoff (1974, p. 31) mentions an even earlier idea of an anonymous Englishman in 1753 about whom virtually nothing is known.

9. In Chardans's *Dictionnaire des trucs* (1960), an amazing book from the publisher Jean-Jacques Pauvert, there is a whole collection of such tricks.

10. Cf. Carl 1871, p. 3f. In the long Latin preliminaries to his book, it is indeed a surprise to find that Gilbert actually uses the term "physiologia nova" (new physiology) for his field of research.

11. Cf. Ullmann 1996, p. 2.

12. For a concise overview, see Teichmann 1996, p. 15f., who includes a useful bibliography on various special fields of electricity.

13. See the collection compiled by Simmen 1967.

14. Emil du Bois-Reymond succeeded in this in1842–1843; he was also one of the most vociferous detractors of Johann Wilhelm Ritter's work (see Schipperges' afterword to the facsimile edition of Ritter's *Fragmente* 1969, p. 8f.)

15. Wetzels 1973, p. 20.

16. As cited in Lenning 1965, p. 119.

17. See the illuminating studies on early romanticism by Tatar (1971) and Wetzels (1973); for the nineteenth century, see Asendorf 1984.

18. Wetzels 1973, p. 1.

19. Here I am indebted to the excellent biography by Richter (1988), which forms the introduction to Ritter's letters to Frommann (pp. 13–84).

20. Cited in Klickowstroem 1929, p. 73.

21. Cf. Richter 2000, p. 11.

22. The nineteenth-century reference work on the electromagnetic telegraph by Schellen (1870) does not mention Ritter at all, not even in the long sections on galvanism.

23. Ostwald's address is published in Ritter 1810 [1984 ed., pp. 321–343]. In his handbook on electrochemistry (1896), Ostwald describes Ritter's research and experiments in detail.

24. Richter 1988, p. 21.

25. Teichmann 1996, pp.48–52, citation p. 48.

26. Cited in Richter 1988, p. 74.

27. Ritter 1810/1984, p. 89.

28. Ostwald in Ritter 1810/1984, p. 328.

29. Ritter 1805, p. 40.

30. Described meticulously in *Annalen der Physik* (Ritter 1801, pp. 447–484).

31. Ibid.; citation in Ritter 1805, p. 40.

32. Karl von Raumer, cited in Richter 1988, p. 71.

33. Published in the *Fragments* (Ritter 1810/1984, pp. 288–320, quotation p. 310).

34. I use this term here, first, because Ritter's concept is remarkably similar to the negative anthropology that Vilém Flusser developed at the end of his life (see Kamper 1999), which at the same time marks its reversal. Second, Ritter remained in close intellectual contact with Schubert and knew of his project to write an anthropology of *physica sacra*.

35. Ritter 1810/1984, p. 256.

36. Ibid., p. 317f.; see also the inaugural thesis by Hartwig (1955), particularly pp. 70–83.

37. Ritter 1810/1984, p. 258.

38. Ritter, Sept. 1803, p. 213f.

39. Ibid., p. 275.

40. Chladni came from a Hungarian family that was forced to emigrate under the reign of the Hapsburgs. When he was only one year old, he was enrolled at the University of Wittenberg where his father was the chancellor (Ullmann 1996, p. 9). Chladni died in Breslau, not far from Ritter's birthplace.

41. Ritter 1805, p. 33f.

42. Ritter wrote this text in 1805 as a reaction to the work of one of his closest friends, the Danish physicist Hans Christian Ørsted, who at the time was investigating Chladni's figures and to whom Ritter was greatly indebted for his knowledge of electromagnetism.

43. Democritus, cited in Mansfeld 1996, p. 289.

44. Cited in Rehm 1973, p. 206.

45. Ritter 1810, item 360.

46. From the afterword by the editors of Ritter 1810, p. 356.

47. See Kittler 1996, p. 290. Additionally, the essay contains a very nice polemic on Soemmering's *Organ der Seele* (dedicated to Immanuel Kant)—cerebral fluid as the unifying principle of the thinking organ.

48. Cf. Schipperges's afterword to Ritter 1910, p. 6; on Soemmering and his telegraph, see Feyerabend 1933, p. 11f.

49. See Aschoff 1974, p. 33f., who does not even acknowledge Ritter's existence.

50. Feyerabend 1933, pp. 12, 19.

51. The sources for Chudy are very sparse and uncertain. Apart from Chudy's own book and the few brief remarks in Aschoff 1984 (pp. 203–206), I rely for the most part on a work by Ede Lósy-Schmidt (1932), which is available only in Hungarian, but appears to me to be a reliable source. Lósy-Schmidt was born in Tirgu Mures in Transylvania, studied engineering in Budapest and Darmstadt, and worked in Wrocław, Katowice, and Kraków as an engineer, mainly on the railways, before returning to Budapest. In addition to working as an engineer, as of 1918 he worked diligently on unearthing and archiving source material relating to the Hungarian history of technology and built up Hungary's first museum of technology. In 1939 the museum moved to Kosice (which today is in Slovakia), where it opened in 1943. Toward the end of the war, the collection was destroyed (Élet és Tudomány Archívum nyitólap/SuliNet nyitólap, 2000). In a conversation with the author in 1997, the composer Mauricio Kagel, who lives in Cologne, mentioned that several years ago he had seen the score of Chudy's opera in an antiquarian bookshop in Vienna, but at that time had not attached any importance to it.

52. Skupin 1986.

53. Chudy n.d. Lósy-Schmidt (1932) reconstructs the publication date of the brochure as 1792; Aschoff (1984), on the other hand, gives 1796. This is certainly too late, for the opera was performed in January 1796 and Chudy refers to it in the brochure's introduction as a vague idea (p. 3f.). Thus the text must have been published between 1792 and 1795.

54. Lósy-Schmidt 1932, p. 8.

55. Lósy-Schmidt created some minimalistic music compositions from Chudy's codes to illustrate how they sound as music.

56. Chudy n.d., p. 12f.

57. Cited in Aschoff 1984, p. 96, who translated this passage of Bacon's text of 1623 from the Latin into German. Here I follow Aschoff broadly, as my intention is not to rewrite this subject but rather to offer an important additional aspect.

58. Aschoff 1984, p. 99f. quotation p. 100.

59. Ibid, p. 143; for further information on Hoffmann, see Wichert 1984. The reference here to the lottery is literal, as lotteries were already common in the eighteenth century. With this suggestion, Hoffmann wanted to reduce the possibility of cheating. Since he speaks of "dissemination" not "transmission," of the information, he was likely thinking of many recipients of his televisualisation of the lottery numbers (p. 90f.)

60. According to Aschoff's calculation (1984, p. 163).

61. Ibid., p. 140f.

62. *Über den Mechanismus der menschlichen Sprache nebst der Beschreibung der Sprachmaschine* [On the Mechanism of Human Language Including a Description of the Speech Machine] was published in 1791 in Vienna; see also Simmen 1967, p. 56f.

63. Friedrich Cramer in *Was heißt "wirklich"?*, 2000, p. 82.

64. From the afterword of the editors of Ritter 1810/1984, p. 364.

65. Goethe 1885, vol. 10, p. 49f.

66. Ibn al-Haytham 1989, vol. 1, p. 51f.

67. In later, much less extreme experiments, both Gustav T. Fechner and Joseph A. F. Plateau sustained severe damage to their eyes.

68. Wetzels 1973, citation p. 101; see also Ostwald in Ritter 1810/1984, p. 328.

69. The dissertation actually appeared in 1818, but bore the date of the following year. Shortly after (1823), Purkyně published it again as the first part of his *Beobachtungen und*

Versuche zur Physiologie der Sinne. Officials of the Hapsburg monarchy's administration germanized his name to Purkinje, the name under which he is known in the English-speaking world and which is commonly found in archives.

70. Purkyně 1819, p. 10f.

71. Ibid., pp. 8, 5.

72. Müller 1826.

73. Rothschuh 1957, pp. 218, 220.

74. Ibid., p. 8.

75. Karger-Decker 1965, p. 174f.

76. Purkyně 1819, p. 7, 49.

77. This is not just a figure of speech: David Brewster's study of the kaleidoscope was published in the same year as Purkyně's doctoral thesis. On this device, which makes symmetrical images out of heterogeneous pieces of material, see also Arber 1960, p. 103.

78. Purkyně 1819, p. 29.

79. Ibid., p. 43f.

80. Purkyně 1819, p. 161 f; emphasis mine. The arts scholar and curator Jaroslav Andel from Prague has also drawn attention to this remarkable statement.

81. Ibid., p. 55f.

82. Purkyně 1825, p. 120f., citations pp. 124, 125.

83. Purkyně 1819, p. 166.

84. For a good overview, see Lindberg 1987.

85. See Zielinski 1989/1999, p. 45.

86. Purkyně 1819, p. 167f.

87. Ibid, pp. 170, 173f. At this point, and only here, I should like to point out that, with respect to Purkyně, Crary (1990) labors under considerable misapprehensions, due partly to the fact that he is acquainted only with Purkyně's work from secondary literature and third-hand sources, which he uses to illustrate his theses. Crary interprets the vast scope of the phenomena investigated by Purkyně solely with reference to the afterimage (pp. 102–104), which only occupies a marginal position in Purkyně's work. For this discussion, and in general on Purkyně's work, see Andel 2000, particularly p. 336.

88. Now in print again: see Goethe 1997.

89. Cf. Karger-Decker 1965, p. 190f.

90. Purkyně n.d., p. 117.

91. Cf. Exner 1919, p. 1, and Hoskovec 2000, p. 33.

92. Cf. Sacks 1997, p. 144.

93. See Ginzburg 1995, p. 35.

Chapter 7

1. Lombroso 1896, first published by Meckelenburg in Berlin, 1887.

2. For an excellent overview, see Rheinberger and Hagner (1993) on the "Experimentalisierung des Lebens" [Experimentalization of Life]. On the history of physiological concepts, see Rothschuh 1957.

3. Cited in Timothy Lenoir's essay in Rothschuh 1957, p. 53.

4. See Leps 1992, p. 23f.

5. Gilman 1977, p. 6.

6. For the biographical details of Lombroso's life I rely on Kurella 1910, Simson 1960 (here p. 155), and Colombo's excellent *"Cronologia Lombrosiana"* in Colombo 1975, pp. 41–53.

7. Kurella (1910, p. 13) thinks that the family's name was originally Lumbroso and that their ancestors were Spanish Jews who were forced to flee to North Africa (*lumbroso*

in Spanish means "bright, luminous"). Lombroso's mother's maiden name was Zefora Levi.

8. Simson 1960, p. 156.

9. See also Leps 1992, p. 60f.

10. *Irrenanstalt* is what the Austrian state called mental institutions in Italy. Cf. "Unterhalt der Irren in den Irrenanstalten den lombardisch-venetianischen Staaten," *Medicinischen Jahrbücher des kaiserl. königl. österr. Staates* [Medical Yearbooks of the Imperial Austrian State], 6 vols. (1821), vol. 4, p. 6f.

11. Translated from the German edition: Lombroso 1887, p. 3. The English translation, entitled *The Man of Genius,* was published in 1891 in London by W. Scott and in New York by C. Scribner's Sons.

12. Lombroso 1887, p. 4.

13. Lombroso 1894c, p. 15f.

14. Lombroso 1894b, part I, see particularly chapter 2, p. 35f.

15. Lombroso and Laschi 1891, vol. 1, p. vi.

16. *Gli anarchici* [The Anarchists] was originally published in Bologna 1894. It has been translated into French (Lombroso 1896).

17. Both Bataille and d'Espezel worked as librarians in the Cabinet des Médailles of the National Library in Paris.

18. Lombroso also wrote a lengthy treatise on graphology, which is undated. The copy I used has a sticker marked "Ex Libris Jonas Cohn" that shows fragments of an antique sculpture and the motto "Das Alte, Wahre, faß es an" [The old, the true, take hold of it].

19. Lombroso's enormous influence was not confined to the standard literature on *Kriminalität und abweichendes Verhalten* [Criminality and Deviant Behavior] (e.g., Hans Joachim Schneider 1983). His publications and ideas affected innumerable publications on pathology in the wide area of research from criminal to sexual deviations. An extreme example is the work of Erich Wulffen, department head of a section of the Saxony Ministry of Justice, particularly his *Psychologie des Verbrechers* [Psychology of a Criminal],

2 vols., (Berlin: P. Langenscheidt, 1922); "Der Sexualverbrecher: Ein Handbuch für Juristen, Verwaltungsbeamte und Ärzte" [The Sexual Offender: A Handbook for Jurists, Administrators, and Doctors] (Berlin: P. Langenscheidt, 1923); "Das Weib als Sexualverbrecherin: Ein Handbuch für Juristen, Polizei- und Strafvollzugsbeamte, Ärzte und Laienrichter" [The Woman as Sexual Offender: A Handbook for Jurists, Police and Prison Officials, Doctors and Lay Judges]. (Berlin: P. Langenscheidt, 1925). (All titles appeared in the book series *Enzyklopädie der modernen Kriminalistik*). A director of the police records department at the police headquarters in Berlin, Hans Schickert, based his criminal psychological study on both Lombroso and Wulffen: *Das Weib als Verbrecherin und Anstifterin* [The Woman as a Criminal and Instigator] (Bonn: A. Marcus & E. Weber, 1919).

20. Cf. Broeckmann 1995/1996, chapter 4, "Criminal Anthropology: A Semiology of Indexicality."

21. Peter Strasser 1984, p. 7; see also his essay in the catalogue of the 1994 Paris exhibition *L'âme au corps* [The Soul in the Body]: "Cesare Lombroso: L'homme délinquant ou la bête sauvage au naturel" (p. 352f.), which is a reprint and first appeared in the catalogue edited by Clair, Pichler, and Pircher (1989).

22. For a detailed account, particularly with reference to England, see Leps 1992, p. 17f.

23. Strasser 1984, p. 13, whose arguments I follow here.

24. Broeckmann 1995/1996, chapter 4 discusses the three congresses at length and particularly the significance accorded to photography.

25. Lombroso 1899/1983.

26. His collection of skulls, photographs, tattooed skin, conserved body parts, graphological documents, etc., provided the basis for the Museo di Antropologia Criminale di Cesare Lombroso in Turin, which unfortunately closed several years ago (see Colombo 1975). A few exhibits are on show in the Museo Criminologico in Via del Gonfalone in Rome, where Lombroso is still prominently presented as the father of criminal anthropology, in close spatial proximity to Gramsci, and with the same amount of interest as that devoted to the Red Brigades. The Museum of Anthropology in Florence also possesses a few small parts of his collection.

27. Lombroso 1894, p. 174.

28. See the excellent essay "The Criminal as Nature's Mistake, or the Ape in Some of Us" in Gould 1977, and also Gould 1984, pp. 122–143.

29. Quoted in Strasser 1984, p. 41.

30. This proposition, an extension of Foucault's argumentation, is endorsed by the studies of Strasser, Leps, and Broeckmann.

31. The study edited by Schneider (1983) discusses Lombroso's approach in a science-historical context and its proximity to Freud's explanation of crime as "Rückschlagsphänomen auf onto- oder phylogenetische psychische Entwicklungsstufen" [phenomenon of a throwback to onto- or phylogenetic stages of mental development] (vol. 1, p. 106).

32. I use the German translation of 1894, p. iii. The German title is *Das Weib als Verbrecherin und Prostituirte* [Woman as a Criminal and a Prostitute], and the edition published in Hamburg has a title page with an ornament of thorns from which hang heavy iron chains with handcuffs.

33. This conclusion does not stop Lombroso and Ferrero from including detailed case studies, very similar in presentation to Krafft-Ebing's *Psychopathia sexualis,* in order to describe perverse acts and behavior.

34. Lombroso and Laschi 1891, vol. 1, p. 222. In *Genie und Irrsinn* [Genius and Madness] Lombroso states the connection with inferiority in a slightly different way: "If we seek to penetrate deeper into the nature of genius with the help of autobiographies and our observations to achieve clarity about what it is that separates the genius from common people, we shall find that this almost always lies in greater or lesser sensitivity, which in the case of a genius can reach pathological proportions" ([1887], p. 19).

35. All citations from Lombroso and Ferrero 1894, p. vi.

36. Lombroso in the preface, ibid., p. vii.

37. Ibid.

38. Using statistics, Galton investigated whether members of the royal family mentioned in weekly prayers lived longer on average (Draaisma 2000, p. 69). On Galton, see also Gould 1987, p. 75f.

39. Lombroso 1887, p. 23.

40. Lombroso and Laschi 1891, vol. 2, plates V–VI, Figs. 1–6.

41. Ibid., vol. 1, diagram II, p. 156f. "Orographic" relates to mountains and things associated with or induced by their presence.

42. Ibid., vol. 1, p. 47f.

43. Kurella 1910, p. 2.

44. Lombroso 1887, p. 23.

45. Ibid., p. 22.

46. From Lombroso's text "L' influenza della civiltà e dell' occasione sul genio" of 1883, cited in Kurella 1910, p. 86.

47. Thanks to an initiative of the Cinématheque in Prague, a restored copy was screened in December 2001 during a symposium that paid tribute to Jan Evangelista Purkyně.

Chapter 8

1. Cited in Johansson 1983, p. 26.

2. My sources here are the Ph.D. thesis by Johansson (in English) and several original texts in Russian from various archives in St. Petersburg and Moscow. Biographical details are given in Baumgarten 1924; the introduction in Gastev 1966, which gives the date of his death as early as 1938; Hielscher 1978, p. 247; Traub 1976; and Köster's afterword in Gastev 1999.

3. Cited in Hielscher 1978, p. 247.

4. Cited in Traub 1976, p. 151.

5. Johansson 1983, pp. 67, 69, 68, 71.

6. Gastev 1966, chapter on "How One Should Work."

7. Gastev 1923/1978, p. 238.

8. The tenor is here very reminiscent of Gustav T. Fechner, a cofounder of psycho-physics. In a short work entitled *Über das höchste Gut* (1923), he celebrates the synthesis of desire and discipline as the ultimate.

9. Translated into English from the German (Köster 1999), n.p.; Johansson (1983) also includes many citations of other Gastev poems translated into English.

10. Quoted in Johansson 1983, p. 101.

11. In 1939, Witkiewicz committed suicide in Poland, in the tiny part of that country not yet occupied by either the Nazis or the Russians, as he testified on paper. On Witkiewicz, see Harten and Stanislawski 1980, pp. 30–86.

12. Akademie der Künste 1983, p. 53. The following citations from the opera are also from this catalogue.

13. See Malevich 1962.

14. Cited in Hielscher 1978, p. 234.

15. Cf. the excellent study on the "fourth dimension" by Henderson 1983, here especially p. 3f.

16. Grau 1988, p. 129.

17. Ullmann 1996, p. 49.

18. Cf. the essay by Gellius N. Povarov in Trogemann, Nitussou, and Ernst 2001, pp. 47–50. In addition, the book's introduction is useful for the history of mathematics, computer science, and calculating machines in Russia.

19. Lévy 1998, p. 911f.

20. For details, see Zielinski 1989/1999.

21. See "Der Rosingsche Fernseher," *Zeitschrift für Schwachstromtechnik,* no. 7 (1911), p. 172f.; Ruhmer 1911; Abramson 1981. In 1998, Maria Barth did some research on

my behalf on Rosing in the St. Petersburg archives. The information I cite here is taken from her unpublished report, which is based on a number of original Russian sources. A particularly rich source of information for her is an essay by A. P. Kupaygorodskaya, "Boris L. Rosing (1869–1933): The Final Years," published in the anthology *Outstanding Figures of Russian Science in the Nineteenth and Twentieth Centuries: Historic Essays,* vol. 3, St. Petersburg 1996, pp. 73–95 (in Russian).

22. Cf. Abramson 1981, p. 579f.

23. See the introduction to the German edition of Taylor 1913, by Rudolf Roesler, p. xii.

24. Citation from the 1911 original, available on the Web courtesy of Eric Eldred (<http://melbecon.unimelb.edu.au/het/taylor/sciman.htm> [September 14, 2004]).

25. Taylor 1913, p. 5f.

26. Fülöp-Miller 1926, p. 283.

27. Gastev 1923/1978, p. 237.

28. Lenin 1961, p. 145.

29. Baumgarten 1924, p. 3.

30. Ibid., pp. 112, 115.

31. Gastev cited in Fülöp-Miller 1926, p. 287.

32. Fülöp-Miller 1934, p. 343.

33. Ibid., p. 276.

34. Cited in Baumgarten 1924, p. 111f.

35. Meyerhold, cited in Bochow 1997, p. 65.

36. See the excellent essay by Chadarevian (1993) on this concept and field of research.

37. Braune and Fischer 1892, p. 409.

38. The text was completed in October 1894 and published by Hirzel in Leipzig in early 1895.

39. Braune and Fischer (1895) describe the practical difficulties they encountered in the course of these experiments, but also how much they enjoyed themselves. For example, lacking a room at their disposal that could be darkened sufficiently during the daytime, they did all their experiments at night. Because they were interested in taking shots of locomotion that was as relaxed and "natural" as possible, the test persons had to make trial "walks" to get accustomed to the electrical-mechanical corset: "We had the additional pleasure of following all phases of the entire movement process directly with the naked eye" (1895, p. 183).

40. Tramm 1921, p. 86f.

41. Bücher 1899, p. 358.

42. This is the title of a paper by Konstantin Sotonin from the University of Kazan, a collaborator of the scientists in St. Petersburg. See also Mitrofanova 2000, p. 180, whom I asked to research the early neurophysiological scene in St. Petersburg; this publication is a summary of her results.

43. *Music for the Mozart Effect* is the title of a series of CDs that some seventy-five years later, utilizes the findings of such research commercially on a large scale. Fragments of musical compositions, particularly early Mozart pieces, are arranged in therapeutic sound packages, such as "Strengthen the Mind" or "Unlock the Creative Spirit," and marketed worldwide.

44. Mitrofanova 2000, p. 175.

45. Etkind 1996, p. 143.

Conclusions

1. This is the title of a legendary book by Gene Youngblood (1970) from the early years of using computers artistically in cinema.

2. Schulz in a letter to Stanislav Ignacy Witkiewicz, in Schulz 1992, vol. 2, p. 92.

3. The correspondence is published in Schulz 1967.

4. All quotations in Schulz 1967, pp. 17–24.

5. In 2001, the writer Christian Geissler discovered in Drohobycz the remains of the murals that Schulz had painted in the nursery of the Gestapo officer's son. Through this spectacular find's being made public, something again became topical and relevant that had remained buried under decades of suppression and oblivion. Through the Drohobycz of Bruno Schulz run lines connecting it with the wide world of art, science, and the media. The poet Geissler from the far north of Germany is one among many: the filmmaker Peter Lilienthal also drew inspiration from Schulz, as did the art theorist John Berger and the film directors The Quay Brothers. In 2002, Christian Geissler's son Benjamin brought out an exciting documentary film, *Bilder finden* [Finding Pictures], which also provides a profound portrait of Schulz. The film's premiere was at the Center for Jewish History in New York.

6. This is also the position of Lynn Thorndike. Cassirer refers here to the work of James George Frazer in the early twentieth century, first published as *The Golden Bough* in 1922.

7. In Cassirer 1985, p. 31.

8. Ibid.

9. The first essay on this subject is Röller and Zielinski 2001, pp. 282–286.

10. For further details, see Zielinski 2001a, pp. 8–27.

11. On the art of Jan and Eva Švankmajer, see the excellent Švankmajer Catalogue 1998.

12. See Milev 1993 on the development of video in eastern Europe.

13. Hultén 1968.

14. The publication that treats some of the works from the exhibition appeared three years later: Reichardt 1971.

15. See the Oxford English Dictionary for the British origin of the word. The tale *Peregrinaggio di tre giovani, figliuoli del re di Serendippo* originated in sixteenth-century Venice but clearly is strongly influenced by Arab and Persian sources.

16. Marc Adrian, "syspot," in Kelemen and Putar 1971, p. 167.

17. *Postindustrial* is not intended here to mean "the time after," because, obviously, industry is still central. I use the term in the sense that Jean-Luc Godard employs it, referring to the hegemony of the post offices and telecommunications over the distribution of images and sound in Europe after World War II.

18. See the essay by Andreas Broeckmann, "Gesichtswechsel oder: Protobalkanische Entidentifizierungen," in Kovats 2000, pp. 364–372, citation p. 368.

19. nettime 1997 provides a useful overview.

20. A striking aspect about the deep time of media is the high number of professional physicians. To my knowledge, no one has yet written a history of the reciprocal relationships bweeen media and medicine.

21. Horkheimer 1934, p. 28.

22. Lyotard 1987, p. 40.

23. Debord 1978. The citation is also the title of Debord's last film before his cinematographic last will, which was completed after his death (see Debord 1985).

24. For an excellent overview, see Taube 1966.

25. In German, *Manchfaltigem*. Lorenz Oken (1843) used this delightful word to refer to things of different kinds.

26. Klossowski 1998, p. 10f.

Bibliography

Abramson, Albert. "Pioneers of television—Vladimir Kosma Zworykin." *SMPTE {Society of Motion Pictures and Television Engineers} Journal* (July 1981): 579–590.

Akademie der Künste, Berliner Festspiele, eds. *Für Augen und Ohren: Von der Spieluhr zum akustischen Environment.* Berlin: AdK, 1980.

Akademie der Künste, Berliner Festwochen, eds. *Sieg über die Sonne: Aspekte russischer Kunst zu Beginn des 20. Jahrhunderts.* Berlin: Fröhlich & Kaufmann, 1983.

Alberti, Leon Battista. *A Treatise on Ciphers.* 1470. Reprint, trans. Alessandro Zaccagnini, Turin: Galimberti Tipografi, 1997.

Alewyn, Richard, and Karl Sälzle. *Das große Welttheater: Die Epoche der höfischen Feste in Dokument und Deutung.* Reinbek bei Hamburg: Rowohlt, 1959.

Alfieri, Vittorio Enzo, ed. *De Aetna.* Palermo: Gellerio, 1981.

Ammann, Peter J. "The Musical Theorie and Philosophie of Robert Fludd." Ph.D. thesis, University of Zürich, n.d. Partly reprinted in *Journal of the Warburg and Courtauld Institute* (1967): 198–211.

Andel, Jaroslav. "Pittura in Boemia: La nascita dell'artista e del critico moderno: Arte, scienza e scoperta del tempo." In *La nascita dell'impressionismo,* ed. Marco Goldin, pp. 332–339. Conegliano: Linea d'ombra libri, 2000.

Arber, Agnes. *Sehen und Denken in der biologischen Forschung.* Reinbek bei Hamburg: Rowohlt, 1960.

Arecco, Davide. *Il sogno di minerva. La scienza fantastica di Athanasius Kircher (1602–1680).* Padua: Cleup Editrice, 2002. (Especially see the chapter "Utopia magnetocratica," p. 119f.)

Aristoteles. *Poetik.* Trans. and ed. Olof Gigon. Stuttgart: Reclam, 1961.

Artau, Joaquin Carreras Y. *De Ramón Lull a los modernos ensayos de formación de una lengua universal.* Barcelona: Instituto Antonio de Nebrija, 1946.

Aschoff, Volker. *Aus der Geschichte der Nachrichtentechnik.* In N 244, lectures, ed. Rheinisch-Westfälische Akademie der Wissenschaften. Opladen: Westdeutscher Verlag, 1974.

———. *Geschichte der Nachrichtentechnik: Beiträge von ihren Anfängen bis zum Ende des 18. Jahrhunderts.* Berlin: Springer, 1984.

———. *Über die Beschreibung des Echos durch Aristoteles.* Ulm: Fabri, 1993.

Asendorf, Christoph. *Batterien der Lebenskraft: Zur Geschichte der Dinge und ihrer Wahrnehmung im 19. Jahrhundert.* Giessen: Anabas, 1984.

Atmanspacher, Harald, Hans Primas, and Eva Wertenschlag-Birkhäuser, eds. *Der Pauli-Jung-Dialog und seine Bedeutung für die moderne Wissenschaft.* Berlin: Springer, 1995.

Authier, Michel. "Die Geschichte der Brechung und Descartes' 'vergessene' Quellen." In Serres 1998, pp. 445–485.

Baatz, Ursula. *Licht—Seele—Auge: Zur Wahrnehmungspsychologie im 19. Jahrhundert.* In Clair et al. 1989, pp. 357–378.

Bachelard, Gaston. *Die Bildung des wissenschaftlichen Geistes.* Trans. Michael Bischoff. Suhrkamp: Frankfurt am Main, 1984.

Bacon, Francis. *The Advancement of Learning.* 1605. Reprint, London: J. M. Dent & Sons; New York: E. P. Dutton, 1915.

Bacon, Roger. *Vom Stein der Weisen/und von den vornembsten Tincturen des Goldes/Viertols und Aneimonij.* Ed. Joachim Tanckium, University of Leipzig. Eißleben: Jacobi Apels, 1608.

————. *The Opus Majus of Roger Bacon.* Trans. Robert Belle Burke. 2 vols. London: Oxford University Press, 1928.

————. *Roger Bacon's Philosophy of Nature: A Critical Edition, with English Translation, Introduction, and Notes, of* De multiplicatione specierum *and* De speculis comburentibus. Ed. David C. Lindberg. Oxford: Oxford University Press, 1983. Reprint, South Bend, Ind.: St. Augustine's Press, 1998.

Bagrow, Leo. *Die Geschichte der Kartographie.* Berlin: Safari, 1951.

Baldwin, Martha. "Athanasius Kircher and the Magnetic Philosophy." Ph.D. thesis, University of Chicago, 1987.

Baltrusaitis, Jurgis. *Imaginäre Realitäten: Fiktion und Illusion als produktive Kraft. Tierphysiognomik, Bilder im Stein, Waldarchitektur, Illusionsgärten.* Trans. Henning Ritter. Cologne: DuMont, 1984.

————. *Der Spiegel: Entdeckungen, Täuschungen, Phantasien.* Trans. Gabriele Ricke and Ronald Voulié. Giessen: Anabas, 1986.

Barbos, Mario Portigliatti. *Cesare Lombrosos deliquenter Mensch.* In Clair et al. 1989, pp. 587–592.

Barthes, Roland. *L'empire des signes.* Geneva: Editions d'Art Albert Skira, 1970.

————. Barthes, R. "Rhetoriker und Magier." Trans. Jutta Prasse. In *Arcimboldo,* ed. Franco Maria Ricci. Parma: Franco Maria Ricci, 1978.

————. *Die helle Kammer: Bemerkungen zur Photographie.* Trans. Dietrich Leube. Frankfurt am Main: Suhrkamp, 1985.

Bataille, Georges. *Die Aufhebung der Ökonomie: Das theoretische Werk in Einzelbänden.* Ed. Gerd Bergfleth. 2d ed. Munich: Matthes & Seitz, 1985.

Baudry, Jean-Louis. "Ideologische Effekte—erzeugt vom Basisapparat." Trans. G. Custance and S. Zielinski. *Eikon, Internationale Zeitschrift für Photographie und Medienkunst* (Vienna) 5 (1993): 34–43.

Baumgarten, Franciska. *Arbeitswissenschaft und Psychotechnik in Russland.* Munich: R. Oldenbourg, 1924.

Baur-Heinhold, Margarete. *Theater des Barock: Festliche Bühnespiele im 17. und 18. Jahrhundert.* Munich: D. W. Callwey, 1966.

Beck, Theodor. *Beiträge zur Geschichte des Maschinenbaus.* 2d ed. Berlin: Springer, 1900.

Beierwaltes, Werner. "Neuplatonisches Denken als Substanz der Renaissance." In Müller et al. 1978, pp. 1–18.

Beinlich, Horst, Hans-Joachim Vollrath, and Klaus Wittstadt, eds. *Spurensuche: Wege zu Athanasius Kircher.* Dettelbach: J. H. Röll, 2002.

Beke, László, and Miklós Peternàk, eds. *Perspektíva—Perspective.* Exhibition catalogue. Budapest: Mucsarnok/C³, 2000.

Belloni, Gabriella. *Giovan Battista della Porta: Criptologia. Edizione, nota biografica, traduzione, con le studio Cognescenza magica e ricerca scientifica in G.B. della Porta.* Rome: Centro internationale di studi umanistici, 1982.

Belz, Ulysses, ed. *Die Ewigkeit ist ein spielendes Kind auf dem Thron: Beiträge von Wissenschaftlern und Künstlern zur Gegenwart des vorsokratischen Denkens.* Bonn, Leipzig: Handdruckpresse, 2000.

Benjamin, Walter. *Lichtenberg—Ein Querschnitt. Hörstück.* In W. Benjamin, *Drei Hörmodelle.* Frankfurt am Main: Suhrkamp, 1971.

———. *Gesammelte Schriften.* In collaboration with Theodor W. Adorno and Gershom Scholem; ed. Rolf Tiedemann and Hermann Schweppenhäuser; vol. 4, no. 1, ed. Tillman Rexroth. Frankfurt am Main: Suhrkamp, 1972.

Benoît, Paul. *Die Theologie im dreizehnten Jahrhundert: Eine Wissenschaft, die anders ist als alle anderen.* In Serres 1998, pp. 315–349.

Benoît, Paul, and Françoise Micheau. *Die Araber als Vermittler?* In Serres 1998, pp. 269–313.

Berkeley, George. *A New Theory of Vision and Other Writings {1709–1721}.* London: Dent & Sons Ltd., 1910.

Bernoulli, Daniel. "Tiefenzeit: Hutton entdeckt die Geologie." *Die Zeitschrift der Kultur* 10 (October 1997): 1.54–2.24.

Bettino, Mario. *Apiaria universae philosophiae mathematicae in quibus paradoxa et nova pleraque machinamenta a usus eximos traducta & facillimis demonstrationimus confirmata.* Bologna: Jo. Baptista Ferroni, 1642. (Vol. 1 also contains an appendix on Euclid's elements: "Euclidis ex apiariis conditus.")

Beutelspacher, Albrecht. *Geheimsprachen: Geschichte und Techniken.* Munich: C. H. Beck, 1997.

————. "Die Wissenschaft vom Verschlüsseln." *Spektrum der Wissenschaft* 4 (2001): 6–11.

Bini, Daniele, ed. *Astrologia—arte e cultura in età rinascimentale* [Art and Culture in the Renaissance]. Italian/English. Modena: Il Bulino, 1996.

Björnbo, Axel Anthon, and Sebastian Vogl, eds. *Alkindi, Tideus und Pseudo-Euklid: Drei optische Werke.* Leipzig: B. G. Teubner, 1912. (Treatises on the history of the mathematical sciences and their applications, founded by Moritz Cantor, no. 26.3, contains Alkindi [al-Kindî]: "De aspectibus," Tideus: "De speculis" [Pseudo] Euclides: "De speculis.")

Bochow, Jörg. *Das Theater Meyerholds und die Biomechanik.* Berlin: Alexander Wewerka, 1997. (Includes video cassette.)

Boeckmann, Johann Lorenz. *Versuch ueber Telegraphic und Telegraphen, nebst der Beschreibung und Vereinfachung des franzoesischen Telegraphen.* Carlsruhe: Macklots Hofbuchdruckerey, 1794. (Facsimile, Düsseldorf: VDI-Verlag, 1966.)

Boehme, Jacob. *The Signature of All Things and Other Writings.* London: Dent & Sons. 1912.

Boehmer, Heinrich. *Ignatius von Loyola.* Stuttgart: K. F. Koehler, 1941.

Boscovich, Rogerius Josephus. *Theoria philosophiae naturalis.* 1763. English edition, Roger Joseph Boscovich, *A Theory of National Philosophy,* Cambridge: MIT Press, 1966.

Brand, Stewart. *The Clock of the Long Now: Time and Responsibility.* New York: Basic Books, 1999.

Braune, Wilhelm, and Otto Fischer. *Bestimmung der Trägheitsmomente des menschlichen Körpers und seiner Glieder.* Vol. 8 of the *Abhandlungen der mathematisch-physischen Classe der Königl. Sächsischen Gesellschaft der Wissenschaften.* Leipzig: S. Hirzel, 1892. (English edition: *Determination of the Moments of Inertia of the Human Body and Its Limbs,* trans. P. Macquet and R. Furlong, Berlin/New York: Springer, 1988.)

———. *Der Gang des Menschen.* Part I: *Versuche am unbelasteten und belasteten Menschen.* Vol. 21 of *Abhandlungen der mathematisch-physischen Classe der Königl. Sächsischen Gesellschaft der Wissenschaften.* Leipzig: S. Hirzel, 1895. (English edition: *The Human Gait,* Berlin/New York: Springer, 1987.)

Braunmühl, Anton von. *Christoph Scheiner als Mathematiker, Physiker und Astronom.* Bayerische Bibliothek, founded and edited by R. Stoettner and K. Trautmann. 24 vols. Bamberg: Gebr. Buchner, 1891.

Brecht, Bertolt. "Leben des Galilei." In B. Brecht, *Gesammelte Werke* 3, Dramas 3, pp. 1229–1345. Frankfurt am Main: Suhrkamp, 1967.

Brewster, David. *Briefe über die natürliche Magie an Sir Walter Scott.* Trans. Friedrich Wolff. Berlin: Theod. Chr. Friedrich. Enslin, 1833. Reprint, Weinheim: Verlag Chemie, 1984.

Brischar, K. *Athanasius Kircher: Ein Lebensbild.* Würzburg: Selbstverlag, 1877.

Broeckmann, Andreas. "A Visual Economy of Individuals: The Use of Portrait Photography in the Nineteenth-Century Human Sciences." Ph.D. thesis, University of East Anglia, Norwich, England, 1995. Revised version: <http://www.v2.nl/~andreas/phd/> (1996).

Bruno, Giordano. *Heroische Leidenschaften und individuelles Leben.* Ed. and introduction by Ernesto Grassi. Reinbek bei Hamburg: Rowohlt, 1957.

———. *Giordano Bruno—Selected and Introduced by Elisabeth von Samsonow.* Munich: dtv, 1999.

Brunschwig, Jacques, and Geoffrey Lloyd, eds. *Das Wissen der Griechen.* Introduction by Michel Serres. Trans. Volker Breidecker et al. Munich: Fink, 2001.

Bücher, Karl. *Arbeit und Rhythmus.* 2nd ed. Leipzig: B. G. Teubner, 1899.

Burckhardt, Jacob. *Die Kultur der Renaissance in Italien.* Berlin: Deutsche Buchgemeinschaft, 1961.

Butor, Michel. *Die Alchemie und ihre Sprache: Essays zur Kunst und Literatur.* Trans. Helmut Scheffel. Frankfurt am Main: Fischer, 1990.

Carl. *Die electrischen Naturkräfte, der Magnetismus, die Electrizität und der galvanische Strom.* Munich: R. Oldenbourg, 1871.

Cassirer, Ernst. *Symbol, Technik, Sprache.* Hamburg: Felix Meiner, 1985.

Ceram, C. W. *Archaeologie des Kinos.* Reinbek: Rowohet, 1965.

Chadarevian, Soraya de. *Die "Methode der Kurven" in der Physiologie zwischen 1850 und 1900.* In Rheinberger and Hagner 1993, pp. 28–49.

Chardans, Jean-Louis. *Dictionaire des trucs (Les faux, les fraudes, les truquages).* Paris: Jean-Jaque Pauvert, 1960.

Chladni, Ernst Florenz Friedrich. *Entdeckungen über die Theorie des Klanges.* Leipzig: Weidmanns Erben und Reich, 1787.

————. *Kurze Übersicht der Schall- und Klanglehre, nebst einem Anhange die Entwicklung und Anordnung der Tonverhältnisse betreffend.* Mainz: Grosh. Hofmusikhandlung, 1827.

Chudy, Josef. *Beschreibung eines Telegraphs, welcher im Jahr 1787 zu Preßburg in Ungarn ist entdeckt worden.* Ofen: Königliche Universitätsschriften, n.d.

Clair, Jean, Cathrin Pichler, and Wolfgang Pircher, eds. *Wunderblock: Eine Geschichte der modernen Seele.* Vienna: Löcker, 1989.

Clubb, Louise George. *Giambattista Della Porta—Dramatist.* Princeton: Princeton University Press, 1965.

Colombo, Giorgio. *Lo Scienza infelice: Il museo di antropologia criminale di Cesare Lombroso.* Turin: Paolo Boringhieri, 1975.

Coudert, Allison. "Some Theories of a Natural Language from the Renaissance to the Seventeenth Century." In Müller et al. 1978, pp. 56–118.

Cram, David, and Jaap Maat. "Universal Language Schemes in the Seventeenth Century." In *Geschichte der Sprachwissenschaften,* ed. Sylvain Auroux et al. Berlin: de Gruyter, 1999.

Crary, Jonathan. *Techniques of the Observer: On Vision and Modernity in the Nineteenth Century.* Cambridge: MIT Press, 1990.

"Criptografia—Breve storia della criptografia (quinta parte), il 16° secolo." <http://www.sancese.com/Cripto5.html> (July 3, 2001).

Crombie, Alistair Cameron. *Medieval and Early Modern Science.* Vol. 1 and 2. New York: Doubleday, 1959.

Dante Alighieri. *Die göttliche Komödie.* Trans. Ida and Walther von Wartburg. Zürich: Manesse, 1963.

Daxecher, F. "Christoph Scheyner's Eye Studies." *Documenta Ophtalmologica* 81 (1992): 27–35.

Debord, Guy. *Die Gesellschaft des Spektakels.* Trans. Jean-Jacques Raspaud. Hamburg: Edition Nautilus, 1978.

———. *In girum imus nocte et consumimur igni. Wir irren des Nachts im Kreis umher und werden vom Feuer verzehrt.* Berlin: Tiamat, 1985.

Dee, John. *The Mathematicall Praeface to the Elements of Geometry of Euclid of Megara.* 1570. Reprint, ed. and introduction by Allen G. Debus, New York: Science History. Publications, 1975.

———. *John Dee on Astronomy: Propaedeumata Aphoristica (1558 and 1568).* Reprint, Latin and English, ed. and introduction by Wayne Shumaker, Los Angeles: University of California Press, 1978.

———. *Die Monas-Hieroglyphe.* Trans. Agnes Klein. Interlaken: Ansata, 1982. (The Latin original was published under the title of *Monas Hieroglyphica* in 1564 in Antwerp.)

Deleuze, Gilles, and Félix Guattari. *Tausend Plateaus: Kapitalismus und Schizophrenie.* Trans. Gabriele Ricke and Ronald Voullié. Berlin: Merve, 1992. (English edition: *A Thousand Plateaus: Capitalism and Schizophrenia,* Minneapolis: University of Minnesota Press, 1987.)

Denker, W. *Sichtbarkeit und Verlauf der totalen Sonnenfinsternis in Deutschland am 19. Aug. 1887*. Berlin: Ferd. Dümmler, 1887.

Derrida, Jacques. *Dem Archiv verschrieben: Eine Freudsche Impression*. Berlin: Brinkmann+Bose, 1997.

Descartes, René. *Philosophische Schriften*. Hamburg: Meiner, 1996. (With an introduction by Rainer Specht and including "Descartes' Wahrheitsbegriff" by Ernst Cassirer.)

———. *Le Monde ou Traité de la Lumière/Die Welt oder Abhandlung über das Licht*. Trans. and afterword by Matthias Tripp. Weinheim: VCH acta humanoria, 1989.

Didi-Huberman, Georges. *Invention de l'hysterie: Charcot et l'iconographie photographique de la Salpêtrière*. Paris: Editions Maculas, 1982.

Diels, Hermann. "Passages Relating to Empedocles from Diels' *Doxographi Graeci*." *Hanover College Historical Texts Project*, 1998/2001. <http://history.hanover.edu/texts/presoc/emp.htm> (January 19, 2003).

Dotzler, Bernhard J. *Papiermaschinen: Versuch über Communication und Control in Literatur und Technik*. Berlin: Akademie Verlag, 1996.

Draaisma, Douwe. "Francis Galton: Inquiries into Human Faculty and Its Development." In *Klassiker der Psychologie*, ed. Helmut E. Lück et al, pp. 66–71. Stuttgart: Kohlhammer, 2000.

Eco, Umberto. "Kircher tra steganografia e poligrafia." In Lo Sardo 2001, pp. 209–213.

Elson, Louis C. *The Theory of Music as Applied to the Teaching and Practice of Voice and Instruments*. Boston: New England Conservatory of Music, 1890.

Ernst, Wolfgang. "Ist die Stadt ein Museum? Rom zum Beispiel—Bausteine zu einer Archäologie der Infrastruktur." In *Stadt und Mensch: Zwischen Chaos und Ordnung*, ed. Dirk Röller, Frankfurt/Main: Lang, 1996.

Etkind, Alexander. *Eros des Unmöglichen: Die Geschichte der Psychoanalyse in Rußland*. Trans. Andreas Tretner. Leipzig: Gustav Kiepenheuer, 1996.

Exner, Franz. *Zur Kenntnis des Purkinje'schen Phänomens: Aus den Sitzungsberichten der Akademie der Wissenschschaften in Wien, mathemathisch-naturwissenschaftliche.* Klasse, Abtlg. IIa, vol. 128, no. 1, Vienna (1919), pp. 1–14.

Fechner, Gustav Theodor. *Über das höchste Gut.* Stuttgart: Strecker und Schröder, 1923.

Feyerabend, E. *Der Telegraph von Gauß und Weber im Werden der elektrischen Telegraphie.* Berlin: Reichspostministerium, 1933.

Filseck, Karin Moser von. *Kairos und Eros.* Bonn: Habelt, 1990.

Fischer, Otto. *Kinematik organischer Gelenke.* Braunschweig: Friedrich Vieweg und Sohn, 1907.

Fletcher, John, ed. *Athanasius Kircher und seine Beziehungen zum gelehrten Europa seiner Zeit.* Wiesbaden: Harrassowitz, 1988.

Fludd, Robert. *Utriusque Cosmi Maioris scilicet et Minoris, Metap(h)ysica, Physica atque Technica Historia.* 2 vols. Volume 1: *De Macrocosmi Historia in duos tractatus divisa.* Oppenheim: Johann.-Theod. de Bry, 1617. Volume 2: *De Supernaturali, Naturali, Praeternaturali et contranaturali, Microcosmi historia in Tractatus tres distrubuta.* Oppenheim: Johann.-Theod. de Bry, 1619.

————. *Storia metafisica, fisica e tecnica dei due mondi, cioè del maggiore e del minore, ripartita in due tomi, secondo la divisione del cosmo* (1617–1621). Extracts trans. into Italian in *La magia naturale,* 1989.

————. *Philosophia Moysaica.* Gouda: Petrus Rammazenius, 1638.

Flusser, Vilém. "Mittel und Meere: Ein Vortrag." *Spuren* 16 (August 1988): 12–16.

————. *Vom Subjekt zum Projekt: Menschwerdung.* Frankfurt am Main: Fischer, 1998.

Foucault, Michel. *Die Ordnung der Dinge: Eine Archäologie der Humanwissenschaften.* Trans. Ulrich Köppen. Frankfurt am Main: Suhrkamp, 1974. (English edition: *The Order of Things: An Archaeology of the Human Sciences,* New York: Pantheon, 1970.)

————. *Über Sexualität, Wissen und Wahrheit: Dispositive der Macht.* Berlin: Merve, 1978.

————. Discipline and Punish: The Birth of the Prison, trans. A. Sheridan, New York: Vintage.

Frazer, James George. *The Golden Bough: A Study in Magic and Religion.* Toronto: Macmillan, 1922.

Fülöp-Miller, René. *Geist und Gesicht des Bolschewismus: Darstellung und Kritik des kulturellen Lebens in Sowjet-Russland.* Zürich: Amalthea, 1926. (English edition: *The Mind and Face of Bolshevism,* London 1926.)

————. *Macht und Geheimnis der Jesuiten: Eine Kultur- und Geistesgeschichte.* Berlin: Th. Knaur Nachf., 1927.

————. *Die Phantasiemaschine. Eine Saga der Gewinnsucht.* Berlin/Vienna/Leipzig: Paul Zsolnay, 1931.

————. *Führer, Schwärmer und Rebellen: Die großen Wunschträume der Menschheit.* Munich: F. Bruckmann, 1934.

Galilei, Galileo. *Sidereus Nuncius: A Reproduction of the Copy in the British Library.* Alburgh: Archival Facsimiles, 1987. (Original: Venice: Thom. Baglionus, 1610.)

Gastev, Aleksej Kapitanovic. "Geradebiegen des Volkes." 1922. Trans. Karla Hielscher. *Alternative* 122/123 (October/December 1978): 242–246.

————. "Rüstet euch, Monteure!" 1923. Trans. Karla Hielscher. *Alternative* 122/123 (October/December 1978): 236–241.

————. *Kak nado rabotat'.* [How One Should Work: Practical Introduction to the Scientific Organization of Labor]. Moskow: Ekonomika, 1966. (New edition of the original, published in the 1920s)

————. *Ein Packen von Ordern.* Trans. Cornelia Köster. Oberwaldbehrungen: Peter Engstler, 1999.

Giacomo, Salvatore di. *La Prostituzione in Napoli nel secoli XV, XVI e XVII.* Naples: Riccardo Marghieri, 1899; reprint, Naples: Edition Gazetta, 1994.

Gilly, Carlos, and Cis van Heertum, eds. *Magia, Alchimia, Scienza dal 1400 al 1700: L'influsso di Ermete Trismegisto. Magic, Alchemy and Science 15th–18th Centuries: The Influence of Hermes Trismegistus.* Italian and English. Venice: Centro Di, 2002.

Gilman, Sander L., ed. *The Face of Madness: Hugh W. Diamond and the Origin of Psychiatric Photography.* Secaucus, N.J.: Citadel Press, 1977.

Ginzburg, Carlo. *Spurensicherung: Die Wissenschaft auf der Suche nach sich selbst.* Trans. Gisela Bonz and Karl F. Hauber. Berlin: Wagenbach, 1995.

Glissant, Édouard. *Traktat über die Welt.* Trans. Beate Thill. Heidelberg: Das Wunderhorn, 1999.

Godwin, Joscelyn. *Robert Fludd: Hermetic Philosopher and Surveyor of Two Worlds.* London: Thames and Hudson, 1979a.

———. *Athanasius Kircher: A Renaissance Man and the Quest for Lost Knowledge.* London: Thames and Hudson, 1979b.

———. "Athanasius Kircher and the Occult." In Fletcher 1988, pp. 17–36.

Goethe, Johann Wolfgang von. *Sämtliche Werke.* Complete edition with 10 vols. Stuttgart: Cotta, 1885.

———. *Das Sehen in subjektiver Hinsicht, von Purkinje, 1819.* Reprinted in Konersmann 1997, pp. 168–179.

Gorman, Michael John. "The Scientific Counter-Revolution: Mathematics, Natural Philosophy and Experimentalism in Jesuit culture 1580–c. 1670." Ph.D thesis, European University Institute, Florence, 1998.

Göttert, Karl-Heinz. *Magie: Zur Geschichte des Streits um die magischen Künste unter Philosophen, Theologen, Medizinern, Juristen und Naturwissenschaftlern von der Antike bis zur Aufklärung.* Munich: Fink, 2001.

Gouk, Penelope. *Music, Science and Natural Magic in Seventeenth-Century England.* New Haven: Yale University Press, 1999.

Gould, Stephen Jay. *Ever Since Darwin: Reflections in Natural History.* New York: W. W. Norton, 1977.

————. *The Mismeasure of Man.* 1981. London: Penguin, 1987.

————. *Time's Arrow, Time's Cycle: Myth and Metaphor in the Discovery of Geological Time.* 1987. Reprint, London: Penguin, 1991.

————. *Wonderful Life: The Burgess Shale and the Nature of History.* 1989. Reprint, London: Penguin, 1991.

————. *Bully for Brontosaurus: Further Reflections in Natural History.* London: Penguin, 1992.

————. "Ladders and Cones: Constraining Evolution by Canonical Icons." In *Hidden Histories of Science,* ed. Robert B. Silvers, pp. 37–67. London: Granta, 1997.

————. *Illusion Fortschritt: Die vielfältigen Wege der Evolution.* Trans. Sebastian Vogel. Frankfurt am Main: S. Fischer, 1998 (Original: *Full House: The Spread of Excellence from Plato to Darwin.* New York: Crown Publishers, 1996).

————. "Time Scales and the Year 2000." In Umberto Eco, S. J. Gould, Jean-Claude Carrière, and Jean Delumeau, *Conversations about the End of Time,* ed. by Catherine David, Frédéric Lenoir, and Jean-Philippe de Tonnac, pp. 1–44. London: Penguin,

————. *The Lying Stones of Marrakech.* London: Vintage, 2001.

Gould, S. J., and Rosamond Wolff Purcell. *Crossing Over: Where Art and Science Meet.* New York: Three Rivers Press, 2000.

Grafton, Anthony. *Die tragischen Ursprünge der deutschen Fußnote.* Trans. H. Jochen Bußmann. Berlin: Berlin Verlag, 1995.

Graham, A. C., and Nathan Sivin. "A Systematic Approach to the Mohist Optics (ca. 300 B.C.)." In *Explorations of an Ancient Tradition,* ed. Nathan Sivin and Shigeru Nakayama, pp. 105–152. MIT East Asian Science Series. Cambridge: MIT Press, 1973.

Grau, Konrad. *Berühmte Wissenschaftsakademien: Von ihrem Entstehen und ihrem weltweiten Erfolg.* Frankfurt am Main: Harri Deutsch, 1988.

Grosjean, Georges, and Rudolf Kinauer. *Kartenkunst und Kartentechnik vom Altertum bis zum Barock.* Bern: Hallwag, 1970.

Grote, Hans Henning Freiherr, ed. *Vorsicht! Feind hört mit! Eine Geschichte der Weltkriegs- und Nachkriegsspionage.* Dresden: Zwinger, n.d.

Guillermou, Alain. *Ignatius von Loyola.* Reinbek: Rowohlt, 1981.

Guyot, M. *Nouvelles récréations physiques et mathématiques.* 3 vols. Paris: Gueffier, 1786.

Haakman, Anton. *De onderaardse wereld van Athanasius Kircher.* Amsterdam: Meulenhoff, 1991.

Hagen, Albert. *Die sexuelle Osphresiologie: Die Beziehungen des Geruchssinnes und der Gerüche zur menschlichen Geschlechtsthätigkeit.* Charlottenburg [Berlin]: H. Barsdorf, 1901.

Halliwell, James Orchard, ed. *The Private Diary of John Dee and the Catalogue of His Library of Manuscripts, from the Original Manuscripts in the Ashmolean Museum at Oxford and Trinity College Library, Cambridge.* London: John Bowyer Nichols and Son, 1842. Reprint, Largs, Scotland: Banton Press, 1990.

Hammond, John H. *The Camera Obscura: A Chronicle.* Bristol: Adam Hilger, 1981.

Harten, Jürgen, and Ryszard Stanislawski, eds. *Hommage à Stanislaw-Ignacy Witkiewicz.* Düsseldorf: Kunsthalle und Museum Sztuki Lódz, 1980.

Hartwig, Wolfgang. "Physik als Kunst: Über die naturphilosophischen Gedanken Johann Wilhelm Ritters." Inaugural diss., Albert Ludwig University, Freiburg im Breisgau, 1955.

Haskell, Yasmin Annabel. "Didactic Tradition and Modern Science in Giuseppe Maria Mazzolari's *Electricorum: libri VI* (Rome 1767)." In Sassoferrato, *Studi Umanistici Piceni XIX, 1999,* Istituto Internazionale Di Studi Piceni, 1999.

———. *Loyola's Bees: Ideology and Industry in Jesuit Latin Didactic Poetry.* The British Academy Series. Oxford: Oxford University Press, 2003.

Helden, Albert van. "Origine e sviluppo del telescopio." In Miniati 1991, pp. 64–71.

Heilborn, Ernst. *Novalis, der Romantiker.* Berlin: Georg Reimer, 1901.

Heilbron, John. *Electricity in the Seventeenth and Eighteenth Centuries: A Study of Early Modern Physics.* Berkeley: University of California Press, 1979.

Hein, Olaf. *Die Drucker und Verleger der Werke des Polyhistors Athanasius Kircher S. J.: Eine Untersuchung zur Produktionsgeschichte enzyklopädischen Schrifttums im Zeitalter des Barock, unter Berücksichtigung wissenschafts- und kulturgeschichtlicher Aspekte.* Vol. 1. Cologne: Böhlau, 1993.

Helden, Albert van. "Porta, Giambattista della: Catalogue of the Scientific Community." *Galileo Project.* <http://es.rice.edu/ES/humsoc/Galileo/Catalog/Files/porta.html> (July 3, 2001).

Henderson, Linda Dalrymple. *The Fourth Dimension and Non-Euclidean Geometry in Modern Art.* Princeton: Princeton University Press, 1983.

[Heron] Hero of Alexandria. *Pneumatics.* Ed. and trans. Bennet Woodcroft. London: Taylor Walton and Maberly, 1851.

Heydenreich, Hasso. *Das Feuerzeug: Ein Beitrag zur Geschichte der Technik.* Weimar: Stadtmuseum, n.d.

Hielscher, Karla. "Kleine Gastev-Biographie." *Alternative* 122/123 (October/December 1978): 247.

Himmelmann, Nikolaus. *Utopische Vergangenheit: Archäologie und moderne Kultur.* Berlin: Gebr. Mann, 1976.

Hocke, René Gustav. *Die Welt als Labyrinth: Manier und Manie in der europäischen Kunst. Manierismus.* Vol. 1. Reinbek bei Hamburg: Rowohlt, 1957.

———. *Manierismus in der Literatur: Sprach-Alchemie und esoterische Kombinationskunst. Manierismus.* Vol. 2. Reinbek bei Hamburg: Rowohlt, 1959.

———. *Die Welt als Labyrinth: Manierismus in der europäischen Kunst und Literatur.* Special edition. Reinbek bei Hamburg: Rowohlt, 1991.

Hölderlin, Friedrich. *Der Tod des Empedokles.* Ed. Friedrich Beissner. Stuttgart: Reclam, 1973.

Hooke, Robert. *Micrographia or Some Physiological Descriptions of Minute Bodies Made by Magnifying Glasses, with Observations and Inquiries Thereupon.* London: Jo. Martyn, Ja. Allestry, 1663.

Horkheimer, Max [Heinrich Regius, pseud.]. *Dämmerung: Notizen in Deutschland.* Zürich: Oprecht and Helbing, 1934.

Hort, G. M. *Dr. John Dee: Elizabethan Mystic and Astrologer.* London: William Rider and Son, 1922. Reprint, Largs, Scotland: Banton Press, 1991.

Hoskovec, Jiri. "Jan Evangelista Purkinje: Beobachtungen und Versuche zur Physiologie der Sinne (1819–1825)." In *Klassiker der Psychologie,* ed. Helmut E. Lück et al. Stuttgart: Kohlhammer, 2000, pp. 31–35.

Hultén, K., and G. Pontus. *The Machine—As Seen at the End of the Mechanical Age.* New York: Museum of Modern Art, 1968.

Hutton, James. *Theory of the Earth with Proofs and Illustrations.* Vol. 1 and 2. Edinburgh: 1795.

Ibn al-Haytham. *The Optics of Ibn Al-Haytham.* Trans. and introduction by A. I. Sabra. 2 vols. London: Warburg Institute, University of London, 1989.

Isa, Ali Ibn. *Erinnerungsbuch für Augenärzte.* Trans. from Arabic manuscripts by J. Hirschberg and J. Lippert. Leipzig: Veit & Comp., 1904. (Reprinted as vol. 44 in the series Islamic Medicine, Publications of the Institute for the History of Arabic-Islamic Science, Frankfurt am Main: Johann Wolfgang von Goethe University, 1996.)

Johansson, Kurt. "Aleksej Gastev: Proletarian Bard of the Machine Age." Ph.D. diss., University of Stockholm, Dept. of Slavic and Baltic Languages, Stockholm, 1983.

Johnen, Chr. *Geschichte der Stenographie—im Zusammenhang mit der allgemeinen Entwicklung der Schrift und der Schriftkürzung.* Vol. 1. Berlin: Ferdinand Schrey, 1911.

Johnson, Frank H., and Yata Haneda, eds. *Bioluminiscence in Progress.* Princeton, N.J.: Princeton University Press, 1966.

Johnston, Norman. *The Human Cage: A Brief History of Prison Architecture.* New York: Walker, 1973.

Josten, C. H. *Robert Fludd and His Philosophicall Key.* Transcription of the manuscript at Trinity College, Cambridge, with an introduction by Allen G. Debus. New York: Science History Publications, 1979.

Jun'ichiro, Tanizaki. *Lob des Schattens: Entwurf einer japanischen Ästhetik.* Trans. Eduard Klopfenstein. Zürich: Manesse, 1987. (Japanese original: Tokyo 1933.)

Kahn, Alfred. "Die Didaktiker auf dem Gebiete der physikalischen Geographie im XVIII. Jahrhundert in ihren Beziehungen zu Kircher, Riccoli und Varenius." Inaugural diss., Philosophical Faculty, University of Würzburg. Würzburg: Anton Boegler, 1906.

Kahn, Fritz. *Das Leben des Menschen: Eine volkstümliche Anatomie. Biologie, Physiologie und Entwicklungsgeschichte des Menschen.* 5 vols. Stuttgart: Franckh'sche Verlagsbuchhandlung, 1923–1931.

Kamper, Dietmar. *Zur Geschichte der Einbildungskraft.* Reinbek bei Hamburg: Rowohlt, 1990.

————. *Körper-Abstraktionen: Das anthropologische Viereck von Raum, Fläche, Linie und Punkt.* Ed. Vilém_Flusser_Archiv at the Academy of Media Arts Cologne, Cologne: Walther König, 1999.

Karger-Decker, Bernt. *Ärzte im Selbstversuch: Ein Kapitel heroischer Medizin.* Leipzig: Koehler & Amelang, 1965.

Karmarsch, Karl. *Geschichte der Technologie seit der Mitte des achtzehnten Jahrhunderts.* Vol. 1 of *Geschichte der Wissenschaften in Deutschland. Neuere Zeit.* Munich: R. Oldenburg, 1872.

Kaufmann, Thomas DaCosta. *The Mastery of Nature: Aspects of Art, Science, and Humanism in the Renaissance.* Princeton Essays on the Arts. Princeton, N.J.: Princeton University Press, 1993.

Kelemen, Boris and Radoslav Putar. *dijalog sa strojem/dialogue with the machine.* Zagreb: Bit International, Galerije Grada, 1971.

Kemp, Martin. *The Science of Art: Optical Themes in Western Art from Brunelleschi to Seurat.* New Haven: Yale University Press, 1990.

Kerkhoff, Manfred. "Zum antiken Begriff des Kairos." *Zeitschrift für philosophische Forschung* 27 (1973).

Kestler, Johann Stephan. *Physiologia Kircheriana experimentalis, qua summa argumentorum multitudine & varietate naturalium rerum scientia per experimenta physica, mathematica, medica, chymica, musica, magnetica, mechanica combrobatur atque stabilitur.* Amsterdam: Jansson (Johannes) van Waesberge, 1680.

Kircher, Athanasius. *Institutiones mathematicae de aritmetica computu ecclesiast: geometria aliisq. scientiis mathematicis.* Würzburg 1630, unpublished manuscript.

————. *Magnes sive de arte magnetica.* Rome: Hermann Scheus, 1641, third edition 1654.

————. *Ars magna lucis et umbrae.* Rome: H. Scheus, 1646 (or 1645). Revised edition, Amsterdam: Jansson van Waesberge, 1671. (Includes a section on steganography at the end, "Cryptologia nova.")

————. *Musurgia universalis sive Ars magna consoni et dissoni in X libros digesta.* 2 vols. Rome: Francesco Corbelletti, 1650. Reprint, with a foreword and indexes by Ulf Scharlau, Hildesheim: Olms, 1970.

————. *Iter exstaticum II.* Rome: Francesco Corbelletti, 1657.

————. *Polygraphia nova et universalis ex combinatoria arte detecta.* Rome: Varesii, 1663.

————. *Mundus subterraneus, in XII libros digestus.* 2 vols. Amsterdam: Jansson van Waesberge, 1664-1665.

————. *China monumentis qua sacris qua profanis, Nec non variis naturae & artis spectaculis, Aliarumque rerum memorabilium argumentis illustrate.* Amsterdam: Jakob a Meurs, 1667. Reprint, Katmandu, Nepal: Bibliotheca Himalayica, 1979. (Cited as *China illustrata.*)

————. *Ars magna sciendi sive combinatoria.* Amsterdam: Jansson van Waesberge, 1669.

————. *Phonurgia nova sive Conjugium mechanico-physicum artis et naturae paranympha phonosophia concinnatum.* Campidona (Kempten): Rudolph Dreher, 1673. (Summary of book 9 of *Musurgia universalis* of 1650).

————. *Turris Babel.* Amsterdam: Jansson van Waesberge, 1679.

————. *Neue Hall- und Tonkunst.* Trans. Agatho Carione. Nördlingen: Arnold Heylen, 1684. Reprint, Hannover: Edition libri rari, 1983.

[Kircher, Athanasius]. *Selbstbiographie des P. Athanasius Kircher aus der Gesellschaft Jesu.* Trans. Nikolaus Seng. Fulda: Aktiendruckerei, 1901.

Kirchhoff, Jochen. *Giordano Bruno.* Reinbek bei Hamburg: Rowohlt, 1980.

Kittler, Friedrich. "Lakanal und Soemmering: Von der optischen zur elektrischen Telegraphie." In *Wunschmaschine Welterfindung, Eine Geschichte der Technikvisionen seit dem 18. Jahrhundert,* ed. Brigitte Felderer, pp. 286–295. Vienna: Springer, 1996.

Klemm, Friedrich, and Armin Hermann. *Briefe eines romantischen Physikers: Johann Wilhelm Ritter an Gotthilf Heinrich Schubert und Karl von Hardenberg.* Munich: Moss, 1966.

Klickowstroem, Graf Carl von. "Johann Wilhelm Ritter und der Elektromagnetismus." *Archiv für Geschichte der Mathematik, der Naturwissenschaften und der Technik* 9 (1929): 68–85.

Klossowski, Pierre. *Die lebende Münze.* Trans. Martin Burckhardt. Berlin: Kadmos, 1998.

Knobloch, Eberhard. *"Musurgia Universalis:* Unknown Combinatorial Studies in the Age of Baroque Absolutism." *History of Science* 17 (1979): 258–275.

Konersmann, Ralf. *Kritik des Sehens.* Leipzig: Reclam, 1997.

Köster, Cornelia. *Aleksej Gastev: Ein Packen von Ordern.* Ostheim/Rhön: Peter Engstler, 1999.

Kovats, Stephen, ed. *Ost-West Internet/Media Revolution.* Frankfurt am Main: Campus, 2000.

Krebs, Peter. *Die Anthropologie des Gotthilf Heinrich von Schubert.* Inaugural diss., University of Cologne. Cologne: Orthen, 1940.

Krehl, Stephan. *Fuge: Erläuterung und Anleitung zur Komposition derselben.* Leipzig: Göschen, 1908.

Kubler, George. *Die Form der Zeit: Anmerkungen zur Geschichte der Dinge.* Trans. Bettina Blumenberg. Frankfurt am Main: Suhrkamp, 1982.

Kuhn, Thomas S. *Die Struktur wissenschaftlicher Revolutionen.* 2d ed. Frankfurt am Main: Suhrkamp, 1976.

Künzel, Werner. *Der* Oedipus Aegyptiacus *des Athanasius Kircher: Das ägyptische Rätsel in der Simulation eines barocken Zeichensystems.* Berlin: AsiMPaCS, 1989.

Künzel, W., and Peter Bexte. *Allwissen und Absturz: Der Ursprung des Computers.* Frankfurt am Main: Insel, 1993.

———. *Maschinendenken/Denkmaschinen: An den Schaltstellen zweier Kulturen.* Frankfurt am Main: Insel, 1996.

Künzel, Werner, and Heiko Cornelius. *Die Ars Generalis Ultima des Raymundus Lullus: Studien zu einem geheimen Ursprung der Computertheorie.* Berlin: Advanced Studies in Modern Philosophy and Computer Science, 1986.

Kurella, Hans. *Cesare Lombroso als Mensch und Forscher.* Wiesbaden: J. F. Bergmann, 1910.

Leibniz, Gottfried Wilhelm. *Monadologie.* Ed. and trans. Hartmut Hecht. Stuttgart: Reclam, 1998.

Leinkauf, Thomas. *Mundus combinatus: Studien zur Struktur der barocken Universalwissenschaft am Beispiel Athanasius Kirchers S. J. (1602–1680).* Berlin: Akademie Verlag, 1993.

Lenin, Vladimir Ilyich. *Werke.* Vol. 20. Berlin: Dietz, 1961.

Lennig, Walter. *Marquis de Sade, in Selbstzeugnissen und Dokumenten.* Reinbek bei Hamburg: Rowohlt, 1965.

Leps, Marie-Christine. *Apprehending the Criminal: The Production of Deviance in Nineteenth-Century Discourse.* Durham, N.C.: Duke University Press, 1992.

Levi, Eliphas. *Geschichte der Magie.* 2 vols. Vienna: Otto Wilhelm Barth, 1926.

Lévy, Pierre. "Die Erfindung des Computers." In Serres 1998, pp. 905–945.

Liano, Ignacio Gómez de. *Athanasius Kircher: Itinerario del éxtasis o las imágines de un saber universal.* Madrid: Ediciones Siruela, 1985; 2d ed., 2001.

Lichtenberg, Georg Christoph. *Aphorismen, Essays, Briefe.* Ed. Kurt Batt. 2d ed. Leipzig: Dieterich, 1965.

Lindberg, David C. *Auge und Licht im Mittelalter: Die Entwicklung der Optik von Alkindi bis Kepler.* Trans. Matthias Althoff. Frankfurt am Main: Suhrkamp, 1987.

Link, David. "Poesiemaschinen/Maschinenpoesie." Inaugural Diss., Humboldt University Berlin and Academy of Media Arts Cologne, 2004.

Linke, Detlef B. *Kunst und Gehirn: Die Eroberung des Unsichtbaren.* Reinbek: Rowohlt, 2001.

Lippmann, Edmund O. von. *Abhandlungen und Vorträge zur Geschichte der Naturwissenschaften.* Leipzig: Veit & Comp., vol. 1: 1906; vol. 2: 1913.

Lombroso, Cesare. *Genie und Irrsinn, in ihren Beziehungen zum Gesetz, zur Kritik und zur Geschichte.* Trans. A. Courth. Leipzig: Reclam, n.d. [1887].

—————. *Handbuch der Graphologie.* Trans. Gustav Brendel. Leipzig: Reclam, n.d. [ca. 1893].

—————. *Der Verbrecher* [Homo delinquens]—*in anthropologischer, ärztlicher und juristischer Beziehung.* 3 vols. Hamburg: Verlagsanstalt u. Druck. A. G., vorm. J. F. Richter, 1894b.

—————. *Studien über Genie und Entartung.* Trans. Ernst Jentsch. Leipzig: Reclam, n.d. [1894c].

—————. *Les anarchistes.* Paris: Flammarion, 1896.

—————. *Kerker-Palimpseste: Wandinschriften und Selbstbekenntnisse gefangener Verbrecher.* 1899. Trans. Hans Kurella. Reprint, Osnabrück: Reinhard Kuballe, 1983.

—————. "Das Verbrechen in Spanien und seine Geschichte." In C. Bernaldo de Quiros and J. M. L. Aguilaniedo, *Verbrechertum und Prostitution in Madrid,* vol. 3, pp. v–xii, Sexualpsychologische Bibliothek, 1st series, ed. Iwan Bloch. Berlin: Louis Marcus, n.d. [ca. 1910, Lombroso's essay is dated from September 1909].

—————. *Handbuch der Graphologie—mit graphologischen Anmerkungen und 470 Faksimiles.* Leipzig: Reclam, n.d.

Lombroso, C., and G. Ferrero. *Das Weib als Verbrecherin und Prostituierte: Anthropologische Studien gegründet auf eine Darstellung der Biologie und Psychologie des normalen Weibes.* Trans. Hans Kurella. Hamburg: Verlagsanstalt u. Druck. A. G., vorm. J. F. Richter, 1894a.

Lombroso, C., and Rodolfo Laschi. *Der politische Verbrecher und die Revolutionen—in anthropologischer, juristischer und staatswissenschaftlicher Beziehung.* Hamburg: Verlagsanstalt u. Druck. A. G., vorm. J. F. Richter, 1891.

Lo Sardo, Eugene. "The Courtly Machines (Le Macchine cortigiane)." In Lo Sardo 1999, pp. 233–274 (Italian pp. 1–62).

———. *Athanasius Kircher S. J.—Il Museo del Mundo.* Rome: Edizioni de Luca, 2001.

Lo Sardo, E., ed. *Iconismi & Mirabilia da Athanasius Kircher.* Introduction by Umberto Eco and notes by Roman Vlad. Rome: Edizioni dell'Elefante, 1999.

Lósy-Schmidt, Ede. *Chudy Jószef optikai és akusztikai távirója; A mai írógéprendszeru gyorstáviró öse a 18. század végérol.* Budapest: F'ovárosi Ny., 1932.

Lothar, R. *Die Sprechmaschine: Ein technisch-ästhetischer Versuch.* Berlin, 1924.

Loyola, Ignatius of. *Die Exerzitien.* Lucerne: Josef Stocker, 1946.

Lucretius. *De rerum natura: Welt aus Atomen.* Latin and German edition, trans. and afterword by Karl Büchner. Stuttgart: Reclam, 1973.

Ludwig, Hellmut. *Marin Mersenne und seine Musiklehre.* Halle: Buchhandlung des Waisenhauses, 1935.

Lullus, Raimundus. *Ein kleiner Schlüssel (clavicula) des R. L. von Manorca, welcher auch ein Schatzkasten (aptorilt) Dietrich genannt wird, worinnen alles, was zur Alchemey-Arbeit erfordert wird, eröffnet und erkläret wird; und R. Lullus: Codicill (Testaments-Anhang) oder Vademecum (Handbüchlein), worinnen die Urquellen der Alchimie-Kunst, wie auch der verborgenen Weltweisheit gezeiget werden.* Cologne: Arnold Birkmanns Erben, 1563. (Both texts in *Neue Sammlung von einigen alten und sehr rar gewordenen philosophisch und alchymistischen Schriften.* Frankfurt, Leipzig: Kraussischer Buchladen, 1767.)

———. *Ars magna generalis et ultima.* Frankfurt: Cornelius Sutorius, 1596.

Lyotard, Jean-François. "Zeit haben." *Ästhetik & Kommunikation* 67/68 (1987): 40.

Mach, Ernst. *Kultur und Mechanik.* Stuttgart: W. Spemann, 1915. (Photomechanical reprint of further texts by Mach, ed. Joachim Thiele. Amsterdam: E. J. Bonset, 1969.)

———. *Die Prinzipien der physikalischen Optik: Historisch und erkenntnispsychologisch ent-wickelt.* Leipzig: Johann Ambrosius Barth, 1921.

———. *Grundlinien der Lehre von den Bewegungsempfindungen.* Amsterdam: Bonset, 1967 [Reprint of the Leipzig edition 1875.]

(La) Magia naturale nel Rinascimento: Testi di Agrippa, Cardano, Fludd. Introduction by Paolo Rossi, trans. and notes by Silvia Parigi. Turin: UTET, 1989.

Malevich, Kasimir. *Suprematismus: Die gegenstandslose Welt.* Trans. Hans von Riesen, ed. Werner Haftmann. Cologne: DuMont Schauberg, 1962.

Mankiewicz, Richard. *Zeitreise Mathematik: Vom Ursprung der Zahlen bis zur Chaostheorie.* Trans. Sabine Lorenz and Felix Seewöster. Cologne: VGS, 2000.

Mann, Heinz Herbert. "Optische Instrumente." In *Erkenntnis Erfindung Konstruktion: Studien zur Bildgeschichte von Naturwissenschaften und Technik vom 16. bis zum 19. Jahr-hundert,* ed. Hans Holländer. Berlin: Gebr. Mann, 2000.

Mannoni, Laurent, Donata Pesenti Campagnoni, and David Robinson. *Light and Move-ment: Incunabula of the Motion Picture 1420–1896.* In English, French, and Italian. Porde-none: Le Giornate de Cinema Muto, 1996.

Mansfeld, Jaap, ed. and trans. *Die Vorsokratiker.* 2 vols., in Greek and German. Stuttgart: Reclam, 1995–1996.

Marchant, Joanna. "First Light." *New Scientist* (July 2000): 34f.

Marek, Jiri. "Athanasius Kircher und die 'neue' Physik im 17. Jahrhundert." In Fletcher 1988, pp. 37–52.

Marey, Etienne-Jules. *La méthode graphique dans les sciences expérimentales et particulièrement en physiologie et en médicine.* Paris: Masson, 1878.

Martin, Alain, and Oliver Primavesi. *L'Empédocle de Strasbourg (P. Strasb. gr. Inv. 1665–1666).* Berlin: de Gruyter, 1999.

Marx, Karl. *Theorien über den Mehrwert* [vol. 4 of *Das Kapital*]. In Marx Engels Werke (MEW) vol. 26.3., Berlin: Dietz, 1974. (English version online: <http://www.marxists.org/archive/marx/works/1863/theories-surplus-value/>

Mattschoss, Conrad. *Geschichte des Zahnrads: Nebst Bemerkungen zur Entwicklung der Verzahnung von K. Kutzbach.* Berlin: VDI-Verlag, 1940.

Mazzolari, G. M. [Josephus Marianus Parthenius]. *Electricorum, libri VI.* Rome: Generosus Salomoni, 1767.

McQueen, Steve. *Barrage.* Ed. Friedrich Meschede. Cologne: DAAD and Walther König, 2000.

Merkel, Franz R. "Der Naturphilosoph Gotthilf Heinrich Schubert und die deutsche Romantik." Inaugural diss., University of Strassburg. Munich: Oskar Beck, 1912.

Middleton, W., and E. Knowles. "Archimedes, Kircher, Buffon, and the Burning-Mirrors." *Isis* 52 (1961): 533–543.

Milev, Rossen. *Video in Osteuropa.* Sofia (Bulgaria): Balkan Media, 1993.

Miniati, Mara, ed. *Museo di Storia della Scienza Catalogue.* Florence: Istituto e Museo di Storia della Scienza, 1991.

Mitrofanova, Alla. "Conditioned Reflexes, Music and Neurofunction." In *Lab, Jahrbuch für Apparate und Künste,* ed. T. Hensel, H. U. Reck, and S. Zielinski, pp. 171–182. Cologne: Walther König, 2000.

Morello, Nicoletta. "Nel Corpo della terra: Il Geocosmo di Athanasius Kircher." In Lo Sardo 2001, pp. 178–196.

Müller, Johannes. *Über die phantastischen Gesichtserscheinungen: Eine physiologische Untersuchung, mit einer physiologischen Urkunde des Aristoteles über den Traum, den Philosophen und Ärzten gewidmet.* Koblenz, 1826. Reprint, Munich: Werner Fritsch, 1967.

Müller, Kurt, Heinrich Schepers, and Wilhelm Totok, eds. "Magia naturalis und die Entstehung der modernen Naturwissenschaften. Studia leibnitiana." *Zeitschrift für Geschichte der Philosophie und der Wissenschaften* 7 (1978).

Müller-Jahncke, Wolf-Dieter. "Agrippa von Nettesheim: 'De occulta philosophia.' Ein Magisches System." In Müller et al. 1978, pp. 19–29.

Müller-Tamm, Jutta. "Die 'Empirie des Subjektiven' bei Jan Evangelista Purkinje: Zum Verhältnis von Sinnesphysiologie und Ästhetik im frühen 19. Jahrhundert." In

Wahrnehmung der Natur und Natur der Wahrnehmung. Studien zur Geschichte visueller Kultur um 1800, ed. Gabriele Dürbeck et al., pp. 153–164. Dresden: Verlag der Kunst, 2001.

Musil, Robert. *Der Mann ohne Eigenschaften.* 9th ed. Reinbek bei Hamburg: Rowohlt, 1968.

Myers, Charles S. *Mind and Work: The Psychological Factors in Industry and Commerce.* London: University of London Press, 1920.

Nachrichten von einer Hallischen Bibliothek. 8 vols. Halle: Johann Justinus Gebauer, July–December 1751.

Needham, Joseph. *Physics and Physical Technology.* Vol. 4, *Science and Civilisation in China.* Cambridge: Cambridge University Press, 1962.

Neidhöfer, Herbert, and Bernd Ternes, eds. *Was kostet den Kopf? Ausgesetztes Denken der Aisthesis zwischen Abstraktion und Imagination. Dietmar Kamper zum 65. Geburtstag.* Marburg: Tectum, 2001.

nettime, eds. *Netzkritik. Materialien zur Internet-Debatte.* Berlin: Edition I-D Archiv, 1997.

Novalis. *Heinrich von Ofterdingen: Ein nachgelassener Roman.* Berlin: Buchhandlung der Realschule, 1802. Reprint, Stuttgart: Reclam, 1987.

Oken, Lorenz. *Lehrbuch der Naturphilosophie* 1810–1811. 3d ed. Zürich: Friedrich Schultheß, 1843.

Ong, Walter J. *The Presence of the Word: Some Prolegomena for Cultural and Religious History.* Minneapolis: University of Minnesota Press, 1967.

————. *Rhetoric, Romance and Technology: Studies in the Interaction of Expression and Culture.* Ithaca, N.Y.: Cornell University Press, 1971.

Os Jesuítas no Brasil. <www.ars.com.br/cav/si16/si17.htm> (August 22, 2001).

Ostwald, Wilhelm. *Elektrochemie: Ihre Geschichte und Lehre.* Leipzig: Veit & Comp., 1896.

————. *Grundriss der Naturphilosophie.* Leipzig: Reclam, 1908.

Pasolini, Pier Paolo. *Trilogia della vita: Il Decameron, I racconti di Canterbury, Il fiore delle mille e una notte.* Ed. Giorgio Gattei. Bologna: Cappelli, 1975.

Pauli, W. "Der Einfluss archetypischer Vorstellungen auf die Bildung naturwissenschaftlicher Theorien bei Kepler." In *Naturerklärung und Psyche. Studien aus dem C. G. Jung-Institut.* Zürich: Rascher, 1952.

Peuckert, Will-Erich. *Gabalia: Ein Versuch zur Geschichte der magia naturalis im 16. bis 18. Jahrhundert.* Vol. 2 of Peuckert, *Pansophie.* Berlin: Erich Schmidt, 1967.

Pfister, Oskar. *Die psychologische Enträtselung der religiösen Glossolalien und der automatischen Kryptographie.* Vol. 3 of Jahrbuch für psychoanalytische und psychopathologische Forschungen. Leipzig: Deuticke, 1912.

Plötzeneder, Karl. "Giovan Battista Della Porta: Visionär zwischen Magie, Imagination und Mathematik." Thesis, Faculty of Humanities, University of Salzburg, 1994.

Porta, Giovan Battista della. *Magia naturalis sive De miraculis rerum naturalium. Libri IIII.* Naples: Matthias Cancer, 1558. (Cited as *Magia* I; editions in other languages are listed in the bibliography according to their date of publication.)

———. *De furtivis literarum notis, vulgò de zifferis. Libri IIII.* Naples: Joa. Maria Scotus, 1563. (A new, miniature edition was published in 1593.)

———. *L'Arte del ricordare.* Naples: Matthias Cancer, 1566. (This book was published first in one volume together with *De furtivis* of 1563 and later with various editions of *Ars Reminiscendi,* first published in 1602.)

———. *Phytognomonia. Libri VIII.* 1583. Reprint, Naples: Horatius Salvianus, 1588.

———. *De humana physiognomonia. Libri IV.* Naples: Vici Aequensis, 1586. (This work was first published in Italian as *Fisonomia dell'huomo* in 1598 by T. Longo in Naples, under the pseudonym of Giovanni de Rosa, to avoid censorship. Cited here according to the German edition: *Die Physiognomie des Menschen,* ed. Theodor Lessing and Will Rink, vol. 1 of *Schriftenreihe zur Gestaltenkunde "Der Körper als Ausdruck,"* Radebeul, Dresden: Madaus, 1930.)

———. *Magia naturalis. Libri XX.* Naples: Horatius Salvianus, 1589. (Cited as *Magia* II. About thirty editions of this work were published, in a number of different languages.)

————. *De occultis literarum notis seu Artis animi sensa occulte alijs significandi, aut ab alijs significata expiscandi enodandique. Libri IIII.* Montisbeligardi: Jacob Foillet, Lazari Zetzneri, 1593a.

————. *De refractione: Optices parte, libri novem.* Naples: Jacobus Carlinus & Antonius Pace, 1593b.

————. *Pneumaticorum libri tres.* Naples: Carlino, 1601.

————. *Ars reminiscendi.* Naples: Ioan. Bapt. Subt., 1602. Reprint in the *Edizione nazionale delle opere di Giovan Battista della Porta of the Edizione Scientifiche Italiane (ESI),* Naples 1996, ed. Raffaele Sirri. Published in the same volume with *L'Arte del Ricordare* as vol. 3 of the *Collected Works.* (In the following references, the publisher's acronym, ESI, is used for the *Collected Works.*)

————. *Magia naturalis. Libri XX.* Frankfurt: Samuel Hempel, 1607.

————. *De distillatione. Libri IX.* Rome: Camera Apostoilica, 1608.

————. *Elementorum curvilineorum. Libri III.* 3d ed. Rome: Bartholomaeus Zannetus, 1610. (The first edition was published in 1601 by Antonio Pace in Naples.)

————. *De aeris transmutationibus. Libri IV.* Rome: Bartholomaeus Zannettus, 1610. Reprint, ed. Alfonso Paoella, Naples: ESI, 2000.

————. "Libro di fisonomia naturale." Naples 1611a. (Handwritten manuscript in the Biblioteca Nazionale of Naples; in vol. 1., chapters 1–3 are completely missing and chapter 4 is partly missing, for a total of nine double-sided pages; the introduction breaks off after two pages.)

————. *Della magia naturale.* Naples: Jiacomo Carlino, e Costantino Vitale, 1611b. (Cited as *Magia* III.)

————. *Natürliche Magia: Das ist ein ausführlicher und gründlicher Bericht/ von den Wunderwercken Natürlicher Dinge.* 4 books. Magdeburg: Martin Rauschern, 1612.

————. *Natural Magick.* (Trans. anon.) London: Thomas Young, Samuel Speed, 1658. Reprint, Collector's Series in Science, ed. Derek J. Price, 2d ed., The Smithsonian Institution, New York: Basic Books, 1958.

————. *Della chirofisionomia.* (Trans. and biographical preface by Pompeo Sarnelli. Naples: Ant. Bulifon, 1677.

————. *Magia naturalis: oder Hauß-, Kunst- und Wunderbuch. Nach dem vermehrten/ in XX. Büchern bestehenden lateinischen Exemplar/ ins Hochteutsche übersetzt.* Nürnberg: Johann Friedrich Rüdigern, 1719.

————. *La Magia Naturale o Esposizione dei Segreti e delle meraviglie della Natura. Con cenni biografica sull'autore.* Milan: Alberto Fidi (Biblioteca di Scienze Occulte), 1925. (This is a shortened version of the text published in the first edition of 1589.)

————. *Criptologia.* Ed. Gabriella Belloni. Rome: Centro internazionale di studi umanistici, 1982. (Edizione Nazionale dei Classici del Pensiero Italiano, Serie II/37.)

————. *Tabernaria.* Ed. Raffaele Sirri. Naples: de Simone Editori, 1990.

————. *Ars reminiscendi aggiunta: L'arte del ricordare.* 1602. Reprint, trans. Dorandino Falcone da Gioia, ed. Raffaele Sirri. [National edition of Giovan Battista della Porta's works.] Naples: ESI, 1996.

————. *Coelestis physiognomonia e in appendice Della celeste fisonomia.* 1603. Reprint, ed. Alfonso Paoella. Naples: ESI, 1996.

————. *Teatro: Promo Tomo—Tragedie.* Ed. Raffaele Sirri. Naples: ESI, 2000a.

[Porta, Giovan Battista della]. *Claudii Ptolemaei magnae constructionis liber primus, cum Theonis Alexandrini commentariis, Io: Baptista Porta Neapolitano Interprete.* 1605. Reprint, ed. Raffaella de Vivo. Naples: ESI, 2000b.

Purkyně (Purkinje), Jan Evangelista. *Beiträge zur Kenntniss des Sehens in subjectiver Hinsicht.* Prague: Johann Gottfried Calve, 1819.

————. "Beyträge zur näheren Kenntnis des Schwindels aus heautognostischen Daten." In vol. 6 of *Medicinisches Jahrbuch des kaiserlich-königlichen-österreichischen Staates,* part 2, pp. 79–125. Vienna: 1820.

————. "[Purkinje's] Mitteilungen über Scheinbewegungen und über den Schwindel aus den Bulletins der Schlesischen Gesellschaft von 1825 und 1826." In *Physiologische Studien über die Orientierung,* ed. Hermann Aubert. Rostock: Universität Rostock, n.d.

————. *Beobachtungen und Versuche zur Physiologie der Sinne: Neue Beiträge zur Kenntniss des Sehens in subjektiver Hinsicht.* Berlin: C. Reimer, 1825.

————. "Über die Verdienste Berkeleys und die Theorie des Sehens." In *Uebersicht der Arbeiten der Schlesischen Gesellschaft fuer vaterlaendische Kultur.* p. 50f. Breslau: 1828.

Rehm, Else. "Johann Wilhelm Ritter und die Universität Jena." In *Jahrbuch des Freien Deutschen Hochstifts 1973,* Tübingen: Niemeyer, 1973.

Reichardt, Jasia. *The Computer in Art.* London: Studio Vista, Van Nostrand Reinhold, 1971.

Reichert, Klaus. "Von der Wissenschaft zur Magie: John Dee." In *Der Magus: Seine Ursprünge und seine Geschichte in verschiedenen Kulturen,* ed. Anthony Grafton and Moshe Idel, pp. 187–106. Berlin: Akademie Verlag, 2001.

Reiss, Erwin. *Pension Sehblick. Eidetik audiovisueller Medien. Eine Videotopik der Seherkenntnis.* Frankfurt/Main: Peter Lang, 1995.

Repcheck, Jack. *The Man Who Found Time: James Hutton and the Discovery of the Earth's Antiquity.* Cambridge: Perseus Publishing, 2003.

Rheinberger, Hans-Jörg, and Michael Hagner, eds. *Die Experimentalisierung des Lebens: Experimentalsysteme in den biologischen Wissenschaften 1850–1950.* Berlin: Akademie Verlag, 1993.

Richter, Klaus. "Der Physiker Johann Wilhelm Ritter." In *Der Physiker des Romantikerkreises Johann Wilhelm Ritter in seinen Briefen an den Verleger Carl Friedrich Ernst Fromm,* ed. K. Richter, pp. 13–84. Weimar: Hermann Böhlaus Nachf., 1988.

————. *Johann Wilhelm Ritter: Bibliographie.* Erfurt: Akademie Gemeinnütziger Wissenschaften, 2000.

Riordan, Michael, and David N. Schramm. *Die Schatten der Schöpfung: Dunkle Materie und die Struktur des Universums.* Foreword by Stephen W. Hawking. Heidelberg: Spektrum, 1993.

Ritter, Johann Wilhelm. *Beweis, dass ein bestaendiger Galvanismus den Lebensprozess in dem Thierreich begleite: Nebst neuen Versuchen und Bemerkungen über den Galvanismus.* Weimar: Industrie-Comptoir, 1798.

————. "Wirkung des Galvanismus der Voltaschen Batterie auf menschliche Sinnes-werkzeuge." *Annalen der Physik* 7 (1801): 447–484.

————. *Beyträge zur näheren Kenntnis des Galvanismus.* 2 vols. Jena: Friedrich Frommann, 1802.

————. "Nachricht von der Fortsetzung seiner Versuche mit Volta's galvanischer Batterie." In *Magazin für den neuesten Zustand der Naturkunde, mit Rücksicht auf die dazu gehörigen Hilfswissenschaften* 4 (September 1802): 575–661.

————. "Versuche und Beobachtungen über den Galvanismus." In *Magazin für den neuesten Zustand der Naturkunde* 6 (August 1803): 97–126; 7 (September 1803): 181–215.

————. "Anmerkungen zum vorstehenden Schreiben des Hr. D. Oersted" [on Chladni's acoustic figures]. *Magazin für den neuesten Zustand der Naturkunde* 9 (1805): 33–47.

————. *Die Physik als Kunst: Ein Versuch, die Tendenz der Physik aus ihrer Geschichte zu deuten.* Munich: Lindauer, 1806.

————. *Fragmente aus dem Nachlasse eines jungen Physikers: Ein Taschenbuch für Freunde der Natur.* Heidelberg: Mohr und Zimmer, 1810. Reprint, ed. with afterword by Steffen and Birgit Dietzsch, Hanau: Müller & Kiepenheuer, 1984. (Facsimile reprint with an afterword by Heinrich Schipperges, Heidelberg: Lambert Schneider, 1969.)

————. *Elektrische Versuche an der Mimosa pudica L, in Parallele mit gleichen Versuchen an Fröschen, Vorlesungsskript aufgezeichnet von Dr. M. Ruhland,* pp. 245–400. Munich: Königl. Akademie der Wissenschaften, 1811.

————. *Die Begründung der Elektrochemie: Eine Auswahl aus den Schriften des romantischen Physikers.* Ed. Armin Hermann. Frankfurt am Main: Akademische Verlagsanstalt, 1968.

Rivosecchi, Valerio. *Esotismo in Roma Barocca: Studi sul Padre Kircher. Biblioteca di storia dell'arte 12.* Rome: Bulzoni, 1982.

Robinson, David, Stephen Herbert, and Richard Crangle, eds. *Encyclopaedia of the Magic Lantern.* London: The Magic Lantern Society, 2001.

Roessler, Otto E. *Endophysik: Die Welt des inneren Beobachters.* Ed. Peter Weibel. Berlin: Merve, 1992.

————. "Mikrokonstruktivismus." In *Lab, Jahrbuch für Künste und Apparate,* ed. Kunsthochschule für Medien mit dem Verein ihrer Freunde, pp. 208–227. Cologne: Walther König, 1996a.

————. *Das Flammenschwert, oder: Wie hermetisch ist die Schnittstelle des Mikrokonstruktivismus?* Bern: Benteli, 1996b.

Rohr, Moritz von. *Zur Geschichte und Theorie des photographischen Teleobjectivs.* Weimar: Verlag der Deutschen Photographen-Zeitung, 1897.

————. *Zur Entwicklung der dunklen Kammer (camera obscura): Sammlung optischer Aufsätze.* Ed. H. Harting. Berlin: Verlag der Central-Zeitung für Optik und Mechanik, 1925.

————. *Die optischen Instrumente: Brille, Lupe, Mikroskop, Fernrohr, Aufnahmelinse und ihnen verwandte Vorkehrungen.* Berlin: Springer, 1905; 4th ed., 1930.

Rolland, Romain. *Empedokles von Agrigent und das Zeitalter des Hasses.* Trans. Leo Götzfried, including Eduard Saenger's adaptation of the fragments. Leipzig: Reclam, 1918.

Röller, Nils. *Migranten: Edmond Jabès, Luigi Nono, Massimo Cacciari.* Berlin: Merve, 1995.

————. and Siegfried Zielinski. "On the Difficulty to Think Twofold in One." In *Sciences of the Interface,* ed. Hans Diebner, Timothy Druckrey, and Peter Weibel. Tübingen: Genista, 2001.

————. *Medientheorie im epistemischen Übergang: Hermann Weyls Philosophie der Mathematik und Naturwissenschaft und Ernst Cassirers Philosophie der symbolischen Formen.* Weimar: Verlag und Datenbank für Geisteswissenschaften, 2002.

Ronchi, Vasca. *Optics: The Science of Vision.* 1897. Reprint, trans. and ed. Edward Rosen, New York: Dover Publications, 1991. Italian original: *Ottica, scienza della visione,* Bologna: Nicola Zanichelli, 1957.

Rosenbusch, Hans. "Der Okkultismus als Beobachtungswissenschaft." In *Die neue Volkshochschule, Bibliothek für moderne Geistesbildung,* vol. 5, pp. 3–79. Leipzig: Weimann, 1928.

Rothschuh, K. E. "Der Begriff der 'Physiologie' und sein Bedeutungswandel in der Geschichte der Wissenschaft." *Archives internationales d'histoires des sciences* 40 (July–September 1957): 217–225.

Rowland, Ingrid D. *The Ecstatic Journey: Athanasius Kircher in Baroque Rome*. Chicago: University of Chicago Library, 2000.

Ruhmer, E. "Ein bedeutsamer Fortschritt im Fernsehproblem (Der Rosingsche Fernseher)." *Die Umschau,* no. 25 (1911): 508–510.

Rumjantsev, S. "Kommunistische Glocken." *Sowjetische Musik* (Moscow) 11 (1984).

Rumpf. J. *Ueber das Fernrohr.* Vienna: Separatum ex Schr., 1879.

Rybczynski, Zbigniew. "Looking to the Future: Imagining the Truth." In Beke and Péternàk 2000, pp. 357–366.

Sacks, Oliver. "Scotoma: Forgetting and Neglect in Science." In *Hidden Histories of Science,* ed. Robert B. Silvers, pp. 141–187. London: Granta, 1997.

Samsonow, Elisabeth von. "Ars/Techne." In Neidhöfer and Ternes 2001, pp. 347–361.

Sarnelli, Pompeo. "Vita di Gio. Battista della Porta Napolitano." Foreword to Porta 1677.

Sartre, Jean-Paul. *Briefe an Simone de Beauvoir und andere.* Vol. 1: 1926–1939. Ed. Simone de Beauvoir. Reinbek bei Hamburg: Rowohlt, 1986.

Schall, Bild, Optik. *Naturwissenschaft und Technik: Vergangenheit, Gegenwart, Zukunft.* Cologne: Lingen 1991.

Scharlau, Ulf. *Athanasius Kircher (1601–1680) als Musikschriftsteller: Ein Beitrag zur Musikanschauung des Barock.* Studien zur hessischen Musikgeschichte, ed. Heinrich Hüschen, vol. 2. Marburg: n. p., 1969.

————. "Athanasius Kircher und die Musik um 1650: Versuch einer Annäherung an Kirchers Musikbegriff." In Fletcher 1988, pp. 53–68.

Scheiner, Christoph. *Rosa ursina sive sol ex admirando facularum et macularum.* Bracciano: Andreas Phaeus, 1626–1630.

Schellen, H. *Der elektromagnetische Telegraph in den Hauptstadien seiner Entwickelung und in seiner gegenwärtigen Ausbildung und Anwendung, nebst einem Anhang über den Betrieb elektrischer Uhren.* 5th ed. Braunschweig: Vieweg, 1870.

Schmidt-Biggemann, Wilhelm. *Topica Universalis: Eine Modellgeschichte humanistischer und barocker Wissenschaft. Paradeigmata 1, Innovative Beiträge zur philosophischen Forschung.* Hamburg: Felix Meiner, 1983.

Schneider, Hans Joachim, ed. *Kriminalität und abweichendes Verhalten.* 2 vols. Weinheim: Beltz, 1983.

Schneider, Joseph. "Athanasius Kircherus." *Henschels Janus, Zeitschrift für Geschichte und Literatur der Medizin* (Breslau) 2 (1847): 599–608. Reprint, Leipzig: Henschel, 1931.

Schneider, K. *Gotthilf Heinrich von Schubert: Ein Lebensbild.* Bielefeld: Velhagen & Klasing, 1863.

Schott, Caspar. *Technica curiosa sive mirabilia artis.* 1664. Reprint, Hildesheim: Olms, 1977.

——. *Magia optica, Das ist Geheime doch naturmässige Gesicht- und Augenlehr / In zehen unterschiedliche Bücher abgetheilet.* Bamberg: Johann Arnold Cholins, 1671.

Schrage, Dominik. *Psychotechnik und Radiophonie: Subjektkonstruktionen in artifiziellen Wirklichkeiten.* Munich: Fink, 2001.

Schrödinger, Erwin. *1944. What is Life? with Mind and Matter and Autobiographical Sketches.* Cambridge: Cambridge University Press. 2004.

——. *Die Natur und die Griechen: Kosmos und Physik.* Trans. Míra Koffka. Reinbek bei Hamburg: Rowohlt, 1956; (English edition: *Nature and the Greeks,* Cambridge: Cambridge University Press, 1954.)

Schubert, Gotthilf Heinrich von. *Nachtseite der Naturwissenschaft.* Dresden: Arnoldsche Buchhandlung, 1818.

——. *Die Symbolik des Traums: Mit einem Anhange aus dem Nachlasse eines Visionairs und einem Fragment über die Sprache des Wachens.* Leipzig: Brockhaus, 1840.

Schulz, Bruno. *Sklepy Cynamonowe.* Kraków/Wrocław: Wydawnictwo Literacki, 1957.

——. *Die Zimtläden.* Trans. Josef Hahn. Munich: Hanser, 1966.

————. *Die Republik der Träume: Fragmente, Aufsätze, Briefe, Grafiken.* Ed. Mikolaj Dutsch, trans. Josef Hahn and Mikolaj Dutsch. Munich: Hanser, 1967.

————. *Die Wirklichkeit ist Schatten des Wortes: Aufsätze und Briefe.* Vol. 2 of *Bruno Schulz Complete Works.* Ed. Jerzy Ficowski. Munich: Hanser, 1992.

————. *Bruno Schulz—Z listów odnalezionych.* Warszawa: Wydawnictwo Cimera, 1993.

Schwartz, K. "Kircher [Athanasius]". In *Allgemeine Encyklopädie der Wissenschaften und Künste,* ed. J. S. Ersch and J. G. Gruber, section 2, part 36, pp. 266–271. Leipzig: Brockhaus, 1884.

Schwenter[um], M. Daniel[um]. *Deliciae physico-mathematicae oder Mathematische und philosophische Erquickstunden.* Nürnberg: Jeremias Dümleis, 1636.

Segalen, Victor. *Die Ästhetik des Diversen: Versuch über den Exotismus.* Trans. Uli Wittmann. Frankfurt am Main: Fischer, 1994.

Sepibus, Girgio de. *Romani Cellegii Societatus Musaeum celeberrimum.* Amsterdam: Jansson van Waesberghe, 1678.

Serres, Michel, ed. *Elemente einer Geschichte der Wissenschaften.* Trans. Horst Brühmann. Frankfurt am Main: Suhrkamp, 1998.

Shea, William R. "Galileo, Scheiner, and the Interpretation of the Sunspots." *ISIS* 61 (1970): 498–519.

Simmen, René. *Der mechanische Mensch: Texte und Dokumente über Automaten, Androiden und Roboter.* Zürich: Selbstverlag, 1967.

Simon, Gérard. *Der Blick, Das Sein und die Erscheinung in der antiken Optik. Mit einem Anhang: Die Wissenschaft vom Sehen und die Darstellung des Sichtbaren.* Bild & Text series. Munich: Fink, 1992.

Simonovits, Anna. *Dialektisches Denken in der Philosophie von Gottfried Wilhelm Leibniz.* Berlin: Akademie Verlag, 1968.

Simson, Gerhard. *Einer gegen Alle: Die Lebensbilder von Christian Thomasius, Cesare Lombroso, Fridtjof Nansen.* Munich: C. H. Beck, 1960.

Skupin, Frithjof, ed. *Abhandlungen von der Telegraphie oder Signal- und Zielschreiberei in die Ferne, nebst einer Beschreibung und Abbildung der neuerfundenen Fernschreibmaschine in Paris.* Heidelberg: R. v. Decker's Nachdruck, 1986. (Four early texts on telegraphy, 1794–1795.)

Specht, Rainer. *René Descartes.* Reinbek bei Hamburg: Rowohlt, 1966; 8th ed., 1998.

Spiritual Exercises of St. Ignatius of Loyola (The), translated from the autograph by Father Elder Mullan, S. J., http://www.ccel.org/ccel/ignatius/exercises.html (22 August 2004).

Starke, Dieter. *Geschichte der Naturwissenschaften. Erste Anfänge.* Frankfurt am Main: Deutsch, 1999.

Stauder, Wilhelm. *Einführung in die Akustik.* 4th ed. Wilhemshaven: F. Noetzel, 1999.

Steinhart. "Empedokles." In *Allgemeine Encyklopädie der Wissenschaften und Künste,* ed. J. S. Ersch and J. G. Gruber, section 1, part 34, pp. 83–105. Leipzig: Brockhaus, 1840.

Stekel, Wilhelm. *Der nervöse Magen: Hygienische Zeitfragen.* Vienna: Paul Knepler, 1918.

Stengers, Isabelle. "Die Gallilei-Affären." In Serres 1998, pp. 395–443.

Sterling, Bruce. "Die Molkereiprodukte-Theorie der toten Medien." In *Hyperorganismen,* ed. Olaf Arndt, Stefanie Peter, and Dagmar Wünnenberg, pp. 360–372. Hannover: Internationalismus, 2000.

Stolzenberg, Daniel. *The Great Art of Knowing: The Baroque Encyclopedia of Athanasius Kircher.* Stanford: Stanford University Libraries, 2001; Fiesole: Edizioni Cadmo, 2001.

Strasser, Gerhard F. *Spectaculum Vesuvii: Zu zwei neuentdeckten Handschriften von Athanasius Kircher mit seinen Illustrationsvorlagen.* In *Theatrum Europaeum: Festschrift für Maria Szarota,* ed. R. Brinkmann et al., pp. 363–384. Munich: Fink, 1982.

———. *Lingua Universalis: Kryptologie und Theorie der Universalsprachen im 16. und 17. Jahrhundert.* Wiesbaden: Harrassowitz, 1988.

Strasser, Peter. *Verbrechermenschen: Zur kriminalwissenschaftlichen Erzeugung des Bösen.* Frankfurt am Main: Campus, 1984.

————. "Die Bestie als Natur." In Clair et al. 1989, pp. 593–600.

Strathern, Paul. *Mendeleyev's Dream.* London: Hamish Hamilton, 2000.

Švankmajer, Eva/Jan. *Anima-Animus-Animation.* Prague: Slovart, 1998.

Szczesniak, Baleslaw. Athanasius Kircher's *China Illustrata.* In *Osiris* 8 (1952): 385–411.

Székely, Bertalan. *Mozgástanulmányai.* Exh. cat. Budapest: Academy of Fine Arts, 1992.

Taes, Frédéric. "Le Système Della Porta" (1563). In *Histoire de la Cryptographie* (6), <http://users.online.be/tst/h6.html> (2 October 2001).

Tatar, Maria M. *Romantic "Naturphilosophie" and Psychology: A Study of G. H. Schubert and the Impact of His Works on Heinrich von Kleist and E.T.A. Hoffmann.* Ann Arbor, Mich: University Microfilms, 1971.

Taube, Mortimer. *Der Mythos der Denkmaschine.* Reinbek: Rowohlt, 1966. (English edition: *Computers and Common Sense: The Myth of Thinking Machines,* New York: Columbia University Press, 1961.)

Taylor, Frederick Winslow. *Die Grundsätze wissenschaftlicher Betriebsführung* [The Principles of Scientific Management]. Trans. Rudolf Roesler. Munich: R. Oldenbourg, 1913.

Teichmann, Jürgen. *Elektrizität: Elektrostatik, Galvanische Elemente, Elektromagnetismus, Mathematik und Atomismus, Elektron und Röntgenstrahlen.* 3d ed. Munich: Deutsches Museum, 1996.

Theile, Friedrich Wilhelm. "Empfindung." In *Allgemeine Encyklopädie der Wissenschaften und Künste,* ed. J. S. Ersch and J. G. Gruber, section 1, part 34, pp. 110–112. Leipzig: Brockhaus, 1840.

Thompson, Michael. *Rubbish Theory: The Creation and Destruction of Value.* London: Oxford University Press, 1979.

————. "Oblivion, Eternity and Tick-Tock." In *Lab: Jahrbuch für Apparate und Künste,* ed. Kunsthochschule für Medien mit ihren Freunden. Cologne: Walther König, 2002.

Thorndike, Lynn. *A History of Magic and Experimental Science.* Vol. 5, 7, and 8. New York: Columbia University Press, 1958.

Tiemann, Veit. *Symmetrische/klassische Kryptographie: Ein interaktiver Überblick.* Bielefeld: Lehrstuhl für Statistik und Information der Universität Bielefeld, 2000.

Tomlinson, Gary. *Music in Renaissance Magic: Toward a Historiography of Others.* Chicago: Chicago University Press, 1993.

Traber, P. Zacharia[s]. *Vervus opticus sive Tractatus theoricus in tres libros opticam, catoptricam, dioptricam distributus.* Vienna: Johann. Christoph. Cosmer, 1675.

Tramm, K. A. *Psychotechnik und Taylor-System.* Vol. 1, *Arbeitsuntersuchungen.* Berlin: Springer 1921.

Traub, Rainer. "Lenin und Taylor: Die Schicksale der 'wissenschaftlichen Arbeitsorganisation' in der (frühen) Sowjetunion." *Kursbuch* (Berlin) 43 (1976): 146–158.

Trogemann, Georg, Alexander Y. Nitussov, and Wolfgang Ernst, eds. *Computing in Russia: The History of Computer Devices and Information Technology Revealed.* Trans. A. Y. Nitussov. Braunschweig: Vieweg 2001.

Trümpy, Rudolf. "James Hutton und die Anfänge der modernen Geologie." In *Schottische Aufklärung: "A Hotbed of Genius,"* ed. Daniel Brühlmeier, pp. 75–89. Berlin: Akademie Verlag, 1996.

Turing, Alan. "Intelligent Machinery." 1948 report for the National Physical Laboratory. In *Cybernetics: Key Papers,* ed. C. R. Evans and A. D. J. Robertson, Baltimore and Manchester: University Park Press, 1968.

Tyndall, John. *Sound.* 4th ed. London: Longmans, Green and Co., 1883.

————. *Das Licht: Sechs Vorlesungen gehalten in Amerika im Winter 1872–1873.* Trans. Gustav Wiedemann. Braunschweig: Vieweg, n.d.

Ullmann, Dieter. "Zur Frühgeschichte der Akustik: A. Kirchers 'Phonurgia Nova.'" *Wissenschaftliche Zeitschrift der Friedrich-Schiller-Universität Jena, Mathematische-Naturwissenschaftliche Reihe* 27 (1978): 355–360.

————. "Ein akustisches Experiment A. Kirchers und seine Geschichte." In *NTM-Schriftenreihe für Geschichte der Naturwissenschaften, Technik und Medizin* (Leipzig) 17, no. 1: 61–88.

———. *Chladni und die Entwicklung der Akustik von 1750–1860.* Science Networks Historical Studies Series 19. Basel: Birkhäuser, 1996.

Universale Bildung im Barock: Der Gelehrte Athanasius Kircher. Exhibition catalogue, ed. Stadt Rastatt, collaborators Reinhard Dieterle, John Fletcher, Christel Römer. Rastatt: Stadt Rastatt; Karlsruhe: Badische Landesbibliothek, 1981.

Visker, Rudi. "Foucaults Anführungszeichen einer Gegenwissenschaft." In *Spiele der Wahrheit, Michel Foucaults Denken,* ed. F. Ewald and B. Waldenfels, p. 298f. Frankfurt am Main: Suhrkamp, 1991.

Vollenweider, Alice, ed. *Italienische Reise: Ein literarischer Reiseführer durch das heutige Italien.* Berlin: Wagenbach, 1985.

Volta, Ornella. *Erik Satie.* Trans. Simon Pleasance. Paris: Edition Hazan, 1997.

Vonderau Museum Fulda, ed. *Magie des Wissens: Athanasius Kircher (1602–1680).* Petersberg: Imhof, 2003.

Wagner, Andreas. *Denkrede auf Gotthilf Heinrich von Schubert in der öffentlichen Sitzung der Königlichen Bayerischen Akademie der Wissenschaften am 26. März 1861.* Munich: Verlag der Königlichen Akademie, 1861.

Wanderlingh, Attilio. *Napoli nella Storia, duemilacinquecento anni, dalle origini greche al secondo millenio.* Naples: Intra Moenia, 1999.

Was heisst "wirklich"? Unsere Erkenntnis zwischen Wahrnehmung und Wissenschaft. Ed. Bayerischen Akademie der Schönen Künste. Special publication, Bavarian Academy of Art. Waakirchen-Schaftlach: Oreos, 2000.

Watelet, Marcel. *Gérard Mercator—Cosmographe.* Antwerpen: Fonds Mercator, 1994.

Waterhouse, J. "Notes on Early Tele-Dioptric Lens-Systems, and the Genesis of Telephotography." *Photographic Journal* (January 31, 1902): 4–21.

Weibel, Peter. "Freud und die Medien/Freud and the Media. Foto Fake II." German/English. *Camera Austria* 36 (Graz, 1991): 3–21.

Weigl, Engelhard. *Instrumente der Neuzeit: Die Entdeckung der modernen Wirklichkeit.* Stuttgart: Metzler, 1990.

Wessely, Othmar. "Zur Deutung des Titelkupfers von Athanasius Kirchers *Musurgia Universalis* (Romae 1650)." *Römische Historische Mitteilungen* (Vienna) 23 (1981): 385–405.

Westfall, Richard S. "Porta, Giambattista della." Catalog of the Scientific Community. *Galileo Project:* <http://es.rice.edu/ES/humsoc/Galileo/Catalog/Files/porta.html> (July 4, 2001).

Wetzels, Walter D. *Johann Wilhelm Ritter: Physik im Wirkungsfeld der deutschen Romantik.* Berlin: de Gruyter, 1973.

White, Michael. *Isaac Newton: The Last Sorcerer.* London: Fourth Estate, 1998.

Wichert, Hans Walter. "Ein Vorschlag zur optischen Telegraphie aus Westfalen aus dem Jahre 1782." *Technikgeschichte* 51 (1984): 86–93.

Wiedemann, E. "Über Musikautomaten. Beiträge zur Geschichte der Naturwissenschaften." 36. In *Sitzungsberichte der Physikalisch-medizinischen Sozietät in Erlangen,* ed. Oskar Schulz, vol. 46, pp. 17–26. Erlangen: Kommissionsverlag Max Mencke, 1915.

———. "Theorie des Regenbogens von Ibn al Haitham. Beiträge zur Geschichte der Naturwissenschaften." 36. In *Sitzungsberichte der Physikalisch-medizinischen Sozietät in Erlangen,* ed. Oskar Schulz, vol. 46, pp. 39–56. Erlangen: Kommissionsverlag Max Mencke, 1915.

———. "Über die Camera Obscura bei Ibn al Haitham. Beiträge zur Geschichte der Naturwissenschaften. 39. In *Sitzungsberichte der Physikalisch-medizinischen Sozietät in Erlangen,* ed. Oskar Schulz, vol. 46, pp. 155–169. Erlangen: Kommissionsverlag Max Mencke, 1915.

Wiener, Norbert. *Cybernetics, or Control and Communication in the Animal and the Machine.* Cambridge: MIT Press, 1948.

Wilde, Emil. *Geschichte der Optik: Unveränderter Nachdruck der Ausgabe 1838–1843.* Wiesbaden: Martin Sändig, 1968.

Wilding, Nick. "'If you have a secret, either keep it, or reveal it': Cryptography and universal language." In Stolzenberg 2001, pp. 93–104.

Wittels, Fritz. "Hypnose, Suggestion und magisches Denken." In *Die neue Volkshochschule, Bibliothek für moderne Geistesbildung,* vol. 5, pp. 3–40. Leipzig: Weimann, 1928.

Wittgenstein, Ludwig. *Wiener Ausgabe, Studien Texte.* 5 vols. in one. Frankfurt am Main: Zweitausendeins, n.d. (Reprint of the Vienna edition: Springer, 1994–1996).

Wooley, Benjamin. *The Queen's Conjuror: The Science and Magic of Dr. Dee.* London: Harper Collins, 2001.

Worbs, Erich. "Johann Wilhelm Ritter, der romantische Physiker: Seine Jugend in Schlesien." *Schlesien* 4 (1971): 223–237.

Wright, M. R. *Empedocles: The Extant Fragments.* New Haven, Conn.: Yale University Press, 1981; 2nd ed., with additional material, 1995.

Würschmidt, Joseph. "Zur Theorie der Camera obscura bei Ibn al Haitham." In *Sitzungsberichte der Physikalisch-medizinischen Sozietät in Erlangen,* ed. Oskar Schulz, vol. 46, pp. 151–154. Erlangen: Kommissionsverlag Max Mencke, 1915.

Xenakis. "Formalized Music." In *Conferenze e Mostra di Discografica Verdiana. Situazione e perspettive degli studi verdiani nel mondo.* Conference proceedings. Venice: Fondazione Giorgio Cini/San Giorgio Maggiore, 1966.

Yates, Frances A. *The Art of Memory.* London: Pimlico Edition, 1992.

Youngblood, Gene. *Expanded Cinema.* London: Studio Vista, 1970.

Zebrowski, Ernest. *A History of the Circle: Mathematical Reasoning and the Physical Universe.* London: Free Association Books, 1999.

Zedler, Johann Heinrich. *Großes vollständiges Universal-Lexikon aller Wissenschaften und Künste.* Vol. 61. Leipzig, Halle: Zedler, 1749.

Zielinski, Siegfried. *Zur Geschichte des Videorecorders.* Berlin: Volker Spiess, 1985.

———. *Audiovisionen. Kino und Fernsehen als Zwischenspiele in der Geschichte.* Reinbek bei Hamburg: Rowohlt, 1989; updated and revised English edition, *Audiovisions: Cinema and Television as Entr'actes in History.* Trans. Gloria Custance. Amsterdam: Amsterdam University Press, 1999.

———. "Von Nachrichtenkörpern und Körpernachrichten." In *Vom Verschwinden der Ferne. Telekommunikation und Kunst,* ed. Edith Decker and Peter Weibel. Cologne: DuMont, 1990.

————. "Towards an Archaeology of the Audiovisual." Trans. G. Custance. In *Towards a Pragmatics of the Audiovisual: Theory and History,* vol. 1, ed. Jürgen E. Müller. Münster: Nodus, 1994a.

————. "Medienarchäologie: In der Suchbewegung nach den unterschiedlichen Ordnungen des Visionierens." *Eikon* (Vienna) 9 (1994b).

————. "Media Archaeology." In *Digital Delirium,* ed. Arthur and Marielouise Kroker. Montreal: New World Perspectives, 1997a.

————. "Towards a Dramaturgy of Differences." In *Interfacing Realities,* ed. V2. 2d ed. Rotterdam: V2 Organisatie, 1997b.

————. "Zu viele Bilder—Wir müssen reagieren!" In *Godard intermedial,* ed. V. Roloff and S. Winter. Siegen: Stauffenberg, 1997c.

————. "Time Machines." In *Lier en Boog,* Series of Philosophy of Art and Art Theory 15 (2000): 173–183.

————. "For an Expanded Epic Art through Media." Trans. G. Custance. In *Symptomatic—Recent Works by Perry Hoberman.* Bradford: National Museum of Photography, Film and Television, 2001a.

————. "From Territories to Interval: Some Preliminary Thoughts on the Economy of Time/the Time." In *net.condition,* ed. Peter Weibel and Timothy Druckrey. Cambridge: MIT Press, 2001b.

————. "Im Zustand der Schwingung kann es keine Ruhe geben: Ein kurzes Portrait des Physikochemikers Johann Wilhelm Ritter." In *05-03-04: Liebesgrüsse aus Odessa. Für/For/à Peter Weibel,* ed. E. Bonk, P. Gente, and M. Rosen. Berlin: Merve, 2004.

————. "Mouse on Mars—Mixed Another Way: A Semantic Construction of Nine Planets." Trans. G. Custance. In *doku/fiction: Mouse on Mars Reviewed and Remixed.* Catalogue Berlin: Kunsthalle Düsseldorf, Die Gestalten, 2004.

Zielinski, S., and Angela Huemer, eds. *Keith Griffiths—The Presence.* Graz: edition blimp, 1992.

Zigaina, Giuseppe, and Christa Steinle. *P. P. Pasolini oder die Grenzüberschreitung—organizzar il trasumanar.* Neue Galerie Graz catalogue. Venice: Marsilio, 1995.

Credits

Biblioteca Nazionale di Napoli: Figures 4.3 (bottom), 4.7, 4.8, 4.12, 5.1 (bottom), 5.3, 5.4.

Deutsches Museum, Munich: Figures 6.3, 6.4, 6.7, 6.8.

Fülöp-Miller 1926: Figures 8.2, 8.6, 8.7.

Fülöp-Miller 1927: Figure 5.9.

Herzog Anton Ulrich Museum Braunschweig: Figure 5.21.

Istituto e Museo Storia della Scienza, Florence: Figures 5.18, 6.13.

Academy of Media Arts Cologne library: Figures 4.11, 6.15.

Staatsbibliothek Preussischer Kulturbesitz Berlin: Figure 5.8.

Stiftung Weimarer Klassik, Herzogin Anna Amalia Library: Figures 2.3, 2.4, 4.1, 5.5, 5.19, 5.20, 5.22.

University of Cologne library: Figures 5.1 (top), 5.6, 5.10.

University of Münster library: Figure 6.1.

University of Salzburg library: Figures 2.2, 5.13, 5.14, 5.15, 5.16, 5.17, 5.23.

All other illustrations are credited in the figure legends or are from the author's private archive.

INDEX

Index

Nollet, Antoine (1700–1770), 165

Notation, 185

Novalis, Friedrich Leopold Freiherr von Hardenberg (1772–1801), 18, 22, 169, 171

Novikov, Timur (1958–2002), 264

Observer, 30, 136, 138–139

Occultism, occult, 65, 75, 177, 192

Oken, Lorenz (1779–1851), 319n

Opera, 183, 234, 251,

Optics, 83–88, 90, 107, 130, 132, 134, 192,

Organism, 25, 44, 65, 179, 193, 197, 206, 252

Ørsted, Hans Christian (1777–1851), 306n

Oscillographs, 167

Osphresiology, sexual, 61

Ostwald, Friedrich Wilhelm (1853–1932), 170, 172

Paik, Nam June, 272

Paleontological, paleontologist, paleontology, 5–7

Palermo, 37

Panopticon, 128

Paracelsus (Theophrastus [Bombastus] von Hohenheim) (1493–1541), 102, 104, 155

Pasolini, Pier Paolo (1922–1975), 59

Pathology, 209, 212, 219. *See also* Criminal

Pauli, Wolfgang (1900–1958), 298n

Pavlov, Ivan Petrovich (1949–1936), 251

Perception, 46–55, 84, 89, 102, 107, 138, 149, 178, 193, 195, 201–202, 207, 273, 274
subjective, 172–175

Performance, 258, 264, 271, 274

Perfume, 61

Perspectiv, 91

Peternàk, Miklos, 265, 295n

Phenakistoscope, 200

Photography, 31, 85, 205, 212, 215, 216, 217

Physics, 16, 17, 30, 37, 40, 42, 45, 53, 101, 104, 116, 130, 157, 160, 165, 172, 175, 177–178, 190, 194, 235, 236, 245
of the visible, 85

Physiognomy, 62, 66–67, 71, 102, 224

Physiography, 195, 197

Physiology, 53, 193, 195, 196, 198, 201, 203, 206, 242, 245

Phytognomia, 67

Plateau, Joseph Antoine Ferdinand (1801–1883), 200

Pliny, Gaius Plinius Secundus the Elder (23–79), 13, 21

Pneumatic, 66, 105, 125, 129

Polygraphia (polygraphy), 147, 149–151, 155

Pores, theory of, 46–47, 50–51, 53, 55

Porta, Giovan Battista della, (*other than in chapter 4*), 8, 19, 34, 36, 37, 101–102, 104, 113, 128, 135, 136, 155, 164, 187, 194, 195, 208, 268, 272

Prague, 8, 11, 38, 67, 71, 76, 97, 193, 201, 202, 262, 265, 278

Primas, Hans, 111

Primavesi, Oliver, 40, 53

Projection, 86, 89–90, 135, 138, 139, 156, 270, 278
equipment, 136
secret messages, 134

Pseudo-Euclid, 94–96, 136

Psychotechnology (-technik), 239, 248

Psychopathia medialis, 8

Psychopathia sexualis, 219

Ptolemy, 86–88, 95

Purgatorio, purgatory, 123, 127, 136

Purkyně (Purkinje), Jan Evangelista (1787–1869), (*other than in chapter 6*), 36, 206, 263, 265

Pythagoras (ca. 560/70–ca. 480), 9, 105, 107, 127
Pythagorean(s), 43, 58, 109–110, 122, 190
school in Naples, 62, 101

Quay Brothers, The, 275, 318n

Radio Corporation of America (RCA), 237

Reck, Hans Ulrich, 284n

Reflexology, 251

Reichardt, Jasia, 266